Mapping, Monitoring, and Modeling Land and Water Resources

Mapping, Monitoring, and Modeling Land and Water Resources

Advanced Techniques Using Open Source Software

Edited by Pravat Kumar Shit, Pulakesh Das,
Gouri Sankar Bhunia, and Dipanwita Dutta

CRC Press
Taylor & Francis Group
Boca Raton London New York

CRC Press is an imprint of the
Taylor & Francis Group, an **informa** business

MATLAB® is a trademark of The MathWorks, Inc. and is used with permission. The MathWorks does not warrant the accuracy of the text or exercises in this book. This book's use or discussion of MATLAB® software or related products does not constitute endorsement or sponsorship by The MathWorks of a particular pedagogical approach or particular use of the MATLAB® software.

First edition published 2021 by
CRC Press
6000 Broken Sound Parkway NW, Suite 300, Boca Raton, FL 33487-2742

and by
CRC Press
2 Park Square, Milton Park, Abingdon, Oxon, OX14 4RN

© 2021 Taylor & Francis Group, LLC

CRC Press is an imprint of Taylor & Francis Group, LLC

The right of Pravat Kumar Shit, Pulakesh Das, Gouri Sankar Bhunia, and Dipanwita Dutta to be identified as the authors of the editorial material, and of the authors for their individual chapters, has been asserted in accordance with sections 77 and 78 of the Copyright, Designs and Patents Act 1988.

Reasonable efforts have been made to publish reliable data and information, but the author and publisher cannot assume responsibility for the validity of all materials or the consequences of their use. The authors and publishers have attempted to trace the copyright holders of all material reproduced in this publication and apologize to copyright holders if permission to publish in this form has not been obtained. If any copyright material has not been acknowledged please write and let us know so we may rectify in any future reprint.

Except as permitted under U.S. Copyright Law, no part of this book may be reprinted, reproduced, transmitted, or utilized in any form by any electronic, mechanical, or other means, now known or hereafter invented, including photocopying, microfilming, and recording, or in any information storage or retrieval system, without written permission from the publishers.

For permission to photocopy or use material electronically from this work, access www.copyright.com or contact the Copyright Clearance Center, Inc. (CCC), 222 Rosewood Drive, Danvers, MA 01923, 978-750-8400. For works that are not available on CCC please contact mpkbookspermissions@tandf.co.uk

Trademark notice: Product or corporate names may be trademarks or registered trademarks and are used only for identification and explanation without intent to infringe.

Library of Congress Cataloging-in-Publication Data

Names: Shit, Pravat Kumar, editor. | Das, Pulakesh, editor. | Bhunia, Gouri Sankar, editor. | Dutta, Dipanwita, editor.
Title: Mapping, monitoring, and modeling land and water resources : advanced techniques using open source software / edited by Pravat Kumar Shit, Pulakesh Das, Gouri Sankar Bhunia, and Dipanwita Dutta.
Description: First edition. | Boca Raton : CRC Press, 2021. | Includes bibliographical references and index. |
Identifiers: LCCN 2020057790 (print) | LCCN 2020057791 (ebook) | ISBN 9780367486839 (hardback) | ISBN 9781003181293 (ebook)
Subjects: LCSH: Environmental monitoring--India--Case studies. | Environmental mapping--India--Case studies. | Natural resources--Management--Data processing. | Environmental monitoring--Geographic information systems. | Environmental monitoring--Remote sensing. | Environmental management--Computer simulation. | Open source software.
Classification: LCC QH541.15.M64 M35 2021 (print) | LCC QH541.15.M64 (ebook) | DDC 363.7/0630954--dc23
LC record available at https://lccn.loc.gov/2020057790
LC ebook record available at https://lccn.loc.gov/2020057791

ISBN: 978-0-367-48683-9 (hbk)
ISBN: 978-1-032-01992-5 (pbk)
ISBN: 978-1-003-18129-3 (ebk)

Typeset in Times
by Deanta Global Publishing Services, Chennai, India

Dedication

*Dedicated
to
Young Scholars
in the field of
Earth and Environmental Sciences*

Contents

Preface xi
Acknowledgments xiii
Editors xv
Contributors xvii

Chapter 1 Introduction to Part I: Mapping, Monitoring, and Modeling of Land Resources 1

Pravat Kumar Shit and Gouri Sankar Bhunia

Chapter 2 Spatio-Temporal Investigation of Mining Activity and Its Effect on Landscape Dynamics: A Geo-Spatial Study of Beejoliya Tehsil, Rajasthan (India) 5

Brijmohan Bairwa, Rashmi Sharma, Arnab Kundu, and K. K. Chattoraj

Chapter 3 Mapping Areas for Growing Pulses in Rice Fallows Using Multi-Criteria Spatial Decisions 21

Raj Kumar Singh, Chandrashekhar Biradar, Ashutosh Sarker, Atul Dogra, and Javed Rizvi

Chapter 4 Assessing Desertification Using Long-Term MODIS and Rainfall Data in Himachal Pradesh (India) 33

Ravi Teja, N. R. Patel, and Arnab Kundu

Chapter 5 Land Use/Land Cover Characteristics of Odisha Coastal Zone: A Retrospective Analysis during the Period between 1990 and 2017 49

Santanu Roy, Gouri Sankar Bhunia, and Abhisek Chakrabarty

Chapter 6 Evaluating Landscape Dynamics in Jamunia Watershed, Jharkhand (India) Using Earth Observation Datasets 71

Mukesh Kumar, Piyush Gourav, Lakhan Lal Mahato, Arnab Kundu, Alex Frederick, K. M. Sharon, Akash Pal, and Deepak Lal

Chapter 7 Drought Frequency and Soil Erosion Problems in Puruliya District of West Bengal, India: A Geo-Environmental Study 83

Abhisek Santra and Shreyashi S. Mitra

Chapter 8 Effects of Cyclone Fani on the Urban Landscape Characteristics of Puri Town, India: A Geospatial Study Using Free and Open Source Software 103

Anindya Ray Ahmed, Asit Kumar Roy, Suman Mitra, and Debajit Datta

Chapter 9	Land Resource Mapping and Monitoring: Advances of Open Source Geospatial Data and Techniques	121
	Gouri Sankar Bhunia and Pravat Kumar Shit	
Chapter 10	Introduction to Part II: Mapping, Monitoring, and Modeling of Water Resources	145
	P. Das and Dipanwita Dutta	
Chapter 11	Improving Wetland Mapping Techniques Using the Integration of Image Fusion Techniques and Artificial Neural Network (ANN)	149
	Swapan Talukdar, Shahfahad, Roquia Salam, Abdus Samad, Mohd Rihan, and Atiqur Rahman	
Chapter 12	Open Source Geospatial Technologies for Generation of Water Resource Development Plan	165
	Arati Paul and V. M. Chowdary	
Chapter 13	Geo-Spatial Enabled Water Resource Development Plan for Decentralized Planning in India: Myths and Facts	179
	Anand N. Shakya, Indal K. Ramteke, and Pritam S. Wanjari	
Chapter 14	Automatic Extraction of Surface Waterbodies of Bilaspur District, Chhattisgarh (India)	197
	Soumen Brahma, Gouri Sankar Bhunia, S. R. Kamlesh, and Pravat Kumar Shit	
Chapter 15	Valuing Ecosystem Services for the Protection of Coastal Wetlands Using Benefit Transfer Approach: Evidence from Bangladesh	211
	Muhammad Mainuddin Patwary, Md. Riad Hossain, Sadia Ashraf, Rabeya Sultana, and Faysal Kabir Shuvo	
Chapter 16	Identification of Groundwater Prospect of Bilaspur District: A Multi-Criteria Decision Making Approach	225
	Soumen Bramha, Gouri Sankar Bhunia, and S. R. Kamlesh	
Chapter 17	Long-Term Drought Assessment and Prediction Driven by CORDEX-RCM: A Study on a Hydro-Meteorologically Significant Watershed of West Bengal	243
	Shreyashi S. Mitra, Abhisek Santra, Akhilesh Kumar, and Shidharth Routh	
Chapter 18	Use of Open Source Software to Assess Spatio-Temporal Variation of Agricultural Drought at Regional Scale	275
	Debarati Bera and Dipanwita Dutta	

Chapter 19 Snow Cover Monitoring Using Topographical Parameters for Beas River Catchment Area 297

Chetna Soni, Arpana Chaudhary, and Chilka Sharma

Chapter 20 Land Surface Water Resource Monitoring and Climate Change 311

P. Das, V. Pandey, and Dipanwita Dutta

Index 327

Preface

Spatial mapping and modeling land and water resources have become crucial for their preservation and management in the present era. A wide range of challenges and uncertainties exist in the study of earth system dynamics due to the incompletely understood effects of climate change and complex man-made interferences. It is important to use the newest technologies and tools to improve and develop sustainable land and water resource management. In this regard, advanced tools, including remote sensing, geographic information systems, and open source platforms, are acknowledged to be very effective. Land sustainability strategies and water resources management along with effective policies must be based on precise monitoring and assessment, advanced geospatial techniques, and open-source modeling platforms.

This book demonstrates the monitoring, mapping, and modeling of land and water resources using geospatial techniques, open source platforms, and programming languages. It explores state-of-art techniques based on open sources software, R statistical programming, and geographic information systems–based advanced modeling techniques, specifically focusing on the recent trends in data mining and robust modeling for Earth and environmental issues.

The book has been organized into two parts: (I) Land Resources Mapping, Monitoring, and Modeling, and (II) Water Resources Mapping, Monitoring, and Modeling. Each section begins with a short introduction covering an overview of its contents. All chapters focus on the core themes of research and knowledge along with topics that have been little explored. Each chapter provides a review of the current understanding, recent advancements, latest research, and future scope. The book offers huge scope to introduce both novices and scholars to the important issues in land and water resources. It is expected that the book as a whole will provide a timely synthesis of the rapidly growing field of geosciences/remote sensing and will also give the impetus to carry forward new and stimulating ideas that will shape a coherent and fruitful vision for future research.

The contents of this book will be of interest to researchers, professionals, and policymakers in the field of Earth and environmental sciences. Advanced geographic information systems techniques for land and water might be useful to people who want to explore, analyze, and model geographic data with open source data and software. The book will equip readers with the knowledge and skills to understand a wide range of issues manifested in geographic data, including those with scientific, societal, and environmental implications.

We are very thankful to all the authors, who have meticulously completed their chapters at short notice and have put effort into preparing this enriching and pioneering publication. We believe that this will be a very useful book for geographers, ecologists, environmental scientists, water resource managers, and others working in the field of Earth and environment studies, including research scholars, environmentalists, and policymakers.

Pravat Kumar Shit
Pulakesh Das
Gouri Sankar Bhunia
Dipanwita Dutta
Midnapore, West Bengal, India

MATLAB® is a registered trademark of The MathWorks, Inc. For product information, please contact:

The MathWorks, Inc.
3 Apple Hill Drive
Natick, MA 01760-2098 USA
Tel: 508 647 7000
Fax: 508-647-7001
E-mail: info@mathworks.com
Web: www.mathworks.com

Acknowledgments

The preparation of this book has been guided by several geomorphologic pioneers. We are obliged to these experts for providing their time to evaluate the chapters published in this book. We thank the anonymous reviewers for their constructive comments, which led to substantial improvement in the quality of this book. Because this book was a long time in the making, we want to thank our family and friends for their continued support. This work would not have been possible without constant inspiration from our students, knowledge from our teachers, enthusiasm from our colleagues and collaborators, and support from our families. Finally, we also thank our publisher and publishing editor, CRC Press, for their continuous support in the publication of this book.

Editors

Pravat Kumar Shit is an Assistant Professor at the P.G. Department of Geography, Raja N. L. Khan Women's College (Autonomous), Medinipur, West Bengal, India. He received his MSc and PhD degrees in Geography from Vidyasagar University, Medinipur, West bengal, India, and PG Diploma in Remote Sensing and Geographic Information Systems from Sambalpur University, Sambalpur, India. His research interests include applied geomorphology, soil erosion, groundwater, forest resources, wetland ecosystems, environmental contaminants and pollution, and natural resources mapping and modeling. He has published 9 books (7 books for Springer) and more than 60 papers in peer-reviewed journals. He is currently the editor of the GIScience and Geo-environmental Modelling (GGM) Book Series published by Springer-Nature.

Pulakesh Das is currently working at the World Resources Institute India, New Delhi, India. Previously, he was teaching as an Assistant Professor in the Department of Remote Sensing and Geographic Information Systems, Vidyasagar University, Midnapore, West Bengal, India. He received his PhD degree from the Indian Institute of Technology, Kharagpur, India in July 2019. He completed his MSc (2012) in Remote Sensing and Geographic Information Systems and BSc (2010) in Physics from Vidyasagar University, Midnapore, West Bengal, India. His primary research areas include land use forest cover modeling, hydrological modeling, forest cover dynamics and climate change, digital image processing, microwave remote sensing for soil moisture and forest biomass estimation, plant biophysical characterization, etc. He has published more than 13 research articles in reputed peer-reviewed journals.

Gouri Sankar Bhunia received his PhD from the University of Calcutta, India in 2015. His PhD dissertation work focused on environmental control measures of infectious disease using geospatial technology. His research interests include environmental modeling, risk assessment, natural resources mapping and modeling, data mining, and information retrieval using geospatial technology. He is associate editor and on the editorial boards of three international journals in health geographic information systems and geosciences. He has published more than 60 articles in various Scopus indexed journals.

Dipanwita Dutta has been working as an Assistant Professor in the Department of Remote Sensing and Geographic Information Systems, Vidyasagar University, Medinipur, West Bengal, India, since October 2012. She completed her MSc in Geography from the University of Calcutta, Kolkata, West Bengal, India, in 2006 and obtained her second MSc degree in Remote Sensing and Geographic Information Systems under a joint collaboration program of the International Institute for Geo-Information Science and Earth Observation, the University of Twente (The Netherlands), and the Indian Institute of Remote Sensing, Dehradun (India) in 2010. She started her career as a senior research fellow at the Indian Agricultural Research Institute, Pusa, New Delhi. She received her PhD degree from Jamia Millia Islamia, New Delhi in collaboration with the Indian Institute of Technology, Kharagpur, India in 2016. Her broad area of research includes agricultural drought, dryland issues, crop monitoring, land use dynamics, and urban green space, and her research projects have been funded by the University Grants Commission, Department of Science and Technology, Science and Engineering Research Board (Government of India). She has published more than 31 articles and book chapters in reputed international journals and edited book volumes. She is a reviewer of many national and international journals. She was awarded a NUFFIC fellowship for 3 months staying at the International Institute for Geo-Information Science and Earth Observation, The Netherlands, as part of her MSc course. She has been awarded an International Travel Grant by the Department of Science and Technology, Government of India, for visiting the University of Salzburg, Austria. She has also been awarded a World Meteorological Organization-International Centre for Theoretical Physics fellowship for attending an international workshop at the International Centre for Theoretical Physics, Italy.

Contributors

Anindya Ray Ahmed
Department of Geography
Jadavpur University
Kolkata, India

Sadia Ashraf
Environmental Science Discipline
Life Science School
Khulna University
Khulna, Bangladesh

Brijmohan Bairwa
School of Earth Sciences
Banasthali Vidyapith
Tonk, India

Debarati Bera
Department of Remote Sensing and GIS
Vidyasagar University
Midnapore, India

Gouri Sankar Bhunia
Randstad India Pvt Ltd.
New Delhi, India

and

Department of Geography
Nalini Prabha Dev Roy College
Bilaspur, India

and

Department of Geography
Seacom Skill University
Birbhum, India

Chandrashekhar Biradar
International Center for Agricultural Research in Dry Areas (ICARDA)
Cairo, Egypt

Soumen Brahma
Department of Geography
Seacom Skill University
Birbhum, India

Abhisek Chakrabarty
Department of Remote Sensing & GIS
Vidyasagar University
Midnapore, India

K. K. Chattoraj
Department of Geo-Informatics
P.R.M.S. Mahavidyalaya
Bankura University
Bankura, India

Arpana Chaudhary
School of Earth Sciences
Banasthali Vidyapith
Banasthali, India

V. M. Chowdary
Regional Remote Sensing Centre – East
NRSC, ISRO
Kolkata, India

P. Das
World Resources Institute
New Delhi, India

Debajit Datta
Department of Geography
Jadavpur University
Kolkata, India

Atul Dogra
International Centre for Research in Agroforestry (ICRAF)
New Delhi, India

Dipanwita Dutta
Department of Remote Sensing & GIS
Vidyasagar University
Midnapore, India

Alex Frederick
Centre for Geospatial Technologies
Sam Higginbottom University of Agriculture Technology and Sciences
Prayagraj, India

Piyush Gourav
Department of Environmental Science
Sharda University
Greater Noida, India

Md. Riad Hossain
Institute of Disaster Management
Khulna University of Engineering & Technology
Khulna, Bangladesh

S. R. Kamlesh
Government E R R Science PG College
Bilaspur, India

Akhilesh Kumar
Department of Civil Engineering
Haldia Institute of Technology
Haldia, India

Mukesh Kumar
Centre for Geospatial Technologies
Sam Higginbottom University of Agriculture Technology and Sciences
Prayagraj, India

Arnab Kundu
Department of Geo-Informatics
P.R.M.S. Mahavidyalaya
Bankura University
Bankura, India

Deepak Lal
Centre for Geospatial Technologies
Sam Higginbottom University of Agriculture Technology and Sciences
Prayagraj, India

Lakhan Lal Mahato
Centre for Geospatial Technologies
Sam Higginbottom University of Agriculture Technology and Sciences
Prayagraj, India

Shreyashi S. Mitra
Department of Civil Engineering
Haldia Institute of Technology
Haldia, India

Suman Mitra
Department of Geography
University of Calcutta
Kolkata, India

Akash Pal
Centre for Geospatial Technologies
Sam Higginbottom University of Agriculture Technology and Sciences
Prayagraj, India

V. Pandey
Institute of Environment and Sustainable Development
Banaras Hindu University
Varanasi, India

N. R. Patel
Department of Agriculture and Soil
Indian Institute of Remote Sensing
Indian Space Research Organization
Dehradun, India

Muhammad Mainuddin Patwary
Environment and Sustainability Research Initiative
Khulna, Bangladesh

and

Environmental Science Discipline
Life Science School
Khulna University
Khulna, Bangladesh

Arati Paul
Regional Remote Sensing Centre – East
NRSC, ISRO
Kolkata, India

Atiqur Rahman
Department of Geography
Faculty of Natural Sciences
Jamia Millia Islamia
New Delhi, India

Indal K. Ramteke
Maharashtra Remote Sensing Applications Centre
Nagpur, India

Contributors

Mohd Rihan
Department of Geography
Faculty of Natural Sciences
Jamia Millia Islamia
New Delhi, India

Javed Rizvi
International Centre for Research in Agroforestry (ICRAF)
New Delhi, India

Shidharth Routh
Department of Civil Engineering
Haldia Institute of Technology
Haldia, India

Asit Kumar Roy
Department of Geography
Jadavpur University
Kolkata, India

Santanu Roy
Department of Remote Sensing & GIS
Vidyasagar University
Midnapore, India

Roquia Salam
Department of Disaster Management
Begum Rokeya University
Rangpur, Bangladesh

Abdus Samad
Amity Institute of Geoinformatics and Remote Sensing (AIGIRS)
Amity University
Noida, India

Abhisek Santra
Department of Civil Engineering
Haldia Institute of Technology
Haldia, India

and

Dept. of Remote Sensing
Birla Institute of Technology
Mesra Ranchi, India

Ashutosh Sarker
International Centre for Agriculture Research in Dry Areas (ICARDA)
New Delhi, India

Shahfahad
Department of Geography
Faculty of Natural Science
Jamia Millia Islamia
New Delhi, India

Anand N. Shakya
Maharashtra Remote Sensing Applications Centre
Nagpur, India

Chilka Sharma
School of Earth Sciences
Banasthali Vidyapith
Banasthali, India

Rashmi Sharma
School of Earth Sciences
Banasthali Vidyapith
Tonk, India

K. M. Sharon
Centre for Geospatial Technologies
Sam Higginbottom University of Agriculture Technology and Sciences
Prayagraj, India

Faysal Kabir Shuvo
Population Wellbeing and Environment Research Lab (PowerLab)
School of Health and Society
Faculty of Social Sciences
University of Wollongong
Wollongong, Australia

Pravat Kumar Shit
Department of Geography
Faculty of Science
Raja N. L. Khan Women's College
Medinipur, India

Raj Kumar Singh
International Centre for Research in Agroforestry (ICRAF)
New Delhi, India

Chetna Soni
School of Earth Sciences
Banasthali Vidyapith
Banasthali, India

Rabeya Sultana
Environmental Science Discipline
Life Science School
Khulna University
Khulna, Bangladesh

Swapan Talukdar
Department of Geography
University of Gour Banga
Malda, India

Ravi Teja
Department of Agriculture and Soil
Indian Institute of Remote Sensing
Indian Space Research Organization
Dehradun, India

Pritam S. Wanjari
Vasundhara State Level Nodal Agency
Pune, India

1 Introduction to Part I
Mapping, Monitoring, and Modeling of Land Resources

Pravat Kumar Shit and Gouri Sankar Bhunia

CONTENTS

1.1 Introduction 1
1.2 Individual Chapters 2
References..3

1.1 INTRODUCTION

Land use has a global perspective. "Land" applies to the area of the earth that is not filled by oceans, lakes, or rivers. It contains the entire geographical area of continents and islands. Land may legitimately be pointed to as the original root of all material riches. A country's economic stability is strongly connected to its ecosystem's services. Most of the land and its service operations are subject to central trade policies, including food processing and food production. There is also a strong connection to climate change in the way we utilize land and soil. Soil contains large quantities of carbon and nitrogen, which can be emitted into the ecosystem based on how we treat the land (Shit et al. 2015). The global greenhouse gas imbalance could be worsened, with increased emissions, by eliminating tropical forests for cattle grazing and harvesting (Shit et al. 2020). These authors assume that the multipurpose essence of land entails multiple trade-offs favoring one use to the detriment of another. Land appraisal integrates the knowledge from soil surveys, atmosphere, vegetation, and other land factors with the particular application for which the land is measured. There are many considerations that are primarily associated with scale but are also determined by the survey process, the mapping date (a measure of the information's dependability), and the difficulty of the land mapping. In addition, such research permits the key limiting factors for agricultural production to be established and allows decision-makers, including land owners, land use managers, and agricultural support services, to establish a crop management mechanism capable of addressing these limitations, thus increasing productivity (Bhunia et al. 2018). This analysis offers details of the limits to, and possibilities for, land use and thereby informs decisions on the best use of the asset; this understanding is necessary for the planning and growth of land use.

Modeling of land use at a global scale is difficult for three main reasons (e.g., constraint of measurement, scale pertinency of the retrieval models, and conglomeration of land surface and the features of linearity/nonlinearity of the retrieval models) relative to smaller-scale solutions. However, inherently, land cover simulation is an interdisciplinary practice. In order to ensure the development of accurate and usable landscape projections, as well as the application of the resulting model projections to resolve issues of societal significance, we focus on elements of socioeconomics, geography, hydrology, environment, and other sciences. There are actually three kinds of models of spatial land use change: numerical models of estimation, dynamic models of simulation, and models of rule-based simulation. The majority of dynamic models are unable to integrate adequate socioeconomic conditions. Cellular Automata (CA) models can simulate spatial patterns as a rule-based simulation tool but cannot view spatiotemporal land use transition processes and are more difficult to create. Empirical estimation approaches using statistical techniques will, therefore, simulate the relationships between changes in land use and drivers. A scenario model is used that helps land managers to forecast and respond to a broad variety of possible potential trends due to the high degree of complexity involved in forecasting future changes in dynamic socio-environmental environments. For landscape simulation, remote sensing offers an essential source of data. Several techniques focused on geographic information systems (GIS) and remote sensing (RS) approaches have become beneficial for land management in recent years. The integration into GIS of multi-criteria appraisal approaches has emerged as an exciting research field attracting numerous planners and managers. For instance, the concept of fuzzy predicates was implemented via ordered weighted averaging (OWA) into the GIS-based land suitability analysis. A standardized hierarchical land suitability index framework was proposed to provide a strategic environmental evaluation of land use growth for policy making. The biophysical variables influencing land use decision-making by multiple stakeholders have been shown to expose a spatially and temporally explicit multi-scale decision support system. Land use development models help in understanding this dynamic mechanism and can provide useful evidence on potential future configurations for land use. It has been shown that open data initiative programs have a profound effect on scientific research. The development of data and software has expanded to a point where some people also talk of democratizing access to knowledge, especially in developed countries but also increasingly in developing countries. Enabled by the cloud-based Big Data platform, mass analysis of Landsat data has opened new venues for studying long-term ecological and land cover dynamics. A variety of satellite data from multiple sources with a wide range of spatial, spectral, temporal, and radiometric resolutions are now available free of charge to all user categories. Building local capacities for evidence-based policy in this way will promote local context alignment and ultimately contribute to more effective data analyses. The growing availability of free geospatial data and tools provides new possibilities for greater openness and expanded citizen interest in environmental management. To build land resource development (LRD) and water resource development (WRD) plan generation tools, MapWinGIS is used. While a range of free open source (FOS) raster analysis packages are available, such as GRASS, White Box, and Integrated Land and Water Information System (ILWIS), the FOS software SAGA GIS provides

- A one-step procedure for land data visualization
- A simple, one-step process for erudite hydrological and terrain modeling
- A compact package that does not necessitate installation and is thus easy to share

A variety of guidelines for streamlining geospatial web resources, including a web coverage service and a web map service, have been developed by the Open Geospatial Consortium (OGC) for individual users to use in a client–server spatial computing context. The method thus allows spatial data from diversified sources to be integrated together to interpret and create useful knowledge for decision-makers to support their planning activities. This brings many new possibilities, including real-time maps, more regular and affordable data changes, and geographic knowledge exchange among users around the globe.

The first section of the book provides a succinct guide to the present incarnation of the models of land use, their implications for aspects of spatial policy, and the key research aspects in this era using open source software.

1.2 INDIVIDUAL CHAPTERS

There are eight chapters in this section, which address the soil and land use and landscape dynamic. Chapter 2 deals with how landscape change dynamics triggered by mining disturbance have been examined by geospatial techniques, especially using open source software, and how to combat this by alternate land use practices. Chapter 3, written by a group of researchers under the lead of Singh, describes a pilot study to map and quantify suitable areas for growing short duration pulses (lentil) in rice fallows using GIS and geostatistical techniques. In Chapter 4, Teja et al. examine the spatial distribution of desertification using long-term Moderate Resolution Imaging Spectroradiometer (MODIS) and rainfall data in Himachal Pradesh (India). Chapter 5 describes an overview of the decadal changes of land use/land cover (LULC) and their transformations of Odisha coastal zone using temporal Landsat series satellite images (1990–2017). The LULC maps for each year are prepared by the introduction of a hybrid classification system, which involves image interpretation, band ratios (normalized difference vegetation index [NDVI], normalized difference built-up index [NDBI]), and supervised and unsupervised techniques.

Chapter 6, written by a group of researchers under the lead of Kumar et al., discusses long-term transformation of LULC as well as fragmentation of landscape in the Jamunia watershed of Jharkhand state of India using ILWIS, quantum geographical information system (QGIS), and FRAGSTATS and Patch Analysis. Chapter 7 highlights the spatiotemporal distribution of drought and soil erosion problems in Puruliya district, a semi-arid district of the western part of West Bengal, India.

In Chapter 8, Ahmed et al. describe the recent trend and feasibility of open source software applications and datasets, focusing on freely available Sentinel-2A Multi-spectral Instrument (MSI) datasets of pre– and post–Cyclone Fani phases of Puri town, India. Chapter 9 deals with the role of advances in open source geospatial data and techniques for land resource mapping and monitoring, modeling, and sustainable management strategies.

REFERENCES

Bhunia, GS; Shit, PK and Maiti R (2018). Comparison of GIS-based interpolation methods for spatial distribution of soil organic carbon (SOC). *Journal of the Saudi Society of Agricultural Sciences* 17 (2), 114–126.

Shit, PK; Nandi, AS and Bhunia GS (2015). Soil erosion risk mapping using RUSLE model on Jhargram subdivision at West Bengal in India. *Modeling Earth Systems and Environment* 1 (3), 28.

Shit, PK; Pourghasemi, HR; Das, P and Bhunia, GS (2020). *Spatial Modeling in Forest Resources Management*, Springer, 675 pp. DOI: 10.1007/978-3-030-56542-8

2 Spatio-Temporal Investigation of Mining Activity and Its Effect on Landscape Dynamics
A Geo-Spatial Study of Beejoliya Tehsil, Rajasthan (India)

Brijmohan Bairwa, Rashmi Sharma, Arnab Kundu, and K. K. Chattoraj

CONTENTS

2.1 Introduction 5
2.2 Materials and Methods 6
    2.2.1 Study Area 6
    2.2.2 Data Used..6
    2.2.3 Methodology...7
2.3 Results and Discussion 10
    2.3.1 LULC Change Assessment...10
    2.3.2 LULC Gain and Losses and Prediction Modeling 16
    2.3.3 NDVI Change Assessment 16
    2.3.4 Elevation Profile Change Assessment 17
2.4 Conclusion 19
References..19

2.1 INTRODUCTION

Natural resources like soil, minerals, water, forests, and cultivated lands have been the backbone of human existence by providing the necessities of life, such as food, fodder, medicines, and fuel, from the beginning of civilization. Minerals are non-renewable natural resources for global economies, providing human beings with vital raw materials (Chevrel and Bourguignon 2016). Sandstone is found mainly in the Vindhyan and Trans-Aravalli-Vidhyan region of Rajasthan, which is the largest producer of non-metallic and industrial minerals. In this context, Beejoliya is one of the largest mining areas of Rajasthan, where mining activities have been increasing on a large scale from 1970 to the present. Mining activity has been a major contributor to economic development for India, second only to agricultural practices. It provides revenue for governments from direct foreign investments, gross domestic product, foreign exchange, and employment and income for both skilled and non-skilled staff, but these operations inevitably lead to substantial environmental damage.

Most environmental degradation problems are induced by traditional concepts and technologies, which have changed landscape patterns over the Earth's surface. Direct and indirect mining critically impacts our environment, land use, and society; changes happen at local, regional, and global scales, and most areas are stripped of their vegetation, forest cover, and wildlife habitat, while people are displaced. The impact of mining activity on landscape dynamics has been a top research priority since the 1970s. The landscape of arid and semi-arid zones is sensitive to erosional and depositional processes resulting from mining, deforestation, overgrazing of land, sediment transport, and human actions. Spatio-temporal analysis of natural resources was largely neglected before the 1970s, but today, we need to manage natural resources for the survival of future generations. A sustainable evaluation of mining regions is a modern concept for protecting the environment. Several researchers have conducted studies in specific technical fields, such as the natural physical structures, vegetation, land, and changes in landscape pattern that are caused by mining activity, but we focus on environmental sustainability in mining areas due to the awareness of local people and government departments together with the corresponding mitigation of impacts on natural resources.

The advantages of geo-spatial technologies are their ability to monitor and quantify landscape changes at high spatial resolution and high temporal frequency with new improved analytical techniques, and their lower cost compared with ground-based investigation and monitoring. Landsat satellite images are a gift to global society for determining the location and extent of natural landscape features in time and space worldwide. Remote sensing and geographic information systems (GIS) technology gives immense scope for the ongoing sustainable planning of natural resources. Therefore, we need to regularly monitor and characterize the spatial landscape pattern changes caused by mining activity, and this is necessary at different time intervals.

Several researchers have investigated this area (Jhanwar 1996; Sinha et al. 2000; Chauhan 2010) and found that the mining activity in Beejoliya tehsil (Bhilwara) increased from 0.84 to 12.045 km^2 between 1971 and 1984 and further, from 12.045 to 30.839 km^2 between 1984 and 1991. This chapter focuses on providing a comprehensive picture of surface mining activity and long-term impact on the landscape dynamics from 1978 to 2018 through a geo-spatial approach and open source software (SNAP, R, SAGA GIS, QGIS, etc.) in the study site. The major objectives are (a) land use/land cover (LULC) change assessment; (b) Normalized Difference Vegetation Index (NDVI) variability; and (c) evaluation of the dynamics of the influence of mining on topography.

2.2 MATERIALS AND METHODS

2.2.1 STUDY AREA

Beejoliya has existed from medieval times and through the freedom movement; however, the tehsil came into existence in the census on 28 June 2011. Sand stone is purple to reddish brown in color with pale white bands, located in the southeast of the Bhilwara district of Rajasthan (India). Sandstone quarries have developed all around the area with a concentration in the east–west direction. Geographically, the area extends between 25°19′N and 25°N in latitude and between 75°7′E and 75°25′E in longitude (Figure 2.1). The total extent of the area is 630.25 km^2, of which 625.00 km^2 is rural and 5.25 km^2 is urban. The topography of the study area is undulating, and the maximum and minimum elevation is 588 and 359 m, respectively, above mean sea level. The area has a typical warm, semi-arid, monsoon climate. The maximum temperature ranges up to 45°C, and the minimum temperature goes down to 2.9°C. The average yearly amount of rainfall is 895.2 mm. The major crop is maize, followed by sorghum, groundnut, and cotton in the Kharif season, while in the Rabi season, wheat is the predominant crop in the area, followed by gram, mustard, and pulse crops. The total population of this area is 89,483.

2.2.2 DATA USED

In this study, two types of data were used: the satellite-based Landsat for the transformation of the landscape from 1978 to 2018 by mining activity, and elevation models (SRTM, ASTER, and ALOS PALSAR). Furthermore, a toposheet map was used for preparing the baseline profile of topography, which assists the interpretation of different LULC types. The baseline digital elevation model (DEM) was generated from contours based on the toposheet (1:50,000) by the ArcGIS platform. Satellite

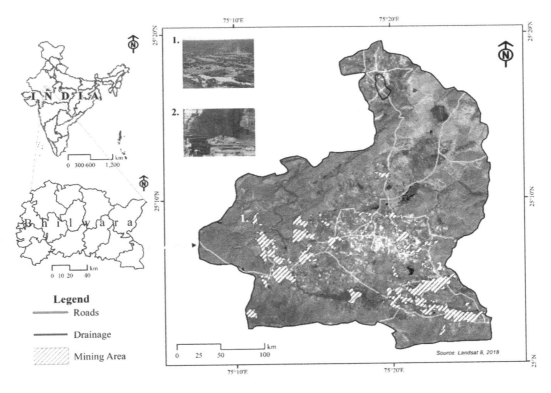

FIGURE 2.1 Location map of the study area.

TABLE 2.1
Descriptions of Satellite Data

Date of Acquisition	Satellite	Sensor	Path and Row	Bands	Resolution (m)
15 December 1978	Landsat 2	MSS	158_43	4	60
15 December 1998	Landsat 5	TM	147_43	7	30
15 December 2018	Landsat 8	OLI and TIRS	147_43	11	30

data were acquired from Landsat MSS (1978), TM (1998), OLI, and TIRS (2018). A brief description of the satellite imageries and elevation models is given in Tables 2.1 and 2.2.

All images were downloaded in December on days clear of atmospheric haze and were processed through ArcGIS, QGIS, and ERDAS Imagine software.

2.2.3 METHODOLOGY

Image processing and spatial analysis are indispensable tools for natural resource analysis and environmental risk assessment. The overall methodology used in this study is shown in Figure 2.2. To show the impact of mining activities on the landscape in the study area, the proposed approach was conceptually developed in eight distinct phases:

- Satellite images and DEM acquisition
- Image pre-processing
- LULC classification

TABLE 2.2
Descriptions of Elevation Data

Digital Elevation	Format	Pixel Size	Data Sources	Year
SRTM	Grid Format	30 M	CGIAR CSI	2000
ASTER	Grid Format	30 M	ASTER	2009
ALOS PALSAR	Grid Format	30 M	ALOS PALSAR	2016

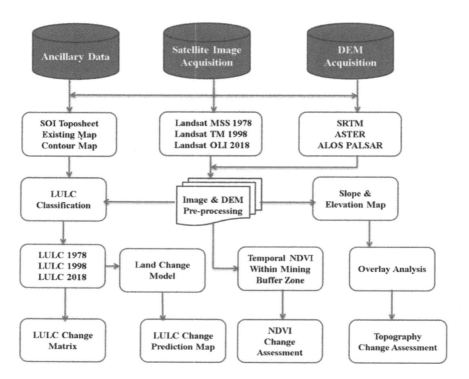

FIGURE 2.2 Brief outline of the workflow.

- Accuracy assessment
- Change matrix calculation
- Gains and losses between 1978 and 1998, and 1998 and 2018
- Derivation of temporal NDVI within the mining buffer zone
- Elevation profile change analysis

Image pre-processing is a key process for enriching the signal properties of satellite images for further processing. Atmospheric and radiometric corrections were executed in QGIS software, and NDVI was computed in ERDAS Imagine. Visible and near-infrared bands were used to produce the standard false color composite (SFCC). All visible and infrared channels were used in the LULC and indices analysis. The LULC classification of an area depends upon available landscape features of that particular area. Also, ArcGIS was used for various stages of classification. Different categories of LULC were generated for a specific period using visual interpretation techniques by GIS. A classification pattern was furnished based on the LULC categories shown in Figure 2.3. For LULC classification, we generated training datasets at 60 and 30 m spatial resolution. Specific bands of the

Mining Activity and Its Effect on Landscape

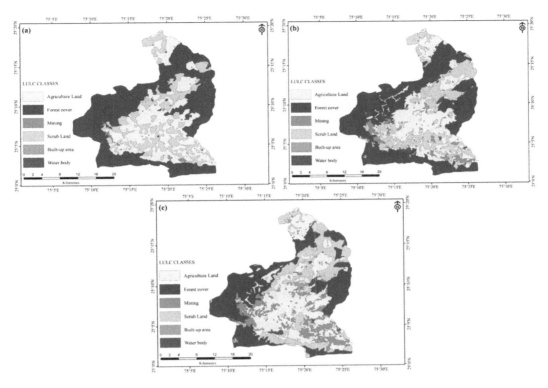

FIGURE 2.3 Area under different classes of LULC in 1978, 1998, and 2018.

Landsat images were used for our applications. LULC accuracy was validated using high-resolution Google Earth images and the toposheet in correspondence with in situ observation. LULC statistics were designed for every single LULC class, and change detection studies were performed in the ArcGIS environment and MS Excel.

Moreover, NDVI was computed within the mining buffer zone to determine changes to vegetation cover. This is a widely used popular index for monitoring vegetation/crop health status propounded by Rouse et al. (1973). Several researchers have employed NDVI in a variety of studies, i.e. the inter-relationship between NDVI and rainfall (Kundu et al. 2018), monitoring desertification (Kundu and Dutta 2011; Kundu et al. 2014, 2015, 2017), and drought assessment (Dutta et al. 2013, 2015; Kundu et al. 2016, 2020; Kundu 2018; Kamble et al. 2019). NDVI is designed using the red (R) and near-infrared (NIR) bands from the Landsat satellite. The basic reason for determining NDVI is that healthy vegetation absorbs most of the visible light that falls on it and reflects a large portion of the NIR light, whereas sparse vegetation reflects more visible light and less NIR light. The NDVI for a given pixel results in a number that ranges from −1 to +1. NDVI is computed using Equation 2.1:

$$\text{NDVI} = (\text{NIR} - \text{R})/(\text{NIR} + \text{R}) \qquad (2.1)$$

On the other hand, all acquired DEMs (SRTM, ASTER, and ALOS PALSAR) are freely available from the different data providers after free registration. After we obtained the data, we pre-processed them in SAGA GIS and QGIS. All the tiles of the DEMs were saved in Geo-TIFF format and re-projected to UTM coordinate system, zone 43N, on WGS-84 ellipsoid. Different filters (simple filters and fill sinks) were used to eliminate noise/sinks. To eliminate the noise, we applied the simple filter tool, which calculates the new values of the raster cells according to a formula. This means that it recalculates the value of the central cell based on the neighboring cells' values. The main task of filtration of this kind is the maximum possible elimination of noise with the preservation of characteristic features of the slope and elevation. It was also important to assign the correct elevation to the study area.

2.3 RESULTS AND DISCUSSION

2.3.1 LULC CHANGE ASSESSMENT

LULC and elevation features are very important in the analysis of mining extent and landscape dynamics. An LULC investigation was performed at different times to determine the recent situation of the landscape and the bases of conversion in the land cover of the area. The LULC maps show the spatial dispersion of the various LULC classes (Figure 2.4). Six major LULC classes (agricultural land, forest cover, mining, scrubland, built-up area, and water body) were identified by interpretation of satellite images and in situ observation. LULC change statistics were generated for the two time periods between 1978 and 1998 and 1998 and 2018 of Beejoliya tehsil (Table 2.3).

The study showed a diminution of forest cover, which was the dominant feature in the whole tehsil. The most significant forests showed a decline from 49.02% (1978) to 46.60% (1998) and from 46.60% (1998) to 33.51% (2018). Growth of the population and expansion of the mining area are the main causal factors of diminution in the forest cover. The influence of population growth can be clearly observed from Table 2.4, which shows an increase in the built-up and cropland classes from 0.36% and 3.24% to 30.74% and 34.26%, respectively. The water body class was increased by the building of dams and patches of mining activity (from 0.41% to 1.08%).

The modification in the landscape pattern of the area is also credited to several ongoing government projects in Beejoliya tehsil. The mining area has increased due to industrial development and human actions; our investigation indicated a marked upsurge from 0.50% to 8.67% during the time studied. Part of the forest was lost due to the increase in agricultural land from 1978 to 2018 (Figures 2.5 and 2.6) (Tables 2.5 to 2.8).

Accuracy assessment was performed from GE images based on an error matrix. The error matrix is a multi-dimensional table in which the cells contain changes from one class to another class. The columns of an error matrix contain the GE data points, and the rows denote the results of classified land use maps. The columns of the error matrix signify the actual field information (GE data), and the rows of the error matrix correspond to a class in the land use map. The overall classification accuracy was calculated as a diagonal point divided by the total number of points. The kappa

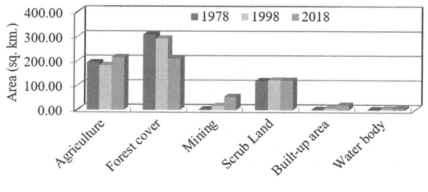

FIGURE 2.4 LULC maps for the year (a) 1978, (b) 1998, and (c) 2018.

TABLE 2.3
LULC Classified Area Statistics (1978 to 2018)

Class	1978	%	1998	%	2018	%
Agriculture	192.77	30.74	181.68	28.97	214.84	34.26
Forest cover	307.37	49.02	292.23	46.60	210.15	33.51
Mining	3.15	0.50	17.90	2.85	54.35	8.67
Scrubland	118.98	18.97	121.66	19.40	120.64	19.24
Built-up area	2.23	0.36	8.28	1.32	20.29	3.24
Water body	2.55	0.41	5.30	0.86	6.78	1.08
Total Area (km^2)	627.05	100	627.05	100	627.05	100

TABLE 2.4
LULC Change Statistics from 1978 to 2018

Classes	1978	1998	Net Change	Change (%)	1998	2018	Net Change	Change (%)
Agriculture	192.77	181.68	−11.09	−5.75	181.68	214.84	33.16	18.25
Forest cover	307.37	292.23	−15.14	−4.92	292.23	210.15	−82.08	−28.09
Mining	3.16	17.90	14.74	466.56	17.90	54.35	36.44	203.58
Scrubland	118.98	121.66	2.68	2.25	121.66	120.64	−1.02	−0.84
Built-up area	2.23	8.28	6.05	271.68	8.28	20.29	12.01	145.14
Water body	2.55	5.30	2.75	107.79	5.30	6.78	1.47	27.76

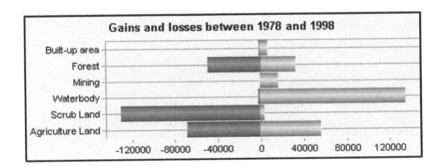

FIGURE 2.5 Gains and losses of LULC categories (in ha) between 1978 and 1998.

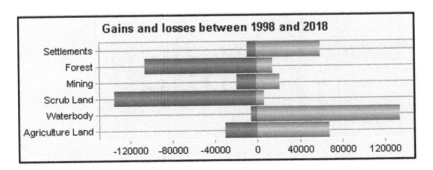

FIGURE 2.6 Gains and losses of LULC categories (in ha) between 1998 and 2018.

TABLE 2.5
Accuracy Assessment

1978		1998							
Classes	Agriculture	Forest Cover	Mining	Scrubland	Built-Up Area	Water Body	Total	User Accuracy (%)
Agriculture	60	5	0	5	3	0	73	82.19
Forest cover	0	43	0	5	0	0	48	89.58
Mining	0	0	27	0	0	1	28	96.43
Scrubland	3	2	0	60	4	0	69	86.96
Built-up area	0	0	3	0	40	0	43	93.02
Water body	0	0	0	0	0	14	14	93.33
Total	63	50	30	70	47	15	275	
Producer accuracy (%)	95.24	86.00	90.00	85.71	85.11	93.33		
Overall accuracy	88.60							

TABLE 2.6
Accuracy Assessment

1998		2018						
Classes	Agriculture	Forest Cover	Mining	Scrubland	Built-Up Area	Water Body	Total	User Accuracy (%)
Agriculture	70	5	0	5	3	0	83	84.33
Forest cover	0	43	0	5	0	0	48	89.58
Mining	0	0	27	0	0	1	28	96.43
Scrubland	3	2	0	60	4	0	69	86.96
Built-up area	0	0	3	0	40	0	43	93.02
Water body	0	0	0	0	0	14	14	100
Total	73	50	30	70	47	15	285	
Producer accuracy (%)	95.89	86.00	90.00	85.71	85.11	93.33		
Overall accuracy	89.12							

TABLE 2.7
Change Matrix between 1978 and 1998

	Classes	Agriculture	Forest Cover	Mining	Scrubland	Built-Up Area	Water Body	1978 Total	Gross Loss
1978	Agriculture	130.52	17.15	5.11	32.82	4.26	2.91	192.77	62.25
	Forest cover	15.28	262.74	1.95	25.84	0.78	0.78	307.37	44.63
	Mining	0.29	0.10	2.66	0.02	0.10	0.00	3.17	0.51
	Scrubland	33.19	12.00	8.06	62.49	1.79	1.42	118.95	56.46
	Built-up area	0.35	0.11	0.13	0.34	1.31	0.00	2.24	0.93
	Water body	2.05	0.13	0.00	0.13	0.05	0.19	2.55	2.36
	1998 Total	181.68	292.23	17.91	121.64	8.29	5.30	627.05	167.14
	Gross gain	51.16	29.49	15.25	59.15	6.98	5.11	167.14	

TABLE 2.8
Change Matrix between 1998 and 2018

	Classes	Agriculture	Forest Cover	Mining	Scrubland	Built-Up Area	Water Body	1998 Total	Gross Loss
1998	Agriculture	154.25	4.66	12.37	4.32	5.28	0.81	181.69	27.44
	Forest cover	20.86	197.07	7.21	65.65	1.15	0.28	292.22	95.15
	Mining	1.15	0.74	15.32	0.00	0.70	0.00	17.91	2.59
	Scrubland	37.21	7.63	19.27	50.55	6.43	0.56	121.65	71.10
	Built-up area	1.26	0.02	0.16	0.09	6.73	0.02	8.28	1.55
	Water body	0.12	0.03	0.01	0.03	0.01	5.10	5.30	0.20
	2018 Total	214.85	210.15	54.34	120.64	20.30	6.78	627.05	
	Gross gain	60.60	13.08	39.02	70.09	13.57	1.67		

statistic was used to evaluate accuracy for linking results from different LULC maps (Table 2.8; Figure 2.3).

$$\text{Overall Accuracy} = \left(\frac{\text{Total number of correct pixels}}{\text{Total number of observed pixels}} \right) \times 100$$

$$\text{Producer Accuracy} = \frac{\text{No. of corrected pixels in each category}}{\text{Total No. of Reference Pixels in Row Category}} \times 100$$

$$\text{User Accuracy} = \frac{\text{No. of corrected pixels in each category}}{\text{Total No. of corrected Pixels in Row Category}} \times 100$$

2.3.2 LULC Gain and Losses and Prediction Modeling

There was a diminution of forest area with each successive increase of mining area. The expansion of agriculture and built-up lands was due to the conversion of forest land to agricultural land and mining areas, respectively (Table 2.4; Figure 2.5).

The rate of forest decline increased from 1.5% during 1978–1998 to 1.88% during 1998–2018. The water body area increased during 1979–1999 by 2.22% due to the construction of dams, but it decreased by 0.21% during 1999–2013. The dry deciduous forest was vastly reduced during 1979–1999 (4.53%) as compared with 1999–2013 (1.25%) (Figure 2.6).

2.3.3 NDVI Change Assessment

NDVI was considered on a temporal scale for analyzing LULC dynamics in the mining area in Beejoliya. The measured NDVI showed that there was a change in vegetation cover from 1978 to 2018. In addition, it indicated that 97.24% of the area was covered by green vegetation in 1978 as against 83.34% vegetation cover in 2018 (Figure 2.7).

In contrast, the non-vegetation categories showed an increase from 2.76% to 16.66% during this period (Table 2.2 and Figures 2.4 and 2.5). Thus, it was obvious that there was a diminution of forest cover in the study site (Table 2.3 and Figures 2.6 and 2.8).

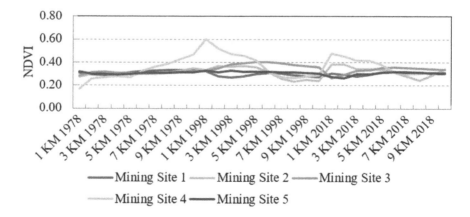

FIGURE 2.7 NDVI change assessment within the mining buffer zone.

Mining Activity and Its Effect on Landscape 17

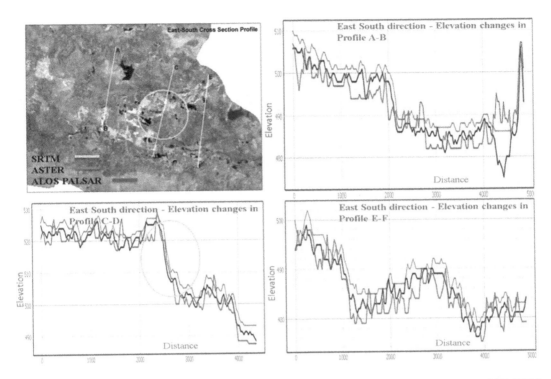

FIGURE 2.8 Topographical profile change assessment from SRTM, ASTER, and ALOS PALSAR DEM across the east–south cross-section profile of Beejoliya tehsil.

The NDVI values for the study area varied mostly from −1 to 0.7. As can be seen, the forest area had the maximum NDVI values, then the non-forest area, followed by habitation, mines, and finally, the water body area. From Figure 2.9, it is clear that most areas in 1978 were covered by forest (Figure 2.2). In the 1998 and 2018 NDVI maps, there was a gradual deterioration of forest area and an increase of mines in the study site. The low NDVI values were recorded from the mines and settlement area. Table 2.4 shows the variation of NDVI values for various classes in 2018.

2.3.4 ELEVATION PROFILE CHANGE ASSESSMENT

For cross-profile elevation analysis, we used three DEM datasets, which provided useful information about several landscape features on a vertical scale. Cross-section comparison in different models allows us to determine the deviation of their elevations from the real values. The obtained values were used to calculate the cross-section elevation profile from DEM. The DEM profile study of terrain variation assessment revealed changes in the landscape due to mining or anthropogenic practices.

In the current study, the nine cross-section profile sets (three profiles for each direction) were separately extracted using SRTM, ASTER, and ALOS PALSAR DEM for the three directions: east–south, west–south, and east–north. Moreover, we extracted nine cross-shore profiles based on the vertical changes on different kinds of landforms. Therefore, the results of vertical differences between these cross-shore profiles signify land loss or land gain from mining activity in the Beejoliya tehsil along with the three directions during 2000–2015 (Figure 2.10).

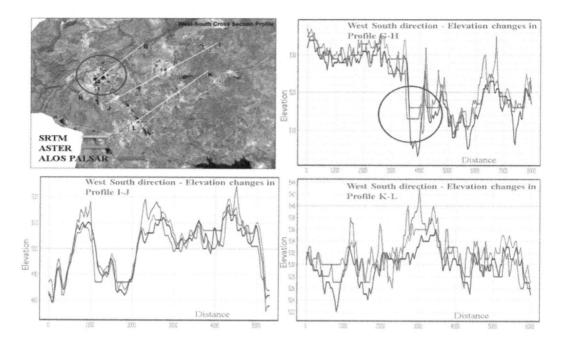

FIGURE 2.9 Topographical profile change assessment from SRTM, ASTER, and ALOS PALSAR DEM across the west–south cross-section profile of Beejoliya tehsil.

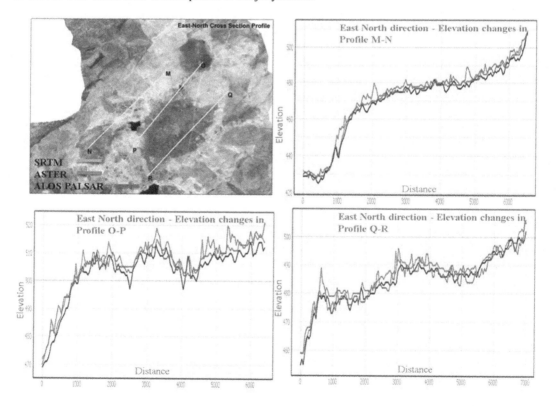

FIGURE 2.10 Topographical profile change assessment from SRTM, ASTER, and ALOS PALSAR DEM across the east–north cross-section profile of Beejoliya tehsil.

2.4 CONCLUSION

The present study is very useful for monitoring changes in landscape pattern and topography due to expansion of the mining area over time. These were effectively mapped and monitored in Beejoliya tehsil by remote sensing and GIS technology as well as open source software. The methodology was premised on the fact that mining-induced land cover and topography changes can be detected based on changes in the landscape characteristics of images obtained at 20-year intervals (1978–1998–2018) obtained with the same acquisition geometry. LULC change features were very beneficial for detecting inter-decadal mining-induced changes along with water bodies and surrounding mining areas. The LULC matrix showed significant influences on agriculture, forest cover, built-up land, and scrubland and also on mining areas in the study area due to increased mining operations (Table 2.3). Similar results were found in the topography profile of the surrounding mining area. The results of change detection showed that the area under forest cover was decreased in both protected and reserved forests and that forest cover was being converted to agricultural land, mining, and scrubland. Due to the changes in agriculture and forest land due to mining activity, the NDVI pattern within the buffer zone of the mining area (Figure 2.9) also showed a diminished intensity of vegetation over time. Furthermore, the current study may be advantageous to detect areas of vegetation that are at risk due to mining activity. Hence, it may be concluded that the expansion of mining operations in the area was caused by LULC change between 1978 and 2018. Thus, it can be concluded that there was a strong relationship between the loss of landscape features and the growth in mining area. Besides, it is necessary to adopt a sustainable approach to mining in the area to reduce the declining land cover dynamics of the tehsil. The indirect impact of open-cast mining in terms of shifting agricultural activities to other places due to its influence on soil fertility needs further study. Thus, this result will help in designing the mining or landscape management plan and conservation of the study area, which is a dynamic step in ecological planning.

REFERENCES

Chauhan, S. S. (2010) Mining, development and environment: a case study of Bijola mining area in Rajasthan, India. *Journal of Human Ecology*, 31(1):65–72.

Chevrel, S., Bourguignon, A. (2016) Application of optical remote sensing for monitoring environmental impacts of mining: from exploitation to post mining. *Land Surface Remote Sensing: Environment and Risks*, 191–220.

Dutta, D., Kundu, A., Patel, N. R. (2013) Predicting agricultural drought in eastern Rajasthan using NDVI and standardized precipitation index. *Geocarto International*, 28(3):192–209.

Dutta, D., Kundu, A., Patel, N. R., Saha, S. K., Siddiqui, A. R. (2015) Assessment of agricultural drought in Rajasthan (India) using remote sensing derived vegetation condition index (VCI) and standardized precipitation index (SPI). *Egyptian Journal of Remote Sensing and Space Sciences*, 18(1):53–63.

Jhanwar, M. L. (1996) Application of remote sensing for environmental monitoring in Bijolia mining area of Rajasthan. *Journal of the Indian Society of Remote Sensing*, 24(4):255–264.

Kamble, D. B., Gautam, S., Bisht, H., Rawat, S., Kundu, A. (2019) Drought assessment for kharif rice using standardized precipitation index (SPI) and vegetation condition index (VCI). *Journal of Agrometeorology*, 21(2):182–187.

Kundu, A. (2018) An Appraisal of Drought Dynamics in Bundelkhand Region (India) using Geo-Spatial Techniques. Ph.D. Thesis, Faculty of Engineering and Technology, Sam Higginbottom University of Agriculture, Technology and Sciences (collaboration with Indian Institute of Remote Sensing, ISRO), India. pp. 1–234.

Kundu, A., Dutta, D., Patel, N. R., Saha, S. K., Siddiqui, A. R. (2014) Identifying the process of environmental changes of Churu district, Rajasthan (India) using remote sensing indices. *Asian Journal of Geoinformatics*, 14(3):14–22.

Kundu, A., Denis, D. M., Patel, N. R., Dutta, D. (2018) A geo-spatial study for analysing temporal responses of NDVI to rainfall. *Singapore Journal of Tropical Geography*, 39(1):107–116.

Kundu, A., Denis, D. M., Patel, N. R., Mall, R. K., Dutta, D. (2020) *Geoinformation Technology for Drought Assessment in Techniques for Disaster Risk Management and Mitigation*. Srivastava, P. K., Singh, S. K., Mohanty, U. C., Murty, T. (Eds.), John Wiley & Sons, Inc., 171–180. ISBN: 9781119359203.

Kundu, A., Dutta, D. (2011) Monitoring desertification risk through climate change and human interference using remote sensing and GIS techniques. *International Journal of Geomatics and Geosciences*, 2(1):21–33.

Kundu, A., Dwivedi, S., Dutta, D. (2016) Monitoring the vegetation health over India during contrasting monsoon years using the satellite remote sensing indices. *Arabian Journal of Geosciences,* 9(2):1–15.

Kundu, A., Patel, N. R., Saha, S. K., Dutta, D. (2015) Monitoring the extent of desertification processes in western Rajasthan (India) using geo-information science. *Arabian Journal of Geosciences,* 8(8):5727–5737.

Kundu, A., Patel, N. R., Saha, S. K., Dutta, D. (2017) Desertification in western Rajasthan (India): an assessment using remote sensing derived rain-use efficiency and residual trend methods. *Natural Hazards*, 86(1):297–313.

Rouse, J. W., Haas, R. H., Schell, J. A. Deering, D. W. (1973) Monitoring vegetation systems in the Great Plains with ERTS. In *3rd ERTS Symposium*, NASA SP-351 I: 309–317.

Sinha, R. K., Pandey, D. K., Sinha, A. K. (2000) Mining and the environment: a case study from Bijolia quarrying site in Rajasthan, India. *The Environmentalist*, 20(3):195–203.

3 Mapping Areas for Growing Pulses in Rice Fallows Using Multi-Criteria Spatial Decisions

Raj Kumar Singh, Chandrashekhar Biradar, Ashutosh Sarker, Atul Dogra, and Javed Rizvi

CONTENTS

3.1 Introduction 21
3.2 Materials and Methods 23
    3.2.1 Study Area 23
    3.2.2 Data and Methodology 23
    3.2.3 Potential Areas for Identification of Pulses 24
    3.2.4 Identification of Assessment Factors 24
    3.2.5 Fuzzy Set Modeling 24
    3.2.6 Analytical Hierarchy Process (AHP) Approach 24
    3.2.7 Calculation of Suitability Index Using Multiple-Criteria Decision-Making (MCDM) 26
    3.2.8 Generating Land Suitability Maps 27
3.3 Results and Discussion 28
3.4 Conclusion 30
References 30

3.1 INTRODUCTION

The human population is growing, and a substantial increase in total arable land area is not anticipated. The intensification and diversification of agriculture will be more a matter of increasing agricultural production in the future. There is still much unexplored potential for agricultural intensification. One such opportunity is the intensification of the cropping system in rice fallow areas after harvesting the first-season or often before second-season crops. The two main concerns about the rice fallow areas are: (i) the population is continuously increasing, thereby requiring a simultaneous increment in the available amount of food grains, and (ii) the continuation of cereal cropping requires crop rotation or diversification for sustainable agriculture (Paroda et al., 1994). Eastern India is the dominant rice cultivation area in India, producing more than 63.3% of the country's total harvest. Rice is mainly produced during the Kharif (rainy) season, and a large part of these agricultural areas remains fallow during the dry Rabi and Zaid seasons due to the lack of appropriate irrigation facilities. Approximately 78.8% of the rice-growing area in the eastern region is rainfed, and rice is grown in this area mostly during the monsoon (June–September) (Kar & Kumar, 2009). The rice cropping area remains fallow between harvest and the next sowing, hence being called rice fallow or Rabi fallow. The existing rice fallow area after the rice harvest (11.65 Mha) is almost equivalent to the net sown area of Punjab, Haryana, and western Uttar Pradesh – the site of the green revolution in India (Varghese et al., 2018). An estimation of fallow area carried out by the International Crops Research Institute for the Semi-Arid Tropics (ICRISAT) indicated that about 82% of the total rice fallow area lies in the eastern region: Assam (0.54 Mha), Bihar and Jharkhand (2.20 Mha), Madhya Pradesh and Chhattisgarh (4.38 Mha), Odisha (1.22 Mha), Uttar Pradesh (0.35 Mha), and West Bengal (1.72 Mha) (Subbarao, 2001). Other limitations are lack of irrigation, cultivation of long-duration varieties, waterlogging, lack of variety of winter crops, and some socio-economic factors, such as stray cattle, etc. (Ali & Kumar, 2009). Rotation of crops is considered an effective alternative and preferable to fallow areas to expedite agricultural intensification (growing another crop during the fallow period) and crop diversification (mainly short-duration lentils and oil seeds).

Therefore, there is a need for precise estimation of such rice fallows and crop site suitability for subsequent winter crops, i.e., pulses. Many tools are available for creation of the required parameters according to the need for land suitability analyses. Such information can be used to identify both opportunities and constraints. Geoinformatics offers various options for carrying out numerous tasks using spatial and attribute data. Also, a geographic information system (GIS) provides the required functionality for spatial decision-making processes (multi-criteria evaluation [MCA]) for land suitability analysis. GIS also has a unique capacity for the integration and analysis of multi-format and multi-source datasets. These data can be manipulated and analyzed for information that is useful for a specific application, such as land-use suitability analysis (Malczewski, 2004). GIS technology is being increasingly employed by different sectors to create resource databases and to find solutions to problems. The Analytical Hierarchical Approach (AHP)–based MCA is an effective approach to integrate the influence of multiple factors in decision-making processes. It considers the relative importance of each theme, while it also accounts for the importance of each factor within a theme.

Paddy is the dominant crop in West Bengal among cereals, pulses, oil seeds, and fibers. The dominant crop among pulses is lentil (Musur), with Maskalai, Khesari, Gram, Arhar, Matar, and Mung (PRFNS, 2018). Pulses occupy 197,000 ha (2.23%) of the total cropped area in West Bengal; the highest productivity is observed for Nadia, South-24 Parganas, Burdwan, Birbhum, Bankura, Purba Medinipur, and Paschim Medinipur district. Medium-productivity districts include Uttar Dinajpur, Malda, Murshidabad, North-24 Parganas, Howrah, and Purulia, and low-productivity districts are Darjeeling, Jalpaiguri, Cooch Behar, Dakshin Dinajpur, and Hoogly (Aktar, 2015). According to the Agricultural Statistics (ASAG, 2018), in India, pulses were cultivated on approximately 29 Mha and recorded the highest ever production of 25.23 Mt during 2017–18. The total area occupied by lentils is about 150,000 ha, and the highest ever production was 150,000 tons during 2017–18 (PRFNS, 2018). According to the annual report of 2017–18, approximately 98% of the lentil-covered area was in the states of Madhya Pradesh (38.35%), Uttar Pradesh (31.15%), West Bengal (11.17%), Bihar (9.65%), Jharkhand (4.47%), Rajasthan (2.20%), and Assam (1.14%). In West Bengal, the total lentil production was 0.15 Mt during 2017–18 (Annual Report, 2018).

Various studies have applied this technique for watershed prioritization (Chowdary et al., 2013); prescribing the suitable areas and structures for rainwater harvesting (Behera et al., 2019); assessment of drought hazard (Pandey & Srivastava, 2019); site suitability analysis for agricultural land use (Pramanik, 2016); crop suitability (Mustafa et al., 2011); land suitability analysis for sugarcane cultivation (Jamil et al., 2018); land suitability analysis for agricultural crops (Bera et al., 2017), etc. Singha et al. (2020) employed the AHP-MCA technique to prescribe the crop rotation selection for rice–jute in the Kharif season and potato–lentil in the Rabi season based on the soil nutrient availability in Hoogly district and recommended about 25% paddy fields for replacement with jute during Kharif and about 8% of potato to replace lentil in the Rabi season. The cultural and socio-economic conditions and livelihood for the majority of the population in West Bengal are dependent on agricultural practices. The identification of suitable areas for lentil production is useful for farmers, agriculture and water resource managers, and policy developers to develop suitable action plans. The present study uses the AHP-based MCA technique to assess the suitability of the lentil crop for West Bengal.

3.2 MATERIALS AND METHODS

3.2.1 STUDY AREA

The research was carried out in one of the agriculture-dominated states of India, West Bengal (Figure 3.1). The total population of the study is more than 90 million, and the majority of the population are dependent on agriculture and allied sectors. The study area is located in the eastern region and lies between 85°48′15.78″E and 89°53′37.61″E longitude and 21°26′59.19″ N and 27°14′17.84″ N latitude (Figure 3.1). In the study area, the elevation varies from 0 to 3636 m, and most of the area has less than 5% slope. The annual average rainfall varies from 1200 to 1600 mm in the different parts of the state. The northern districts receive much higher rainfall, 200–400 cm, while the southwestern districts are relatively drier. The annual average minimum temperature in the state varies between 10 and 14 °C, while the maximum temperature varies between 23 and 27 °C.

3.2.2 DATA AND METHODOLOGY

The soil resource database of India published by the National Bureau of Soil Survey and Land Use Planning (NBSS & LUP) was used as a base map for soil characteristics identification. The soil moisture data were taken from Soil Moisture Active Passive (SMAP), which provides a variety of geophysical fields at 3 h time resolution on the global 9 km modeling grid (Reichle et al., 2015).

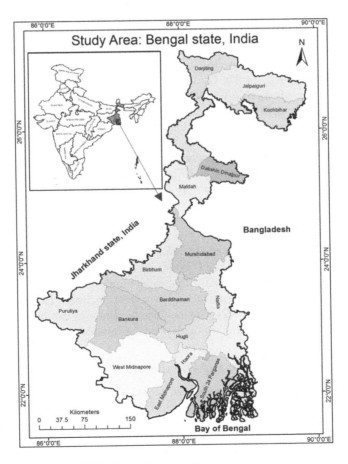

FIGURE 3.1 District-level map of West Bengal state, India.

Climate data taken from the website www.worldclim.org, which provides generic grid data at the highest resolution at 30 arc second (1 km) for current conditions (1950–2000), were used for minimum temperature, maximum temperature, and rainfall. Also, elevation and slope were extracted from the Shuttle Radar Topography Mission (SRTM) digital elevation model (DEM) at 30 m spatial resolution. Crop management data were collected from a review of literature and theses, and furthermore, some crop management data were collected from field visits.

3.2.3 Potential Areas for Identification of Pulses

The vast rainfed rice area of eastern India remains fallow in the Rabi season due to lack of optimum water and crop management. Rabi fallow areas can be considered potential areas for growing pulses in the region. The fallow area has been spatially mapped using Moderate Resolution Imaging Spectroradiometer (MODIS) time series data (Zhang et al., 2017) in terms of start and end dates of fallow period, duration, and yearly crop intensity. The potential area was estimated using fallow dynamics with the following criteria: start date from November, end date up to March (i.e., a duration of 90 to 120 days). These thresholds were applied to existing dynamics, and finally, a spatial map of potential areas was generated.

3.2.4 Identification of Assessment Factors

The first step in tracing suitable areas for the crop is to identify influencing factors. Out of the holistic set of factors that influence the growth and production of crops, those that are easiest to acquire are considered, based on published literature (emphasizing climate, soil, and topographic factors used for crop suitability analysis) (Akinci et al., 2013; Nisar Ahamed et al., 2000). Hence, only the most relevant factors, i.e., soil characteristics, climate conditions, and topography characteristics, were used in the suitability analysis. The land suitability evaluation criteria were classified based on Food and Agriculture Organization (FAO, 1976) crop requirement standards. According to the data availability and the actual condition of the study area, the parameters selected to evaluate the land suitability for lentil production are: (1) soil nutrients and physical condition: pH, soil organic carbon (OC), cation exchange capacity (CEC), salinity, soil moisture, soil texture, and depth; (2) climate conditions: precipitation, minimum and maximum temperature; (3) topographic factors: elevation, slope, and drainage density (Table 3.1). A multi-criteria evaluation process was employed for integrating these factors to assess the land suitability for lentil production (Figure 3.2).

3.2.5 Fuzzy Set Modeling

In comparison to the classical set with a rigid boundary, fuzzy set modeling was used to compute a transitional or partial membership between 0 and 1, set through the membership function (Zadeh, 1965). The membership function is denoted with a value ranging from 0 to 1, which represent non-membership and full membership, respectively, and the intermediate values measure the degree of closeness to unity (Burrough et al., 1992). The membership function can be assessed using various mathematical models, such as parabolic, sigmoid, inverted sigmoid, and linear functions.

3.2.6 Analytical Hierarchy Process (AHP) Approach

The most important step in suitability assessment is to estimate the weightage of each parameter. The weight can be determined by many approaches, although AHP (Malczewski, 2004; Saaty, 1980) has been effectively used. In the AHP method, a hierarchical model consisting of objectives, criteria, and alternatives is used for every problem (Malczewski, 2004). The AHP approach is used to assign differential weights to various parameters or factors based on their importance. The hierarchical model allows weights to be assigned even to the sub-criteria or sub-class categories.

TABLE 3.1
Fuzzy Membership for Categorical/Qualitative Data

Factors	Sub-Criteria	Ratings	Membership	Area (ha)
	Soil Properties			
Texture	Loamy	1	1	4756.05
	Clayey	2	0.8	2857.19
	Sandy	3	0.4	92.95
	Loamy skeletal	5	0.2	123.27
	Clay skeletal	4	0.1	339.65
Depth	Very deep	1	1	125.53
	Deep	2	0.8	6950.71
	Moderately deep	3	0.4	648.57
	Shallow	4	0.0	443.40
Moisture	0.4–0.5	1	1	19.25
	0.3–0.4	2	0.8	382.70
	0.2–0.3	3	0.4	4932.60
	0.1–0.2	4	0.1	3064.47
pH	Neutral (6.5–7.5)	1	1	2535.92
	Slightly acidic (5.5–6.5)	2	0.8	3611.55
	Moderately acidic (4.5–5.5)	3	0.6	554.72
	Strongly acidic (<4.5)	4	0.4	219.61
	Slightly alkaline (7.5–8.5)	5	0.1	860.33
OC	>5.0%	1	1	13.95
	2.0–5.0%	2	0.8	207.21
	1.0–2.0%	3	0.6	399.93
	0.7–1.0%	4	0.4	623.56
	0.5–0.7 %	5	0.2	2183.06
	0.5%	6	0.0	3911.45
CEC	>30	1	1	418.27
	20–30	2	0.8	643.16
	10–20	3	0.4	3046.14
	<10	4	0.1	3409.36
Salinity	Negligible (1–2)	1	1	7155.53
	Slight (2–4)	2	0.8	24.20
	Moderate (4–8)	3	0.6	359.46
	High (8.15)	4	0.1	171.41
	Topography			
Flooding	Nil	1	1	5422.88
	Moderate	2	0.6	2147.76
	Occasional	3	0.4	380.10
	Severe	4	0.0	217.47
Drainage density	Good	1	1	2436.99
	Moderately good	2	0.8	1640.85
	Poor	3	0.4	3577.41
	Excessive	4	0.0	512.96

Fuzzy Scale for Continuous Data

	Min.	Optimum (Min.)	Optimum (Max.)	Max.	Fuzzy functions
Elevation (m)	0	10	200	674	Sigmoid symmetric
Slope (%)	0	1	4	100	Sigmoid symmetric

(Continued)

TABLE 3.1 (CONTINUED)
Fuzzy Membership for Categorical/Qualitative Data

Factors	Sub-Criteria	Ratings			Membership	Area (ha)
		Climate				
Precipitation (mm)	700	700		1400	3616	Sigmoid symmetric
Temperature (°C)	10	11		12	14.7	Sigmoid symmetric
Temperature (°C)	23	24		26	27.5	Sigmoid symmetric
Net radiation (w/m^2)	98				107	Linear

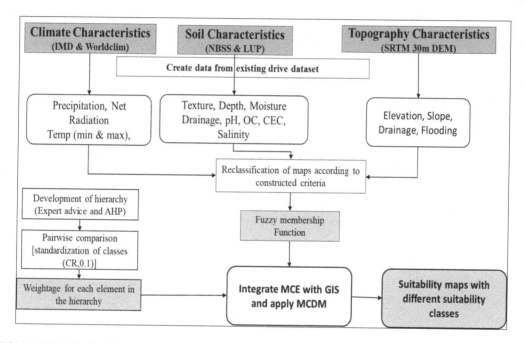

FIGURE 3.2 Methodology for agro-climatic suitability using multi-criteria approach.

A pair-wise comparison matrix of criteria is carried out to assign the relative importance of factors based on Saaty's scale, where the relative importance is judged based on expert opinion. To ensure consistency in defining weights, the consistent ratio (CR) parameter is computed with an acceptable maximum value of 0.10 (Nisar Ahamed et al., 2000; Saaty & Vargas, 2012; Saaty, 1980). If the consistency ratio is above 0.1, the estimated weights are considered to be inconsistent, and the given judgments need to be improved. In this case, CR was 0.014, 0.031, 0.008, and 0.016 for the matrix criteria of soil physical property, soil chemical property, climate, and topography, respectively (Table 3.2).

3.2.7 Calculation of Suitability Index Using Multiple-Criteria Decision-Making (MCDM)

This calculation integrates the response of underlying factors in the decision-making process. It examines the relative importance of multiple conflicting criteria to evaluate the final expected outcome. It includes an evaluation (or aggregation) stage to combine the information from the factors

TABLE 3.2
Pair-Wise Comparison of Criteria and Factors

Topography Characteristics

	Elevation	Slope	Drainage	Flooding	Weights
Elevation	1	1/3	1/2	2	0.16
Slope		1	2	4	0.43
Drainage			1	3	0.32
Flooding				1	0.09

Soil Characteristics

	Texture	Depth	Moisture	pH	OC	CEC	Salinity	
Texture	1	1/3	1/2	2	1/4	4	3	0.11
Depth		1	2	3	1/2	6	5	0.23
Moisture			1	3	1/3	5	4	0.16
pH				1	1/5	3	2	0.07
OC					1	7	6	0.35
CEC						1	1/2	0.03
Salinity							1	0.05

Climate Characteristics

	Tmin	Tmax	Rainfall	Net Radiation	Weights
Tmin	1	1/2	2	3	0.28
Tmax		1	3	4	0.47
Rainfall			1	2	0.16
Net Radiation				1	0.10

Final Comparison for Criteria

	Climate	Topography	Soil	Weights
Climate	1	2[a]	2	0.50
Topography		1	1	0.25
Soil			1	0.25

[a] 2 represents that climate is twice as important as topography with regard to the suitability.

and constraints, such as weighted linear combination (WLC) (Eq. 3.1). After the weights are estimated, the MCE (Multi-Criteria Evaluation) module is used to combine the factors and constraints in the form of a WLC. The final suitability index (SI) is computed as follows:

$$\text{Suitability index (SI)} = \sum_{i=1}^{n} w_i \cdot \mu_i(x) \qquad (3.1)$$

where w_i is the weight of the factor i, μ_i is the membership grade for factor I, and n is the number of factors. The value of SI is between 0 and 1, where 0 represents not suitable and 1 indicates most suitable.

3.2.8 GENERATING LAND SUITABILITY MAPS

A DEM was obtained from SRTM 30 m data, and slope and elevation were obtained from it. The soil parameter was extracted and scaled from 0 to 1 using fuzzy criteria. The distribution of soil,

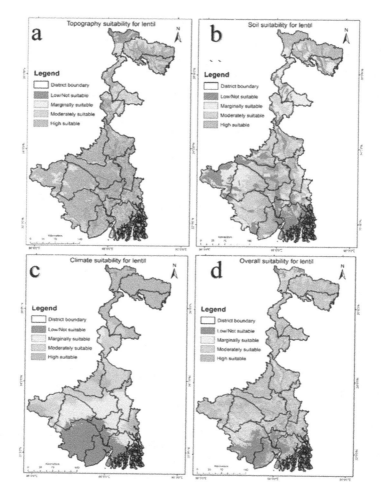

FIGURE 3.3 Suitability maps: (a) topographic, (b) soil, (c) climate, and (d) overall lentil.

climate, topography, and overall suitability map is represented in Figure 3.3a–d. The overall SI was evaluated through a statistical approach, and the data were found to have a mean of 0.684, standard deviation of 0.13, minimum of 0.031, maximum of 0.934, quartile 3 of 0.796, and quartile 1 of 0.665. The suitability classification was based on the FAO framework (FAO, 1976), where S1 is highly suited without any limitations; S2 is moderately suitable with certain limitations; S3 is marginally suited with severe limitations; and N is unsuited for lentil cultivation in order to plant other crops. The classification of suitability was allocated on the basis of classification and data statistics as represented in Table 3.3.

3.3 RESULTS AND DISCUSSION

This study took many parameters into account in determining the suitability of lentil cultivation in West Bengal state. Three main criteria (topography, soil, and climate) were considered with a number of sub-parameters. A pair-wise comparison matrix was employed to derive the weights following the AHP technique. The significance of individual parameters was denoted by the weights for lentil suitability mapping and utilized in MCE. The topographic suitability was generated using several factors, i.e., elevation, slope, drainage, and flooding (Figure 3.3a). The distribution of area

TABLE 3.3
Suitability Areas and Their Distribution

Sl. No.	Suitability	Area (ha)	Area (%)
1	**Topography suitability**		
	Highly suitable	3217.08	38.08
	Moderately suitable	1008.11	11.93
	Marginally suitable	2167.40	25.66
	Low/not suitable	2054.56	24.32
2	**Soil suitability**		
	Highly suitable	2125.50	23.91
	Moderately suitable	2177.39	24.50
	Marginally suitable	2493.04	28.05
	Low/not suitable	2093.16	23.55
3	**Climate suitability**		
	Highly suitable	2102.91	24.88
	Moderately suitable	2150.22	25.44
	Marginally suitable	2088.74	24.71
	Low/not suitable	2109.94	24.96
4	**Overall suitability**		
	Highly suitable	1845.98	21.84
	Moderately suitable	3453.50	40.86
	Marginally suitable	1607.89	19.02
	Low/not suitable	1544.87	18.28

showed that 50.32% was in the range of highly and moderately suitable for lentil production, and 24.96% was unsuitable (Table 3.3). Slope, with high membership 0.43, was the most effective topographic factor, while elevation, with membership 0.16, was the least effective factor (Figure 3.3a). The highly suitable districts were identified in the western part of the study area as Purulia, Bankura, Paschim Medinipur, Bardhaman, and Birbhum, whereas moderately suitable areas were identified for Nadia, Malda, Murshidabad, Uttar Dinajpur, and Jalpaiguri. In contrast, marginally suitable areas were distributed in small pockets of Purba Medinipur, South and North 24 Parganas, Hoogly, and Cooch Behar district, whereas the highly topographic rugged district of Darjeeling was observed to be the least suitable.

The soil suitability map was generated using eight factors, i.e., texture, depth, moisture, pH, OC, CEC, salinity, and soil erosion (Figure 3.3b). The area distribution showed that 48.41% of the total area was highly or moderately suitable for lentil production, and 23.55% was unsuitable (Table 3.3). The highly and moderately suitable areas were mostly observed in the central southern part of the study area. Such areas were seen for Purba Medinipur, Paschim Medinipur, Hoogly, Bardhaman, Birbhum, North 24 Parganas, Nadia, Murshidabad, Jalpaiguri, and Darjeeling district. Marginally and least suitable areas were mostly observed for a few parts of Paschim Medinipur, South 24 Paraganas, Malda, and Cooch Bihar, and the dominant regions of Howrah, Purulia, Uttar Dinajpur, and Dakshin Dinajpur. Climatic parameters such as minimum and maximum temperature, rainfall, and net radiation were used to determine the climate suitability for the lentil production. The climate-dependent land suitability map showed that 50.32% of the total area was highly or moderately suitable for lentil production (Table 3.3), and 24.06% was unsuitable. The highly and moderately suitable areas were observed in the central and northern regions, respectively. On the other hand, the southern region was relatively unsuitable.

The maximum contribution was assigned to climate (0.5) followed by equal contributions for topography (0.25) and soil (0.25) (Table 3.2). Within climate, the dominant influencing factor was maximum temperature (0.47) followed by minimum temperature (0.28), rainfall (0.16), and incoming solar radiation (0.10). In topographic data, the major influence was assigned to slope (0.43) followed by drainage density (0.32), elevation (0.16), and flooding (0.09). The OC (0.47) was the dominant factor considered for lentil production, followed by soil depth (0.23), moisture (0.16), and texture (0.11). The weight was estimated based on past literature on crop suitability, Sensitivity analysis is used to determine the level of importance of each criterion and therefore attempts to reduce the subjectivity of weights (Elsheikh et al., 2013; Olaniyi et al., 2015).

The overall lentil suitability was estimated by integrating 16 factors. The weight of each factor was calculated by AHP, and a WLC model was utilized for the final land SI for lentil production. The highly suitable areas for lentil production were identified for the districts in the central, northern, and eastern regions, while the least suitable areas were estimated for the districts in the southern region. Districts such as Purba Medinipur, Paschim Medinipur, South 24 Parganas, and Howrah were mostly identified as least suitable. The most suitable areas were identified for Birbhum, followed by Jalpaiguri, Nadia, Purulia, and Bardhhaman. However, parts of the Hoogly and Darjeeling districts were identified under the least suitable zones. The area distribution showed that 61% of the study area was suitable for growing lentils, in the range from most suitable (21%) to moderately suitable (40%), followed by marginal (19%) to least suitable (18%). The results are in accordance with the actual production observed in the region; i.e., major lentil planting areas were found in highly and moderately suitable zones. A similar study, land suitability for lentils based on soil nutrients, was done for the Hooghly district of West Bengal and found that nearly 67% was suitable for growing lentils, in the range from most suitable (12.4%) to moderately suitable (54.6%), followed by substantially unsuitable (30%) to least suitable (3%) (Singha, et al., 2018). Similarly, agricultural land suitability was determined using almost the same techniques in Malda district, and 50.13% of the area was estimated to be in the range of highly suitable to moderate suitability, followed by marginal suitability (34.58%) to limited suitability (15.83%) (Mistri & Sengupta, 2020).

3.4 CONCLUSION

A lentil suitability map was produced using the fuzzy, AHP, MCDM, and GIS technologies in West Bengal state. The fuzzy set offers an outstanding framework for translating numerical data of different magnitudes into membership feature grades in the range of 0–1, where 1 represents 100% of the suitable property. AHP is an efficient way of systematically assessing the weight of various variables and monitors the existence of competing parameters by calculated consistency. The fusion of fuzzy and AHP methods with GIS ensures an accurate combination to apply to an analysis of land suitability. Following the previous literature, the maximum influence was assigned to climate, followed by topography and soil characteristics. The dominant contribution was assigned to maximum temperature within the climate group, OC within the soil group, and slope within the topographic group. The result showed that nearly 61% of the total area is suitable for lentil production, 21% most suitable and 40% moderately suitable, followed by marginal (19%) and least suitable (18%). The highly and moderately suitable areas were observed in the central, northern, and western districts: Birbhum, Jalpaiguri, Nadia, Bardhhaman, Purulia Bankura, and Cooch Bihar. The least suitable areas were observed in Purba and Paschim Medinipur. The results provide useful methods to improve the productivity of land use and the management of lentil production in West Bengal state. The output maps and generated data would be useful for agriculture and water resource managers to prescribe appropriate management plans and develop suitable policies for improved land utilization and agricultural productivity.

REFERENCES

Akinci, H., Özalp, A. Y., & Turgut, B. (2013). Agricultural land use suitability analysis using GIS and AHP technique. *Computers and Electronics in Agriculture*. https://doi.org/10.1016/j.compag.2013.07.006.

Aktar, N. (2015). Agricultural productivity and productivity regions in West Bengal. *The NEHU Journal, XIII*(2), 49–61.

Ali, M., & Kumar, S. (2009). Major technological advances in pulses: Indian scenario. In *Milestones in Food Legumes Research*, eds. Masood Ali and Shiv Kumar. Indian Institute of Pulses Research, Kanpur, India, 1–20.

Annal Report. (2018). Annual report of 2017–18. Directorate of Pulses Development, Vindhyachal Bhavan, Government of India. Accessed 19 October 2020. http://dpd.gov.in/Annual%20Report%202017-18.pdf.

ASAG. (2018). Agricultural statistics at a glance 2018. Directorate of Economics and Statistics, Government of India. Accessed 10 August 2020. http://agricoop.gov.in/sites/default/files/agristatglance2018.pdf.

Behera, M. D., Biradar, C., Das, P., & Chowdary, V. M. (2019). Developing quantifiable approaches for delineating suitable options for irrigating fallow areas during dry season – a case study from Eastern India. *Environmental Monitoring and Assessment*. https://doi.org/10.1007/s10661-019-7697-4.

Bera, S., Ahmad, M., & Suman, S. (2017). Land suitability analysis for agricultural crop using remote sensing and GIS – a case study of Purulia district. *IJSRD-International Journal for Scientific Research & Development, 5*(6), 999–1004.

Burrough, P. A., Macmillan, R. A., & van Deursen, W. (1992). Fuzzy classification methods for determining land suitability from soil profile observations and topography. *Journal of Soil Science*. https://doi.org/10.1111/j.1365-2389.1992.tb00129.x.

Chowdary, V. M., Chakraborthy, D., Jeyaram, A., Murthy, Y. V. N. K., Sharma, J. R., & Dadhwal, V. K. (2013). Multi-criteria decision making approach for watershed prioritization using analytic hierarchy process technique and GIS. *Water Resources Management*. https://doi.org/10.1007/s11269-013-0364-6.

Elsheikh, R., Mohamed Shariff, A. R. B., Amiri, F., Ahmad, N. B., Balasundram, S. K., & Soom, M. A. M. (2013). Agriculture Land Suitability Evaluator (ALSE): a decision and planning support tool for tropical and subtropical crops. *Computers and Electronics in Agriculture, 93*, 98–110. https://doi.org/10.1016/j.compag.2013.02.003

FAO. (1976). A framework for land evaluation. FAO Soils Bulletin No. 32. In *A Framework for Land Evaluation*, FAO and Agriculture Organization of the United Nations, Rome, ISBN 92-5-100111-1.

Jamil, M., Ahmed, R., & Sajjad, H. (2018). Land suitability assessment for sugarcane cultivation in Bijnor district, India using geographic information system and fuzzy analytical hierarchy process. *GeoJournal*. https://doi.org/10.1007/s10708-017-9788-5.

Kar, G., & Kumar, A. (2009). Evaluation of post-rainy season crops with residual soil moisture and different tillage methods in rice fallow of eastern India. *Agricultural Water Management*. https://doi.org/10.1016/j.agwat.2009.01.002.

Malczewski, J. (2004). GIS-based land-use suitability analysis: a critical overview. *Progress in Planning*. https://doi.org/10.1016/j.progress.2003.09.002.

Mistri, P., & Sengupta, S. (2020). Multi-criteria decision-making approaches to agricultural land suitability classification of Malda District, Eastern India. *Natural Resources Research*. https://doi.org/10.1007/s11053-019-09556-8.

Mustafa, A. A., Singh, M., Sahoo, R. N., Ahmed, N., Khanna, M., & Sarangi, A. (2011). Land suitability analysis for different crops: a multi criteria decision making approach using remote sensing and GIS. *Water Technology, 3*(12), 1–84.

Nisar Ahamed, T. R., Gopal Rao, K., & Murthy, J. S. R. (2000). GIS-based fuzzy membership model for cropland suitability analysis. *Agricultural Systems*. https://doi.org/10.1016/S0308-521X(99)00036-0

Olaniyi, A. O., Ajiboye, A. J., Abdullah, A. M., Ramli, M. F., & Sood, A. M. (2015). Agricultural land use suitability assessment in Malaysia. *Bulgarian Journal of Agricultural Science*.

Pandey, V., & Srivastava, P. K. (2019). Integration of microwave and optical/infrared derived datasets for a drought hazard inventory in a sub-tropical region of India. *Remote Sensing, 11*(4). https://doi.org/10.3390/rs11040439.

Paroda, R. S., Woodhead, T., & Singh, R. B. (1994). *Sustainability of Rice-Wheat Production Systems in Asia*. RAPA Publication (FAO).

Pramanik, M. K. (2016). Site suitability analysis for agricultural land use of Darjeeling district using AHP and GIS techniques. *Modeling Earth Systems and Environment*. https://doi.org/10.1007/s40808-016-0116-8.

PRFNS. (2018). Pulses revolution from food to nutritional security, Ministry of Agriculture & Farmers Welfare, Government of India. Accessed 10 August 2020. https://farmer.gov.in/SucessReport2018-19.pdf.

Reichle, R. H., Ardizzone, J. V, Kim, G.-K., Luchesi, R. A., Smith, E. B., & Weiss, B. H. (2015). Soil moisture active passive (SMAP) mission level 4 surface and root zone soil moisture (L4_SM) product specification document. *GMAO Office Note*.

Saaty, T. L., & Vargas, L. G. (2012). *Prediction, Projection and Forecasting: Applications of the Analytic Hierarchy Process in Economics.* Springer, Netherlands.

Saaty, Thomas, & Process, A. H. (1980). *The Analytical Hierarchy Process.pdf. Priority Setting. Resource Allocation.* MacGraw-Hill, New York International Book Company. https://doi.org/10.1002/jqs.593.

Singha, C., & Swain, K. C. (2018). Soil profile based land suitability study for jute and lentil using AHP ranking. *International Journal of Bio-Resource and Stress Management.* https://doi.org/10.23910/ijbsm/2018.9.3.1869.

Singha, C., Swain, K. C., & Swain, S. K. (2020). Best crop rotation selection with GIS-AHP technique using soil nutrient variability. *Agriculture (Switzerland).* https://doi.org/10.3390/agriculture10060213.

Subbarao, G. V. (2001). *Spatial Distribution and Quantification of Rice-Fallows in South Asia: Potential for Legumes.* Andhrapradesh, India: International Crops Research Institute for the Semi-Aric-I Tropics (ICRISAT), ISBN 92-9066-436-3.

Varghese, N., Dogra, A., Sarker, A., & Hassan, A. A. (2018). The lentil economy in India. In *Lentils: Potential Resources for Enhancing Genetic Gains.* https://doi.org/10.1016/B978-0-12-813522-8.00009-1.

Zadeh, L. A. (1965). Fuzzy sets. *Information and Control.* https://doi.org/10.1016/S0019-9958(65)90241-X.

Zhang, G., Xiao, X., Biradar, C. M., Dong, J., Qin, Y., Menarguez, M. A., Zhou, Y., Zhang, Y., Jin, C., Wang, J., Doughty, R. B., Ding, M., & Moore, B. (2017). Spatiotemporal patterns of paddy rice croplands in China and India from 2000 to 2015. *Science of the Total Environment.* https://doi.org/10.1016/j.scitotenv.2016.10.223.

4 Assessing Desertification Using Long-Term MODIS and Rainfall Data in Himachal Pradesh (India)

Ravi Teja, N. R. Patel, and Arnab Kundu

CONTENTS

4.1 Introduction 33
4.2 Study Area 35
4.3 Materials and Methods 36
    4.3.1 Methodology 37
    4.3.2 MODIS NDVI Analysis 37
    4.3.3 RUE 37
    4.3.4 Residual Image 38
    4.3.5 Statistical Significance Test 38
    4.3.6 Significance Level 39
4.4 Results and Discussion 39
    4.4.1 Relationship between NDVI and Rainfall 39
    4.4.2 RUE 39
    4.4.3 RUE Trend 39
    4.4.4 Residual Trend and Human-Induced Desertification 40
    4.4.5 Regression Slope between Residual NDVI and Time (Year) 40
    4.4.6 Comparison between RUE and RESTREND 43
4.5 Conclusions 44
4.6 Scope for Further Research 45
Bibliography 45

4.1 INTRODUCTION

Desertification is characterized by the United Nations (UNCCD, 1994) as "land degradation in arid, semiarid and sub-humid areas, resulting from various factors including climatic variations and human activities". Desertification and land degradation in weaker zones with dry and semi-dry climate are brought about by natural and human-prompted procedures (Dupre, 1990), influencing an enormous aspect of the Earth's surface. Climate-originated desertification is unpreventable and tough to control. Nevertheless, it is possible to reduce human-induced desertification. This needs information about areas where desertification and the vegetation degradation process are driven by anthropogenic factors. The present study attempts to investigate the areas where human-induced factors are resulting in desertification. The regions influenced by desertification incorporate the most extreme classification of deserts, known as the cold deserts, lying generally in the mountainous regions of the world, for instance, Himachal Pradesh; the various desert areas in Lahul and Spiti district and Kinnaur district have an area of 7589 and 3400 km², respectively. Various investigations have indicated that the apparent desertification can largely be ascribed to variations in rainfall instead of human-induced land degradation (Tucker et al., 1991; Nicholson et al., 1998; Prince et al., 1998; Anyamba and Tucker, 2005; Kundu, 2010). Since the International Convention on Desertification of the United Nations came into force in 1996, the need to quantify land degradation and desertification measures has significantly expanded. While standard ground evaluation strategies for undertaking such estimations are ineffective or costly, it is generally considered that satellite and airborne remote sensing frameworks offer impressive potential. Satellite observations of land, sea, and air, particularly during natural and human-induced hazards, have become vital for protecting the worldwide climate, reducing disasters, and accomplishing good results (Navalgund et al., 2007). Earth observation satellites offer noteworthy contributions to the evaluation and monitoring of desertification, especially by providing the spatial facts necessary for local-scale examinations of the relations between environmental change, land degradation, and desertification activities. Several studies reveal that the integration of remote sensing and non-remote sensing data in geographic information systems (GIS) can be a useful tool for analyzing desertification processes. Di (2003) added a thorough examination of the modern evolution of remote sensing for desertification at both the local and the worldwide level. This assessment furthermore highlighted the example and basic headway in the field of advancement. Hanan et al. (1991) noted that remote sensing has been effectively applied to the estimation of desert extension and to the appraisal of components that cause desertification. Remote sensing systems have been discovered as a significant tool in desertification measures by various researchers (Kundu, 2010; Kundu and Dutta, 2011; Kundu et al., 2015, 2017). Hill et al. (2008) examined the utilization of traditional methods to monitor desertification and land degradation in the Mediterranean area, adopting a disorder-based strategy to understanding the directions of various locales. Meanwhile, Hellden and Tottrup (2008) applied such strategies in wide zones subject to desertification monitoring, from Senegal to Mongolia. The majority of past studies have emphasized the need for continuous, long-term satellite sensing to capture the high inter-annual variation of dry-land environment services and to recognize the role of human activities and climate variation in vegetation efficiency.

A wide range of geospatial approaches has emerged over the past few decades, which intricately measures desertification by estimating changes of net primary productivity (NPP) over a long time. Earth observation satellite information was effectively utilized to monitor the extent of green vegetation (as a proxy for NPP) over periods of several decades, which provided recognizable proof of degraded vegetation assets on regional scales. These methodologies have been supplemented with extra components of Mann-Kendall (MK) time series examination and applied at local scales dependent on persistent multi-yearly time series from worldwide environmental observing satellites (e.g., Moderate Resolution Imaging Spectroradiometer (MODIS) and National Oceanic and Atmospheric Administration [NOAA]). Recently, efforts were made worldwide to analyze human-induced land degradation or desertification by taking into account the long-term trend in rain-use efficiency (RUE) (Prince et al., 1998; Wessels et al., 2007; Kundu, 2010; Kundu et al., 2017) and the residual of the vegetation–climate association (Evans and Geerken, 2004; Hermann et al., 2005; Cao et al., 2006; Kundu, 2010; Kundu et al., 2017) over the spatial domain.

As per Nicholson et al. (1990) and Davenport and Nicholson (1993), the $\sum$ normalized difference vegetation index ($\sum$NDVI)/rainfall proportion is viewed as a helpful intermediary for RUE in dry areas with a yearly rainfall of 200 to 1000 mm. Similarly, Symeonakis et al. (2003) employed the RUE technique to evaluate landscape degradation in dry land with the assistance of time-integrated MODIS NDVI and rainfall data. In their examination, Bai et al. (2005) found that qualities for RUE determined from NDVI, which are easy to acquire, correlate with those determined from estimations of NPP, which are difficult to get. RUE is a proportion among NPP and NDVI with rainfall (NPP/rainfall or $\sum$NDVI/Rainfall). Wessels et al. (2007) utilized this approach to distinguish the human-initiated and climate-incited land degradation on the premise of NDVI examination. RUE is a local indicator of productivity and land degradation, since it may very well be obtained from remote sensing (for example, NDVI) and rainfall information (Nicholson et al., 1998; Kundu, 2010; Kundu et al., 2017).

Assessing Desertification

Herrmann et al. (2005) showed the trends of vegetation dynamics and their relationship to climate using the residual trend (RESTREND) technique. Wessels et al. (2007) applied this RESTREND technique to show the negative patterns in the contrasts between the observed $\sum$NDVI and the $\sum$NDVI predicted by the rainfall, utilizing regressions considered for every pixel. They demonstrated that, accounted for by precipitation, the RESTREND technique recognized regions with negative residual patterns that had positive $\sum$NDVI patterns. These zones had lower $\sum$NDVI than predicted by the rainfall –$\sum$NDVI association and in this manner, may have faced a reduction in the production per unit rainfall. The negative trend means that the desertification resulted from human activities. In the Global Assessment of Land Degradation and Improvement (GLADA) Report, Bai and Dent (2008) also used the RESTREND method to identify land degradation and progress in Argentina using remote sensing. In their study, Kundu et al. (2017) described desertification over the western part of Rajasthan (India) using RESTREND based on remote sensing. Another study, by Matin et al. (2020), revealed the human-induced land degradation and transformations using the RESTREND tool in the western part of the Ganga river basin (India).

Himachal Pradesh is a distinctive overlap zone of farming and animal husbandry. This region has noticeable desertification features and a multifaceted driving component, typically like the other cold Himalayan deserts. This study envisages the use of publicly available coarse but more advanced long-term time series of satellite-based vegetation indices from MODIS to understand vegetation dynamics and the resultant desertification process.

4.2 STUDY AREA

Himachal Pradesh is completely rugged, with elevations extending from 350 to 6975 m over the mean sea level. Its area is between 30°22′40″ N and 33°12′40″ N latitude and between 75°45′55″E and 79°04′20″ E longitude (Figure 4.1). The total area is 55,673 km². The area under forest is 37,033 km². The total population is 6.8 million (Census of India, 2011). The density of the population is 123 per square kilometer. Being a sub-Himalayan state, it has a varying climate, which changes with elevation (350–6500 m). It has four climatic zones. The Shivalik hill zone climate shifts from hot and sub-humid tropical (350–650 m) in the southern low tracts; it involves about 30% of the geographical zone and about 55% of the cultivated region of the state. The mid hill zone

FIGURE 4.1 Agro-climatic zones of Himachal Pradesh.

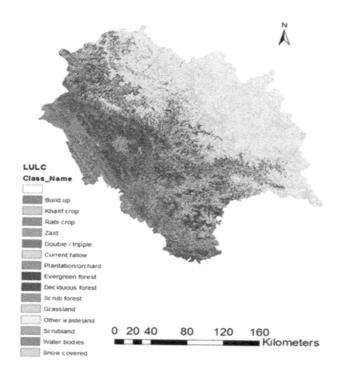

FIGURE 4.2 LULC classification of Himachal Pradesh.

ranges from 651 to 1800 m above mean sea level, having a gentle, calm climate. It possesses about 10% of the total topographical region and about 30% of the cultivated zones of the state. The high hill zone lies from 1801 to 2200 m above sea level with a humid, calm climate and elevated fields. This area covers about 25% of the geographical regions and about 10% of the cultivated zone of the state. The cold dry zone includes Lahul and Spiti and Kinnaur districts and Pangi tehsil of Chamba district, lying around 2200 m above sea level. It involves about 35% of the geographical and 5% of the entire cultivated region of the state. The average temperature in summer ranges between 28 and 32 °C. The rainy season starts toward the end of June. Substantial downpours happen in July and August. The normal rainfall is 1278 mm. Out of all the areas, Dharamsala in Kangra gets the highest rainfall: 3400 mm. Spiti is the driest region of the state (rainfall less than 50 mm).

The explanation is that it is encased by high mountains and dense snowfall on all sides. The rivers that course through this state are Beas, Yamuna, Chenab (Chander Bhaga), Sutlej, Raavi, and Parbati. A number of significant lakes are also located here. All these channels are snow-occupied and perennial. Moreover, the natural lakes and enormous drops accessible in the river courses have massive potential for hydroelectric power generation, reducing costs (Figure 4.2).

4.3 MATERIALS AND METHODS

Data were acquired from two sources: first, MODIS NDVI data and second, NOAA Climate Prediction Center (CPC) Rainfall data. The data were taken from the MODIS onboard Terra: 12 years (2001–2012) of 250 m 16-day MODIS NDVI maximum value composites over Himachal Pradesh. On the side of weather and flood analysis actions at the U.S. Agency for International Development (USAID) and the United States Geological Survey (USGS), another framework was formed and placed into activity at the CPC of NOAA to deliver continuous investigations of day-by-day rainfall with $0.5° \times 0.5°$ resolution over south Asia (70°E–110°E; 5°N–35°N). The

Assessing Desertification

pre-processing of NDVI and CPC rainfall must be done before re-projecting into Albers Conical Equal Area projection with WGS-84 datum. This projection was picked to maintain a comparable pixel size regarding NDVI.

4.3.1 Methodology

A flow diagram for the overall methodology is shown in Figure 4.3.

4.3.2 MODIS NDVI Analysis

It has often been seen that human actions accelerate desertification cycles, and desertification also influences people. If the effect of the human-induced desertification process can be distinguished from climate-induced desertification, the areas of human-induced desertification can be identified. For this purpose, two methods, RUE and RESTREND, were performed (Figure 4.4).

4.3.3 RUE

$$\text{RUE} = \sum \text{NDVI}/R_{total} \quad (4.1)$$

where ΣNDVI is the annual integrated NDVI obtained from 16-day MODIS NDVI composites and R_{total} is an annual total of daily CPC-derived rainfall. To obtain ΣNDVI, first, all pre-processed NDVI images (2001–2012) were normalized with this [(NDVI × 100) + 100] formula. At that point, all normalized NDVI images were stacked together. The fortnights of the growing season

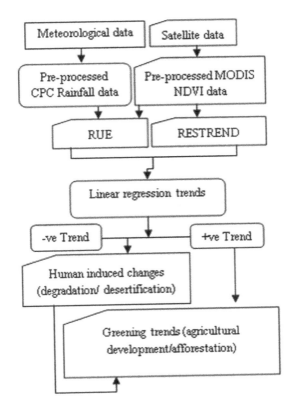

FIGURE 4.3 Flowchart of methodology.

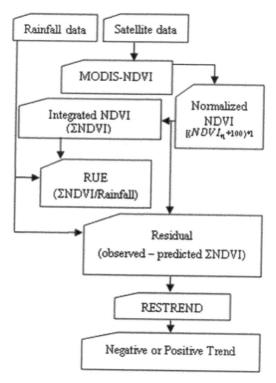

FIGURE 4.4 Flowchart of vegetation trend.

(June–October) were taken into account in this study. After stacking, the fortnights of the growing season were integrated using the integrated NDVI equation: [Integrated NDVI = $16/2(\text{NDVI}t_1 + 2 \times \text{NDVI}t_2 + 2 \times \text{NDVI}t_3 \ldots + 2 \times \text{NDVI}t_n)$]. All layers of rainfall were stacked and afterward subset for the study region. Finally, the RUE equation was applied and an image was generated (Eq. 4.1).

4.3.4 Residual Image

The residual trend means the long-term trend of residual NDVI. Residual NDVI denotes contrasts between observed and predicted NDVI, which depend upon the rainfall. The predicted NDVI was determined by a regression model, which depends on the relationship between growing season NDVI and rainfall. The predicted NDVI shows that the climatic effect on growing season NDVI through observed NDVI is a combination of both climate and anthropogenic components. When the predicted NDVI is deducted from the observed NDVI, the residual value represents the effect of anthropogenic factors on NDVI. The relationship between residual NDVI and time (year) shows the pattern of residual NDVI. This analysis of trend identification was carried out using MK time series analysis and Sen's slope estimator. If the slope of the trend is negative, at that point it shows loss of vegetation because of anthropogenic causes, and a positive slope in residual trend means expansion in vegetation growth because of technical advancement, soil preservation, afforestation, and so on.

4.3.5 Statistical Significance Test

The starting point of a statistical test is to define a null hypothesis (H_0) and an alternative hypothesis (H_1). For example, to test for a trend in a time series, H_0 would be that there is no trend in the data, and H_1 would be that there is an increasing or decreasing trend. The test statistic is a means of comparing H_0 and H_1. It is a numerical value computed from the data series that is being tested.

Assessing Desertification

4.3.6 Significance Level

The significance level means measuring whether the test statistic is very different from the (critical) values that would typically occur under H_0. For example, for $\alpha = 0.05$, the critical test statistic value is the value that would be exceeded by 5% of test statistic values obtained from randomly generated data. If the test statistic value is greater than the critical test statistic value, H_0 is rejected. A possible interpretation of the significance level might be: (1) $\alpha > 0.1$ – little evidence against H_0; (2) $0.05 < \alpha < 0.1$ – possible evidence against H_0; (3) $0.01 < \alpha < 0.05$ – strong evidence against H_0; (4) $\alpha < 0.01$ – very strong evidence against H_0. Therefore, the critical test statistic values for significance levels of $\alpha = 0.1$, $\alpha = 0.05$, and $\alpha = 0.01$ are the 90th, 95th, and 99th percentile values, respectively, of test statistic values from the generated (resampled) time series.

4.4 RESULTS AND DISCUSSION

4.4.1 Relationship between NDVI and Rainfall

The NDVI of the growing season is completely reliant on the volume of rainfall received. In the present study, growing period NDVI and rainfall have been used for monitoring the desertification process. Since desertification is a long-term process, past long-term data from 12 years (2001–2012) were used in this study to show the gradual changes in vegetation. Two methods, RUE and RESTREND, were used for monitoring desertification in this area. Both these methods are based on long-term CPC rainfall and NDVI.

4.4.2 RUE

The ratio between NDVI and rainfall was calculated for each year. Since there is an incredible inconsistency in rainfall from year to year, just looking at one year cannot achieve the expected outcome in assessing desertification. That is the reason why time series RUE was done to screen for long-term vegetation degradation. The MK test was used to perform the trend analysis, and Sen's slope estimator was used to find the slopes of RUE and RESTREND (Figure 4.5).

The slope of the study area has been sub-divided into five classes indicating emphatically negative to unequivocally positive patterns. It can be seen that in the southwestern part of the study region, the slope tends to be positive, and it diminishes towards the northeastern side. The slope of RUE essentially relies on the state of integrated NDVI and rainfall. A negative slope of RUE shows the degradation of vegetation, and a positive slope shows the opposite. The positive pattern on the southwestern side demonstrates that the NDVI of this zone is due to an endless supply of rainfall during the growing season. Though the central and southern parts receive maximum rainfall, the relation observed between RUE and time (year) does not show a strong trend. The negative slope of this region depicts that RUE is not significant with time. The basic cause for this is the high snowfall of this region, resulting in degradation of vegetation or vegetation fully covered by snow, and some human-induced desertification resulting from various anthropogenic factors (Figure 4.6).

4.4.3 RUE Trend

It was observed from the RUE trend that some parts of Kangra and Kinnaur are showing negative trends, while most parts of Lahul and Spiti are showing strong negative trends. On the other hand, the trend has become strongly positive in the central part of the state; most parts of Kangra, Una, Hamirpur, Mandi, Bilaspur, and Sirmaur and some parts of Solan, Chamba, and Shimla are showing positive trends (Figure 4.7).

This indicates the effect of high rainfall, due to which RUE has increased with time. A stable trend in RUE was observed in the maximum area, which denotes no significant change with time.

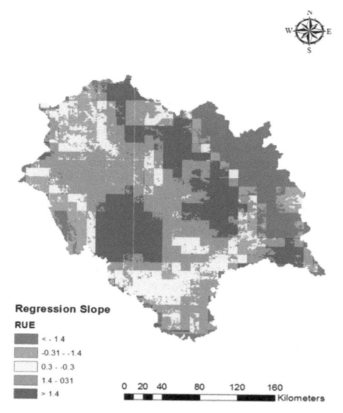

FIGURE 4.5 RUE slope map.

4.4.4 Residual Trend and Human-Induced Desertification

The residual trend has been utilized by numerous researchers for checking human-caused desertification. The residuals represent the distinction between observed and predicted NDVI. The predicted NDVI was determined utilizing linear regression between rainfall and NDVI. The relationship between NDVI and rainfall was investigated by a significance test. The Student's t-test was carried out for the test of significance. A p-level >0.1 was taken as a "not significant" class. Most regions of this area fall into the class with p-level <0.01; just a few areas on the western and southern sides are demonstrating p-level <0.1. This suggests degradation is comparatively less (Figure 4.8).

Because of the regression conditions, the integrated NDVI for every year was predicted. The predicted NDVI shows that the NDVI that should be seen in a specific year is an effect of rainfall. So, the trend of predicted NDVI speaks to the climate instigated biomass creation. When it was deducted from observed NDVI, the resultant residuals of every year show the human-initiated biomass production as far as residual NDVI. MK pattern investigation was performed to show the pattern in NDVI residual. The pattern from the year 2001–2012 was assessed, considering the residual NDVI as a dependent variable and time (year) as an independent variable.

4.4.5 Regression Slope between Residual NDVI and Time (Year)

The regression slope of the residual trend is ordered by the degree of the slope, regardless of whether it is positive or negative and how strong it is. The maximum area of this region is described by

Assessing Desertification

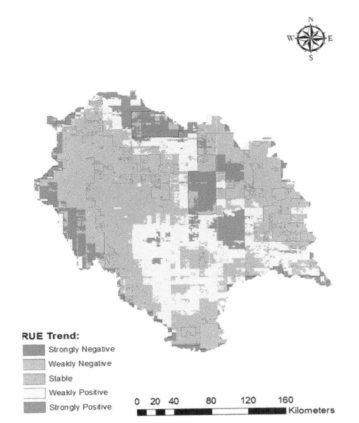

FIGURE 4.6 RUE trend.

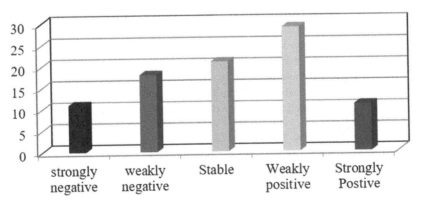

FIGURE 4.7 Slope (RUE).

a strongly positive slope where the contrast between observed and predicted NDVI is near zero. This shows that there is no anthropogenic action. The southern part of Chamba, the central part of Kangra, the northwestern part of Mandi, the northeastern part of Hamirpur, the northern part of Bilaspur, some parts of Kullu, Kinnaur, and Sirmaur show a strongly negative slope. That indicates the occurrence of human-induced desertification. Desertification, cutting of wood for fuel, overgrazing, and so on are the major variables that set off anthropogenic vegetation degradation. On the other hand, some parts of Kangra, most parts of Lahul and Spiti, and Kinnaur are showing a strong

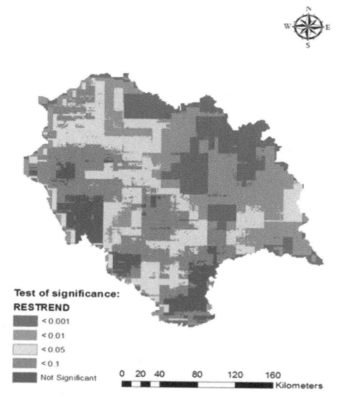

FIGURE 4.8 Rainfall and NDVI significance test.

positive slope. Here also human activities took place, but some other effective factors play a key role here, i.e., lower population, less demand, less overgrazing, high snowfall, maximum area covered by snow throughout 365 days, etc. Some parts of Kangra are showing a positive slope, i.e., anthropogenic activities are overshadowed by technological progress in agriculture, forestation, plantation, etc., helping to increase green vegetation. The positive slope in the central part of Kangra represents the effect of the lake situated nearby on agriculture. The residual slope shows that climatic influences have not played a major role in vegetation degradation of the central-western part of the study area (Figure 4.9).

It was discovered that roughly 11% of the total study area has undergone profoundly human-inflicted desertification. Around 28% of the study region is encountering moderate human-caused desertification (Figure 4.10).

Then again, human-induced positive changes in vegetation have resulted in 26% of the area being weakly positive, and due to some natural climatic effects, 20% of the area is showing strongly positive. The remaining zones (15%) are demonstrating a stable condition or no anthropogenic effects on vegetation (Figure 4.11).

Figure 4.12 shows the year-wise residual NDVI through the integrated observed and predicted NDVI. Two specific regions have been chosen to show the residual. The first shows a positive residual trend. The observed NDVI is indicating a growing trend from 2001 to 2012. Between the years 2008 and 2011, the observed NDVI has abruptly increased. There may be some human-induced growth, such as a plantation, irrigation, or agricultural development, that has increased the NDVI growth. The line graph of the observed NDVI shows a positive trend. On the other hand, the predicted NDVI is showing an unstable condition (Figure 4.12).

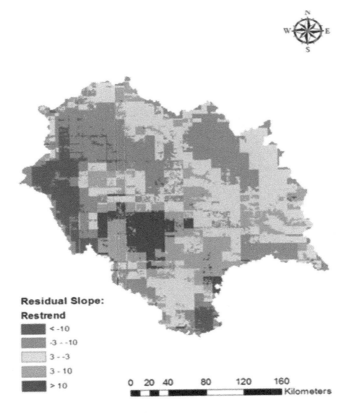

FIGURE 4.9 RESTREND slope map.

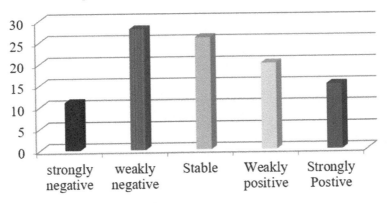

FIGURE 4.10 Regression slope (RESTREND).

4.4.6 COMPARISON BETWEEN RUE AND RESTREND

It was observed that the RUE trend and the residual trend are overall indicating different trends. The residual and RUE trends of an area with a positive trend is shown in Figure 4.13.

The residual trend was observed as being more strongly positive than RUE. The integrated growing season NDVI is also showing a stable and positive trend, in agreement with the RUE and residual trends. This denotes that there is no effect of desertification or human-induced desertification, but it may change its course in the future based upon the frequency of anthropogenic activities.

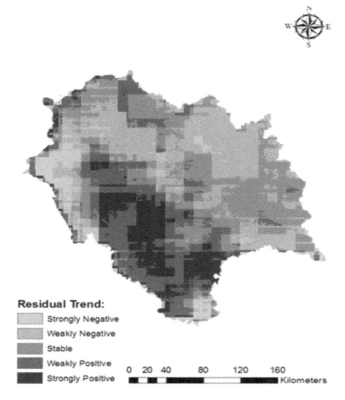

FIGURE 4.11 Residual trend.

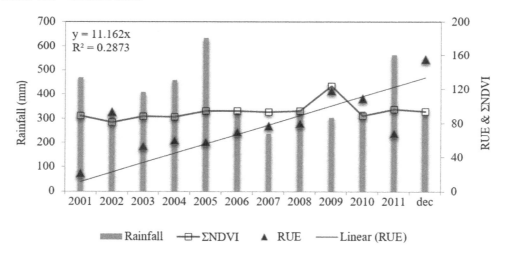

FIGURE 4.12 Rainfall, RUE, and ΣNDVI relationship.

4.5 CONCLUSIONS

Environmental change and human actions are the two principal driving components in desertification, and the determination of their respective roles is vital, both hypothetically and pragmatically, for preventing and controlling desertification. Our study initially advanced a methodology for finding the vegetation trends vis-à-vis climate change with integrated temporal and spatial relationships. It has been proposed that RUE, the proportion of NDVI to rainfall, can standardize the between-year

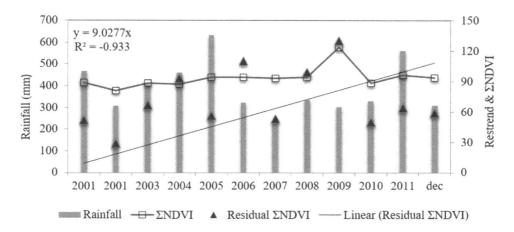

FIGURE 4.13 Rainfall, RESTREND, and ΣNDVI relationship.

variability in NDVI caused by rainfall fluctuation and can therefore provide a record of degradation that is free from the impact of rainfall (Nicholson et al., 1998; Prince et al., 1998). Field tests have indicated that degraded rangelands have decreased RUE (Le Houe'rou, 1984; O'Connor et al., 2001). Moreover, RUE has been proposed as a provincial marker of efficiency and land degradation, since it can easily be obtained from remote sensing evaluations (for example, NDVI) and rainfall information. Several studies have investigated human-initiated factors in desertification by the RUE technique, however, RUE is not an exact indicator, and its application is not very dependable. One of its apparent deficits is that the impacts of climate and human-initiated elements can, without much of a stretch, be confounded when they have a similar temporal propensity.

This study shows that the RESTREND approach is a helpful device for controlling the impacts of rainfall to identify human-instigated land degradation. It has demonstrated that accounting for rainfall, the RESTREND technique recognized regions with negative residual patterns that had positive $\sum$NDVI patterns. These regions had lower $\sum$NDVI values than anticipated by the rainfall–$\sum$NDVI relationship and in this manner may have encountered a decrease in the production per unit rainfall. The negative trends mean that the desertification occurred due to human activities. The RESTREND approach can distinguish regions where there has been a decrease per unit rainfall, yet the specific reason for the negative pattern, for example, overgrazing by domesticated animals, can't be controlled by this strategy alone. It is thus visualized that the RESTREND strategy could fill in as a territorial marker to recognize possibly degraded regions, which could then be examined more closely by utilizing high-resolution remote sensing information and field-based observation. Field reviews will remain a fundamental part of such an observing system, since they provide confirmatory information to the remote sensing items and can follow aspects of land degradation, for example, vegetation species change and soil disintegration, that are not distinguishable with satellite information.

4.6 SCOPE FOR FURTHER RESEARCH

1. Future research can attempt to study the monitoring of climate-induced desertification by aridity index according to data availability.
2. A study can be carried out on human-induced desertification risk mapping, incorporating the population element.

BIBLIOGRAPHY

Anyamba, A. and Tucker, C.J. (2005). Analysis of Sahelian vegetation dynamics using NOAA-AVHRR NDVI data from 1981 to 2003. *Journal of Arid Environments* 63(3), 596–614.

Anyamba, A., Tucker, C.J. and Mahoney, R. (2002). From El Niño to La Niña: vegetation response pattern over east and southern Africa during the 1997–2000 period. *Journal of Climate* 15(21), 3096–3103.

Bai, Z.G., Dent, D.L. (2008). Verification Report on the GLADA Land Degradation Study: Land Degradation and Improvement in South Africa. *Identification by Remote Sensing*. University of South Africa, D. J. Pretorius Department of Agriculture: Stellenbosch, South Africa.

Bai, Z.G., Dent, D.L. and Schaepman, M.E. (2005). Quantitative global assessment of land degradation and improvement: pilot study in North China. Report 2005/6, ISRIC – World Soil Information, Wageningen.

Boer, M. and Puigdefabregas, J. (2003). Predicting potential vegetation index values as a reference for the assessment and monitoring of dry-land condition. *International Journal of Remote Sensing* 24(5), 1135–1141.

Boer, M.M. and Puigdefabregas, J. (2005). Assessment of dry-land condition using remotely sensed anomalies of vegetation index values. *International Journal of Remote Sensing* 26 (18), 4045–4065.

Brief Facts of Himachal Pradesh (2012–13). Economics and statistics department Himachal Pradesh Official Statistics section of the directorate. Baseline survey on assessment of existing knowledge level, awareness and preventive practices of disaster management in Himachal Pradesh draft report Department of Economics & Statistics Government of Himachal Pradesh.

Cao, X., Gu, Z.H., Chen, J., Liu, J. and Shi, P. (2006). Analysis of human induced steppe degradation based on remote sensing in Xilin Cole, Inner Mongolia, China. *Acta Phytoecol Sinica* 30(2), 268–277 (in Chinese with English abstract).

Davenport, M.L. and Nicholson, S.E. (1993). On the relation between rainfall and the normalized difference vegetation index for diverse vegetation types in East Africa. *International Journal of Remote Sensing* 14(12), 2369–2389.

Del Barrio, G., Hill, J. and Peter, D., eds. (1996). The use of remote sensing for land degradation and desertification monitoring in the Mediterranean basin. EUR 16732 EN, Luxembourg: Office for Official Publications of the European Communities.

Di, L. (2003). Recent progress on remote sensing monitoring of desertification. *Annals of Arid Zone* 42(3).

Director Himalayan Forest Research Institute Conifer Campus, Panthaghati SHIMLA-171 009 (H.P.).

Dupre, M. (1990). Historical antecedents of desertification: climatic or anthropological factors? In J.L. Rubio and R.J. Rickson (eds.), Luxembourg: Commission of the European Communities, p. 2–39.

Evans, J.P. and Geerken, R. (2004). Discrimination between climate and human induced dry-land degradation. *Journal of Arid Environments* 57(4), 535–554.

Gabriels, D., Cornelis, W.M., Eyletters, M., and Hollebosch, P., (2008) (Eds.) *Combating Desertification Monitoring, Adaptation and Restoration Strategies*. UNESCO Chair of Eremology, Ghent University, Belgium. pp. 216. ISBN: 978-90-5989-271-2.

Goward, S.N. and Prince, S.D. (1995). Transient effects of climate on vegetation dynamics: satellite observations. *Journal of Biogeography* 22(2/3), 549–563.

Hanan, N.P., Prince, S.D. and Hiernaux, P.H.Y. (1991). Spectral modelling of multicomponent landscapes in the Sahel, *International Journal of Remote Sensing* 12(6), 1243–4259.

Hellden, U. (1991). Desertification time for an assessment. *Ambio* 20(8), 372–383.

Hellden, U. and Tottrup, C. (2008). Regional desertification: a global synthesis. Global Planetary Change 64(3–4), 169–176.

Herrmann, S.M., Anyamba, A. and Tucker, C.J. (2005). Recent trends in vegetation dynamics in the African Sahel and their relationship to climate. *Global Environmental Change* 15(4), 394–404.

Herrmann, S.M. and Hutchinson, C.F. (2005). The changing contexts of the desertification debate. *Journal of Arid Environments* 63(3), 538–555.

Hickler, T., Eklundh, L., Seaquist, J.W., et al. (2005). Precipitation controls Sahel greening trend. *Geophysical Research Letters* 32(4), 214–215.

Hill, T.M. Bateman, H.G. II, Aldrich, J.M. and Schlotterbeck, R.L. (2008). Effects of using wheat gluten and rice protein concentrate in dairy calf milk replacers. *Prof. Anim. Sci.* 24(5), 465–472.

Hostert, P., Röder, A. and Hill, J. (2003). Coupling spectral unmixing and trend analysis for monitoring of long-term vegetation dynamics in Mediterranean rangelands. *Remote Sensing of the Environment* 87(2–3), 183–197.

Hui, Y.J., Chen, C., Zhao, S.Z. and Zhang, C.Y. (1996). Techniques of cultivating grasses and shrubs for controlling grassland desertification. *Grassland of China* 4, 19–23 (in Chinese with English abstract).

India Brand Equity Foundation ("IBEF") and ICRA Management Consulting Services Limited (IMaCS) www.ibef.org.

IPCC (2007). *Climate Change 2007: Climate Change Impacts, Adaptation and Vulnerability*. Summary for policymakers. Inter-Governmental Panel on Climate Change.

Jin, H.L., Dong, G.R., Su, Z.Z. and Sun, L.Y. (2001). Reconstruction of the spatial patterns of desert/loess boundary belt in North China during the Holocene. *Chinese Science Bulletin* 46, 969–975.

Kendall, M.G. 1975. *Rank Correlation Methods*, 4th ed., Charles Griffin, London.

Kundu, A. (2010). Desertification Monitoring and Zonation Modeling Using Remote Sensing and GIS Techniques. M. Sc. Thesis (unpublished), Faculty of Geo-Information Science and Earth Observation (ITC), University of Twente, Netherlands. pp. 1–82.

Kundu, A. and Dutta, D. (2011). Monitoring desertification risk through climate change and human interference using remote sensing and GIS techniques. *International Journal of Geomatics and Geosciences* 2(1), 21–33.

Kundu, A., Dutta, D., Patel, N.R., Saha, S.K. and Siddiqui, A.R. (2014) Identifying the process of environmental changes of Churu district using remote sensing indices. *Asian Journal of Geoinformatics* 14(3), 14–22.

Kundu, A., Patel, N.R., Saha, S.K. and Dutta, D. (2015) Monitoring the extent of desertification processes in western Rajasthan (India) using geo-information science. *Arabian Journal of Geosciences* 8(8), 5727–5737.

Kundu, A., Patel, N.R., Saha, S.K. and Dutta, D. (2017) Desertification in western Rajasthan (India): an assessment using remote sensing derived rain-use efficiency and residual trend methods. *Natural Hazards*, 86(1), 297–313.

Lamprey, H.F. (1975). Report on the Desert Encroachment Reconnaissance in Northern Sudan, 21 October–10 November 1975. UNESCO/UNEP, Nairobi.

Le Houérou, H.N. (1984). Rain-use efficiency: a unifying concept in arid-land ecology. *Journal of Arid Lands* 7, 213–247.

Matin, S., Sullivan, C.A., Finn, J.A., hUallach, D.Ó., Green, S., Meredith, D. and Moran, J. (2020). Assessing the distribution and extent of High Nature Value farmland in the Republic of Ireland. *Ecological Indicators* 108, 1–11.

Navalgund, R.R., Jayaraman, V. and Roy, P.S. (2007). Remote sensing applications: An overview. *Current Science* 12(93), 1747–1766.

Nicholson, S.E. and Farrar, T. (1994). The influence of soil type on the relationships between NDVI, rainfall and soil moisture in semi-arid Botswana. Part I. NDVI response to rainfall. *Remote Sensing of Environment* 50(2), 107–120.

Nicholson, S.E., Tucker, C.J. and Ba, M.B. (1998). Desertification, drought, and surface vegetation: an example from the West African Sahel. *Bulletin of the American Meteorological Society* 79(5), 1–15.

Nicholson, S.E., Davenport, M.L. and Malo, A.R. (1990). A comparison of the vegetation response to rainfall in the Sahel and East Africa, using Normalized Difference Vegetation Index from NOAA AVHRR. *Climatic Change* 17(2–3), 209–241.

Nie, H.G., Wang, M., Sun, H., et al. (2005). Dynamic analysis on changes of desertification in desert-loess boundary belt during the last 10 years: A case study on Yulin area in north Shaanxi province. *North Western Geology* 38(2), 86–93 (in Chinese with English abstract).

O'Connor, T.G., Haines, L.M. and Snyman, H.A. (2001). Influence of precipitation and species composition on phyto-mass of a semi-arid African grassland. *Journal of Ecology*, 89(5), 850–860.

Prince, S.D. (2002). Spatial and temporal scales of measurement of desertification. In: Stafford-Smith, M., Reynolds, J.F. (eds.), *Global Desertification: Do Humans Create Deserts?* Dahlem University Press, Berlin, pp. 23–40.

Prince, S.D., Brown de Colstoun, E. and Kravitz, L. (1998). Evidence from rain use efficiencies does not support extensive Sahelian desertification. *Global Change Biology* 4(4), 359–374.

Röder, A. and Hill, J., eds. (2005). Proc. 1st International Conference on Remote Sensing and Geoinformation Processing in the Assessment and Monitoring of Land Degradation and Desertification (RGLDD), Trier, Germany, 7–9 Sep 2005.

Sen, P.K. (1968). Estimates of the regression coefficient based on Kendall's tau. *Journal of the American Statistical Association* 63, 1379–1389.

Stellmes, M., Sommer, S., and Hill, J. (2005). Use of the NOAA AVHRR NDVI-Ts feature space to derive vegetation cover estimates from long term time series for determining regional vegetation trends in the Mediterranean. Remote Sensing and Geoinformation Processing in the Assessment and Monitoring of Land Degradation and Desertification, Trier, Germany, 231–238.

Sun, J.G., Ai, T.H., Wang, P., et al. (2008). Assessing vegetation degradation based on NDVI-climate variables feature space. *Geomatics and Information Science of Wuhan University* 33 (6), 8–11 (in Chinese with English abstract).

Symeonakis, E. and Drake, N. (2004). Monitoring desertification and land degradation over sub Saharan Africa. *International Journal of Remote Sensing* 25(3), 573–592.

Symeonakis, E., Calvo, A. and Arnau, E. (2003). A study of the relationship between landuse changes and land degradation in SE Spain using remote sensing and GIS. *Proceedings of the International Conference of GIS and Remote Sensing in Hydrology, Water Resources and Environment, Three Gorges Dam site*, China, 16–19 September 2003.

Tucker, C.J., Newcomb, W.W., Los, S.O., Prince, S.D. (1991). Mean and inter-annual variation of growing season normalized difference vegetation index for the Sahel 1981–1989. *International Journal of Remote Sensing* 12(6), 1133–1135.

UNCCD (1994). United Nations Convention to combat desertification in countries experiencing serious drought and/or desertification, particularly in Africa. A/AC.241/27, Paris.

UNCED (1992). Managing Fragile Ecosystems: Combating Desertification and Drought. United Nations Conference on Environment and Development.

UNEP (1994). United Nations Convention to Combat Desertification in those countries experiencing serious drought and/or desertification, particularly in Africa. UNEP, Geneva.

Vegetation Index User's Guide (MOD13 Series) Version 2.00, May 2010 (Collection 5) Ramon Solano, KamelDidan, Andree Jacobson and Alfredo Huete.

Wang, J., Price, K.P. and Rich, P.M. (2001). Spatial patterns of NDVI in response to precipitation and temperature in the central Great Plains. *International Journal of Remote Sensing* 22, 3827–3844.

Wessels, K.J., Prince, S.D., Frost, P.E. and van Zyl, D. (2004). Assessing the effects of human-induced land degradation in the former homelands of northern South Africa with a 1km AVHRR NDVI time-series. *Remote Sensing of Environment* 91(1), 47–67.

Wessels, K.J., Prince, S.D., Malherbe, J., et al. (2007). Can human-induced land degradation be distinguished from the effects of rainfall variability? A case study in South Africa. *Journal of Arid Environments* 68(2), 271–297.

Xin, Z.B., Xu, J.X. and Zheng, W. (2007). The effects of climate change and human activities on vegetation cover of the Loess Plateau. *Science in China (Series D)* 37(11), 1504–1514 (in Chinese with English abstract).

Xu, D.Y., Kang, X.W., Liu, Z.L., et al. (2009). The relative roles of climate change and human activities in the process of desertification in the Ordos region. *Science in China (Series D)* 39(4), 516–528 (in Chinese with English abstract).

5 Land Use/Land Cover Characteristics of Odisha Coastal Zone
A Retrospective Analysis during the Period between 1990 and 2017

Santanu Roy, Gouri Sankar Bhunia, and Abhisek Chakrabarty

CONTENTS

- 5.1 Introduction 50
- 5.2 Study Area 50
- 5.3 Data Used 51
- 5.4 Methodology 52
  - 5.4.1 Pre-processing of Satellite Image 52
  - 5.4.2 Land Use Land Cover (LULC) Map Preparation 52
  - 5.4.3 Land Use/Land Cover Change Analysis 53
  - 5.4.4 Accuracy Assessment 54
  - 5.4.5 Estimation of Resource Degradation 55
- 5.5 Results 55
  - 5.5.1 Spatio-temporal Distribution of LULC Categories 55
    - 5.5.1.1 Agricultural Fallow Land 56
    - 5.5.1.2 Cultivated Land 56
    - 5.5.1.3 Aquaculture 57
    - 5.5.1.4 Fallow/Open Land 57
    - 5.5.1.5 Beach/Sand Dune (with Vegetation) 58
    - 5.5.1.6 Industrial Built-up Area 58
    - 5.5.1.7 Mudflat 60
    - 5.5.1.8 Mangrove 60
    - 5.5.1.9 Marshy Land 60
    - 5.5.1.10 Open Vegetation 60
    - 5.5.1.11 Rural and Urban Settlement 61
    - 5.5.1.12 Scrub Land 61
    - 5.5.1.13 Water Bodies (Streams/Canals/Ponds/Lakes) 61
    - 5.5.1.14 Ocean 62
  - 5.5.2 Estimation of Land Degradation Characteristics 62
- 5.6 Discussion 65
- 5.7 Conclusion 67
- Acknowledgments 68
- References 68

5.1 INTRODUCTION

Coastal areas are highly productive biological areas and essential components of the regional biosystem. Shoreline relocation, habitat transition, biodiversity transition, and environmental degradation are currently largely protected by coastal zone surveillance. The mixed and genetically diversified ecosystems of these areas are abundant and are the primary sources of carbon sinks and oxygen. Throughout the estuary and coastal areas, which have been impacted by climate change and urbanization, there are various types of degradation, such as diminishing wetlands, emissions from the atmosphere, and erosion by other exogenetic process (Tian et al., 2016). Long-term and comprehensive land use management (land use/land cover [LULC]) reform is crucial to address these issues, as it offers valuable knowledge to demonstrate the history, existing circumstances, and potential of the LULC transition and to consider biogeochemical processes and improvements to the LULC mechanism (Chakraborty et al., 2016).

An earlier study suggested that there have been a tremendous increase in population and evolving pressures in the coastal areas for the last four decades (Barbara Neumann, 2015). The average population density in the coastal zone was 77 people/km^2 in 1990 and 87 people/km^2 in 2000, projected to reach 99 people/km^2 in 2010 (UNEP, 2007), which will gradually increase in 2020. Satellite remote sensing has played a vital role in the mapping and monitoring of Earth's surface and sub-surface features (assets). It provides a synoptic view of spatial and temporal information on Earth's surface, which is a significant and cost-effective tool for mapping natural resources (Anderson et al., 1976; Nayak, 2004). Images captured by the Landsat satellite image sensor, namely, Multispectral Scanner (MSS), Thematic Mapper (TM), and Enhanced Thematic Mapper+ (ETM+), are used globally for LULC study (Baby, 2015). The benefit of Landsat images with suitable spectral properties offer better identification of LULC and dynamic behaviors associated to reference data collected from onsite observation during in-situ survey (Kawakubo et al., 2011). Change detection studies of land use have been attempted by several researchers in recent decades to understand the process and trend of land use changes (Rawat, 2013). There is some literature available for LULC classification, such as visual image interpretation, image classification (supervised, unsupervised, and object based), band ratio, etc. (Anderson et al., 1976; Nayak, 2002; SAC, 2012). The Ministry of Rural Development, Government of India (GoI), prepared a land use policy based on the context of land resource management towards sustainable development. The National Land Utilization Plan (NLUP) could be applied at the national, regional, district, and village levels to fulfill these objectives.

LULC characteristics change due to the various physical, socio-economic, and climatic components that influence the local socio-economic conditions of the coastal region in space and time. The Odisha coast has been recorded extensively as a highly vulnerable area in response to climate change forcing sea levels to increase. In the Chandipur coast of the district of Orissa, Mukhopadhyay et al. (2011) studied coastal erosion and subsequent shoreline changes compared with deviation in sea surface height from 1990 to 2010 with multi-storm data. The present condition of LULC characteristics is evolving with increasing migration and consequent urbanization and the growth in tourism. Challenges such as inaccessible terrain conditions, coastal flooding, severe monsoons, and regular cyclone incidents must be resolved through the initiative of sustainable management of natural resources on the coast of Odisha. Furthermore, traditional living opportunities and wood harvesting–based forestry practices are not favorable to sustainability for either the local people or the environment. The present study describes the LULC characteristics of the Odisha coast and the temporal variation and conversion of land during the period between 1990 and 2017. This study also analyzes the impact on coastal resources due to conversion of LULC classes.

5.2 STUDY AREA

The study area geographically extends from 86°12′42.95″E to 87°29′4.75″E and from 21°47′49.30″N to 19°57′43.05″N. It is bounded by West Bengal in the north, the Bay of Bengal in the east, the

LULC Characteristics of Odisha Coastal Zone

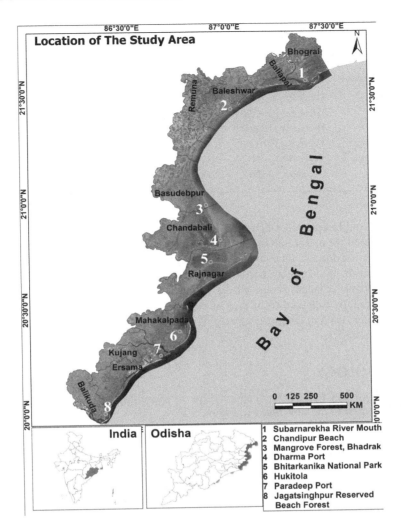

FIGURE 5.1 Location map of the study area.

Puri district of Odisha in the south, and other blocks of more than four districts of the state in the west (Figure 5.1). The study area consists of 12 coastal blocks of 4 districts (Baleswar, Bhadrak, Kendapur, and Jagatsinghpur) of Odisha, namely, Bhograi, Baliapal, Baleshwar, Remuna, Bahanaga, Basudebpur, Chandabali, Rajnagar, Mahakalpada, Kujang, Ersama, and Balikuda. The study area is geomorphologically developed between the flood plain of the Mahanadi–Barmhani–Baitarani river system, the Bhadrak coastal plain, the Budhabalanga flood plain, and the flood plain of the Subarnarekha River. The coastal plain consists of mainly newer alluvial and Pleistocene soil (GSI, 1974).

5.3 DATA USED

The present study was carried out using multi-temporal Landsat series satellite images for assessment of the decadal changes and transformation of LULC features over the Odisha coast. The data preparation phase started with the acquisition of data from six satellites, which were downloaded from USGS Earth Explorer Community (www.ers.cr.usgs.gov). Landsat's Thematic Mapper (TM)

TABLE 5.1
Details of the Satellite Image Used in the Study

Sl. No.	Satellite/Sensor	Date of Acquisition	Path/Row	Spatial Resolution	Data Source
1	Landsat5/TM	December 23, 1990	139/45	30 m	http://earthexplorer.usgs.gov/
		December 23, 1990	139/46		
		December 11, 2009	139/45		
		December 1, 2009	139/46		
2	Landsat8/OLI	December 17, 2017	139/45	30 m	
		December 17, 2017	139/46		

and Operational Land Imager (OLI)/Thermal Infrared Sensor (TIRS) images from 1990, 2009, and 2017 were considered for extraction of LULC characteristics. The temporal resolution of the satellite images is not equal due to data unavailability and processing constraints; therefore, the temporal resolution factor has been taken into account in the land use change analysis calculation. The study area is covered by two adjacent scenes of the Landsat image with path 139 and rows 45 and 46. Secondary data or maps collected from various sources, such as toposheets, census data, and maps (2011), etc., have been used in the present study. The details of the satellite dataset used in this study are presented in Table 5.1.

5.4 METHODOLOGY

The general methodology that has been adopted for the present study is discussed in the subsequent sections and presented in the flow chart in Figure 5.2.

5.4.1 Pre-processing of Satellite Image

Blue and near-infrared (NIR) and short-wave infrared (SWIR) were used to identify features in the coastal zone. The combined picture of the red and NIR bands is graded appropriately for cultivable land, woodland, scrub, and other natural vegetation cover (Chen et al., 2003). The selected spectral bands have been transformed into digital number radiance based on specified equation and calibration coefficients and then converted to top of atmospheric (TOA) reflectance (NASA, 2011). The image of 1990 (geo-coded) was considered as the base image, which has been rectified based on known ground control (GCP) points to match spatially, and the subsequent temporal images (2009 and 2017) have been referenced accordingly while considering the Universal Transverse Mercator (UTM) projection system and the World Geodetic System (WGS 84) as datum with zone 45N. The corrected images were stacked in ERDAS Imagine software v14.0. The LULC classes were collected in shape, scale, surrounding/association, color, texture, and pattern, and combined as a training site. Various features, such as industrial area, built-up area, port/harbor, aquaculture, etc., were extracted from satellite images by visual image interpretation skills.

5.4.2 Land Use Land Cover (LULC) Map Preparation

The LULC map was prepared using a maximum likelihood (MXL) algorithm based on the collection of ground samples and field knowledge of the study area (Foody and Mathur, 2004; Yuan et al., 2009; Jayanth et al., 2015). From a scientific point of view, the MXL algorithm is grouped with maximum likelihood pixels that are continuously upgraded to a certain category based on the multivariate normal circulation values of pixels in the training sets using a probability density function (El Asmar and Hereher, 2011). The clustered signatures of training sets sum up the MXL pixels, which are distributed over an image and assembled into a separate medium vector group with

LULC Characteristics of Odisha Coastal Zone

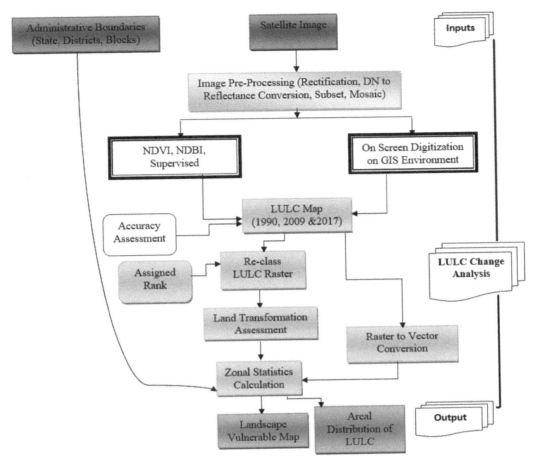

FIGURE 5.2 Flow chart of methodology for LULC mapping and assessment.

covariance matrices. In the Google Earth images, multi-temporal training sites for each LULC class were collected. On average, 250 training sites, which were homogeneously distributed, were chosen, and samples were collected for the study area to identify the LULC. The MXL algorithm produces clustered signature sets as a knowledge engineer that uses the pixels based upon the maximum perspective of a particular class (El Gammal et al., 2010). The training sites in ERDAS Imagine were treated with controlled devices. With the aid of the training sample, the thematic LULC maps were created. Classifiers like the MXL algorithm were processed. In addition, 1990, 2009, and 2017 LULC layers with reported classification schemes were developed with Space Application Center (SAC) attributes (SAC; Nayak, 2002). The details of the classification system are represented in Table 5.2. It is evident that the Normalized Differential Vegetation Index (NDVI; Rouse et al., 1974) and the Normalized Difference Built Up Index (NDBI; Macarof and Statescu, 2017) have been created for the training cycle for the precise identification of areas associated with small urban patches, cultivated land, and internal water bodies.

5.4.3 Land Use/Land Cover Change Analysis

Useful knowledge about land modification or alteration is found in the LULC change identification assessment. Centered on the land use level, the multi-temporal LULC maps (1990, 2009, and 2017) were compared. Approximate improvements in the area and perimeter of LULC groups are

TABLE 5.2
List of Land Use/Land Cover Class Adopted in the Study

Level I	Level II	Level III
Built-Up Land	Human Habitation	Urban Settlement
		Rural Settlement
	Transportation	Port/Harbor/Jetty
	Industrial	
Agricultural Land	Crop Land	
	Agricultural Fallow Land	
Rangeland	Shrub/Grass	
Forest Land	Mixed Forest	
	Open Vegetation	
	Mangrove	
Water	Water Bodies	(Streams/Canals/Pond/Lake/Reservoir/Tank)
	Ocean	
Wetland	Marshy Land	
	Mudflat	
Bare Soil	Fallow Land	
Shore Land	Beach/Sand Dunes	
Other Features	Aquaculture Ponds	

Source: Nayak, S., Use of satellite data in coastal mapping. *Indian Cartographer*, 22, 147–156, 2002; SAC, Coastal Zones of India, Space Application Centre (SAC), ISRO, Ahmadabad, 2012.

required by the relative analysis of character-types (Richards and Jia, 2006). The change detection analysis has been performed by compiling matrix tabulation, expressed as:

$$(\text{Current year} - \text{Initial year})/\text{Initial year}$$

The results of the study include the pixel variations in a common class corresponding to both images, suggesting changes in the state of an object or phenomenon over time (Joshi et al., 2011). Alternatively, the magnitude of change (K) is determined for each class of features to estimate the degree of expansion or area reduction. The equation of K is expressed as

$$K = (f - i/i) \times 100$$

where K is the magnitude of change, f is the initial classified LULC image, and i is the recent classified LULC image. Negative values in the resultant image indicate loss of area of the LULC feature, while positive values in the image refer to gain of area of the LULC feature.

5.4.4 Accuracy Assessment

The accuracy of the categorized image, which can be assessed by normal techniques, is referred to as the confusion matrix or as the error matrix (Mujabar and Chandrasekar, 2013). The confusion matrix collates the pixel or LULC relation of any location on each ground truth with the equivalent pixel value and rating in the classification picture for the same location. The indices of pre-class accuracy and error matrices were determined to distinguish LULC improvements, and the precision of the test was categorized. For this analysis, approximately 100 random sample locations were chosen, points were taken from high-resolution (Google Earth) images, which for each year of research are well dispersed within the study area, and then, their corresponding LULC classes were calculated. In order to test the validity of the LULC classification, these reference points were overlapped on the defined images, and the findings assess the precision of LULC features to post-field testing. Such empirically tested classes measured the pre-class accuracy, derived from the reference images, and determined a distinct range of overall accuracy as well as a Kappa coefficient for each year.

5.4.5 ESTIMATION OF RESOURCE DEGRADATION

The LULC data consist of 16 LULC classes, which have been grouped into five categories (Food Security, Environmental Resources, Economic Stability, Non-Developed Region, and Demographic Status) based on the Draft Land use policy of India (2013) and the Food and Agricultural Organization (FAO) report (2017) for land resource assessment. Individual land use may alter due to natural phenomena and anthropogenic demands, which makes it quite difficult to understand the alteration trends for any region in terms of the nature–mankind cycle. The LULC categorization was performed in order to better understand the degradation of land resources with the aim of sustainable land use management. The rank table for the division is presented in Table 5.3, and the flow chart map for the research methodology is shown in Figure 5.2.

5.5 RESULTS

In the present study, the coastal zone of the study area was classified into 17 classes: agricultural fallow land, cultivated land, aquaculture land, fallow land/open land, beach/sand dune (with vegetation), industrial built-up, mangrove, marshy land, mix-vegetation, mudflat, ocean, open vegetation, port/harbor/jetty, rural settlement, scrub land, urban settlement, and water bodies (streams/canal/ponds/lakes) (Table 5.4 and Figure 5.3). The classification was based on satellite image interpretation, field survey, the condition of the existing study area, and a literature review.

Table 5.5 shows the areal distribution of LULC classes for the Odisha coast between 1990 and 2017. It is evident from Table 5.5 that approximately 50% of the study area is covered with cultivated land, followed by ocean and agricultural fallow land. Industrial built-up, port/harbor/jetty, and beach/sand dune (with vegetation) cover a much lower percentage of the area in the study site. Figure 5.3 shows the percentage LULC distribution in the study site. However, dramatic changes in the LULC classes have been found between 1990 and 2017. The areal statistics showed a decrease of agricultural fallow land, cultivated land, beach/sand dune (with vegetation), marshy land, mangrove, mixed vegetation, and mudflat during the study period.

5.5.1 SPATIO-TEMPORAL DISTRIBUTION OF LULC CATEGORIES

LULC maps for 1990, 2009, and 2017 were prepared, and the spatio-temporal distributions of land use are discussed in the following sub-sections (Figure 5.3).

TABLE 5.3
Table of Assigned Ranks of LULC Classes for Vulnerability Assessment

Sl. No.	Category	LULC Classes	Rank
1	Food Security	Agricultural Land, Agricultural Fallow Land	1
2	Natural Conservation	Beach/Sand Dune (with vegetation), Mangrove, Marshy Land, Mix-Vegetation, Mudflat, Ocean, Open Vegetation, Scrub Land	2
3	Economic Status	Fallow Land/Open Land	3
4	Non-Developed Area	Aquaculture, Industrial Built-up, Port/Harbor	4
5	Demographic Status	Rural Settlement, Urban Settlement	5

TABLE 5.4
Major Land Use/Land Cover Categories Adopted in This Research Work

LULC Class	Description
Agricultural Fallow Land	A piece of land that is usually used for farming is left without crops on it for a certain period, after which it may be used again due to restored fertility
Cultivated Land	Areas where crops are cultivated
Aquaculture Land	Areas dominated by shrimp ponds with fish farming
Fallow Land/Open Land	Exposed soils and open scrub land
Beach/Sand Dune (with Vegetation)	Dunes nearer the ocean and protected with vegetation cover, such as coastal dune grass
Industrial Built-Up	Built-up area used primarily in real estate development and construction industry
Mangrove	Mature halophile plants, which grow in brackish or salt water conditions
Marshy Land	Shallow water on a lakeshore, inhabited by woody herbaceous plants
Mix-Vegetation	Combined type of coastal flora under various headings, such as mangroves, coastal dune vegetation, shore vegetation, algae vegetation, and aquatic weeds
Mudflat	Un-vegetated intertidal area in the coastal ecosystem
Ocean	Coastal oceans are waters that lie above the continental shelf
Open Vegetation	Tidal flora comprises aquatic algal flora in tidal and brackish marshes and is also synonymous with drift lines
Port/Harbor/Jetty	The seaport used to load, and unload freight and boats is grounded for health reasons due to poor weather or atmospheric conditions
Rural Settlement	People live on the land outside towns and cities, mainly dependent on farming and pastoral practices
Scrub Land	Areas covered with weeds/grass/shrubs, etc.
Urban Settlement	Developed dense residential, business, manufacturing, and related infrastructure such as highways, parks, mixed urban, and other public amenities, which may also be called the "Jungle of Concrete"
Water Bodies (Streams/Canals/Ponds/Lakes)	Water networks, canals, lakes, ponds, and active hydrological features

5.5.1.1 Agricultural Fallow Land

Agricultural fallow lands are usually related to cultivated lands that are temporarily uncultivated for one or more seasons or cannot sustain crops that have been traditionally associated with cultivated lands after the planning and growth seasons (Haque et al., 2008). Approximately 6% of the study area comprises agricultural fallow land. In 1990, the estimated area of agricultural fallow land is 400.13 km^2 (6.47%), which decreases to 365.85 km^2 (5.92%) in 2017 (Table 5.5). This LULC class is decreasing at a rate of 8% during 1990–2009 and 0.61% during the period between 2009 and 2017. However, the overall decrease is 34.27 km^2 at a rate of 9.37% from 1990 to 2017. This may be attributed to the effects of salinization of land, chemically affected soil fertility, deposition of sand by cyclones damaging crops, etc. In the study area, the maximum agricultural fallow land is recorded in Chandabali, Mahakalpada, Rajnagar, and Balikuda blocks. The lowest amount of fallow land is observed in Bahanga, Baliapal, Remuna, and Bhorgai blocks (Figure 5.3).

5.5.1.2 Cultivated Land

Cultivable land is primarily distributed in various parts of the coastal plain. Cultivable soil in the sample is usually used to sustain local subsistence trends with paddy and vegetables. Approximately 50% of the study area is covered by cultivable land. The areal coverage of cultivable land is recorded as 3144.57 km^2 (50.87%), 3062.60 km^2 (49.54%), and 2868.66 km^2 (46.41%) in 1990, 2009, and

LULC Characteristics of Odisha Coastal Zone 57

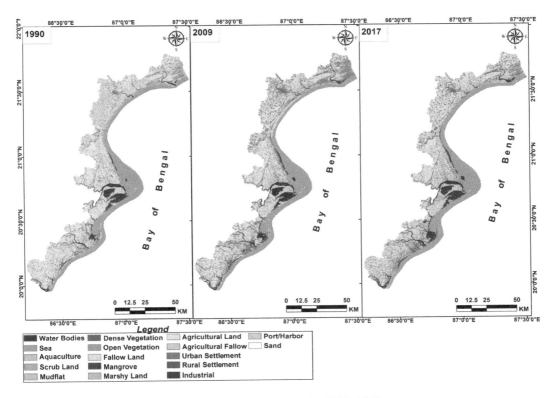

FIGURE 5.3 LULC distribution map of the study area during 1990–2017.

2017, respectively. The area of this land use class shows a decreasing trend of −81.97, −193.94, and −275.91 km² in 1990–2009, 2009–2017, and 1990–2017, respectively (Table 5.6). In the study area, the maximum area of cultivated land is observed in Chandabali, Basudevpur, and Mahakalpada blocks. The minimum area of cultivated land is seen in Bahaga, Bhograi, Ersama, and Kujang blocks. Land under cultivation is decreasing due to rising land loss, invasion of built-up areas, and aquatic agriculture and real estate activities in the region under research. Subsequently, increasing soil instability and sand deposition in the coastal area cause widespread transformation into fallow land.

5.5.1.3 Aquaculture

The aquaculture sector is relevant, and this use of land is increasing in the area being studied, because the shrimp industry and the services involved affect the economy of the local community. The spatial coverage of aquaculture is denoted as 8.75 km² (0.14%) in 1990, increasing to 127.86 km² (2.07%) in 2009, followed by 163.60 km² (2.65%) in 2017 (Table 5.5). In 1990, the area of aquaculture land was very limited for the entire district, but it has now increased dramatically. With the increase of fish farming, the wetland areas are also increased. The rapid growth of aquaculture farming over the last 29 years would appear to be associated with Chandabali, Ersama, and Baleswar blocks (Figure 5.3). A much lower area of aquaculture is observed in Bahanaga, Baliapal, Balikuda, and Kujang blocks. Aquaculture is altering paddy fields along the coastal zone, which may no longer be sufficiently productive in the increasingly saline environment (Mukhopadhyay et al., 2018).

5.5.1.4 Fallow/Open Land

The spatial coverage of fallow/open land in 1990 is 34.45 km², corresponding to 0.56%, and it decreases to 11.05 km² in 2017, or 0.18% of the study area. The results also showed 21.27 km² areal change at a decreasing rate of 61.76% during 1990–2009. However, during 2009–2017, the estimated areal change is 2.13 km², with a decreasing trend of 16.15%, and the calculated overall

TABLE 5.5
Areal Distribution of LULC Features of Odisha Coast, Derived from Satellite Data during Period between 1990 and 2017

	1990		2009		2017	
LULC Class	km²	%	km²	%	km²	%
Agricultural Fallow	400.13	6.47	368.11	5.96	365.85	5.92
Cultivated Land	3144.57	50.87	3062.60	49.54	2868.66	46.41
Aquaculture	8.75	0.14	127.86	2.07	163.60	2.65
Beach/Sand Dune (with Vegetation)	18.37	0.30	38.54	0.62	30.02	0.49
Fallow Land/Open Land	34.45	0.56	13.17	0.21	11.05	0.18
Industrial Built-Up	3.30	0.05	3.58	0.06	4.84	0.08
Mangrove	196.34	3.18	206.26	3.34	267.08	4.32
Marshy Land	117.19	1.90	104.57	1.69	124.45	2.01
Mix-Vegetation	261.90	4.24	211.53	3.42	143.84	2.33
Mudflat	62.72	1.01	41.61	0.67	78.97	1.28
Ocean	1224.05	19.80	1211.27	19.59	1272.99	20.59
Open Vegetation	227.25	3.68	239.36	3.87	260.22	4.21
Port/Harbor	4.53	0.07	6.96	0.11	10.42	0.17
Rural Settlement	163.66	2.65	243.02	3.93	290.67	4.70
Scrub Land	77.20	1.25	72.17	1.17	68.97	1.12
Urban Settlement	14.30	0.23	17.65	0.29	26.69	0.43
Water Bodies (Streams/Canals/Ponds/Lakes)	223.10	3.61	213.26	3.45	193.31	3.13

change rate during the entire study period (1990–2017) is 211.85%. In the study area, maximum fallow land is observed in Balipal, Balikuda, and Mahakalpada blocks, whereas the lowest extent is observed in Bahanaga, Chandabali, and Basudebpur blocks.

5.5.1.5 Beach/Sand Dune (with Vegetation)

Sandy bay, sandy dune, and associated terrace with sparse foliage cover on the surrounding shoreline are part of the coastal face of Odisha. The field is particularly diverse and needs to evolve, in contrast to other groups. The foredunes and sand dunes contain numerous plant types, such as trees, grasses and woody plants, and other coastal plants. This protects beaches and dunes against erosion from wind, storms, and sand on eroding plains. This vegetation covers the beaches and dunes. The effect of this aspect is strong shifts in tidal, aquatic, and anthropogenic behaviors over time. The total area of this category is calculated as 18.37 km² in 1990, 38.54 km² in 2000, and 30.02 km² in 2017, corresponding to 0.30%, 0.62%, and 0.49% of the total study area, respectively (Table 5.5). However, the results showed a positive change of beach and sand dune during the entire study period. It is evident that 20.18 km² positive areal change at a rate of 109.85% is calculated from 1990 to 2000, whereas during 2000 and 2017, there is a negative areal change of 8.52 km² at a rate of 22.12%. However, during the period between 1990 and 2017, there is a positive change at a rate of 38.81% in the study area (Table 5.6). A high rate of deposition is observed in Mahakalpada, Chandabali, Bahanaga, Basudevpur, Baliapal, Bhograi, small pockets of Kujang, and Balikuda in the study area, whereas erosion is observed on the seaward side of Chandabali, Rajnagar, Ersama, Baleshwar, and small pockets of Mahakalpada, Basudevpur, and Bhograi (Figure 5.3).

5.5.1.6 Industrial Built-up Area

The industrial built-up area primarily comprises NTPC Ltd., Indian oil refinery, food processing units, fertilizer industry, construction industry, and real estate development in the study site. The

TABLE 5.6
Differences and Magnitude of Areal Changes of LULC Classes during the Period between 1990 and 2017

LULC Types	Area (km²) in Changes (1990–2009)	K	Area (km²) in Changes (2009–2017)	K	Area (km²) in Changes (1990–2017)	K
Agricultural Fallow	−32.02	−8.00	−2.26	−0.61	−34.27	−9.37
Cultivated Land	−81.97	−2.61	−193.94	−6.33	−275.91	−9.62
Aquaculture	119.11	1361.70	35.74	27.95	154.85	94.65
Beach/Sand Dune (with Vegetation)	20.18	109.85	−8.52	−22.12	11.65	38.81
Fallow Land/Open Land	−21.27	−61.76	−2.13	−16.15	−23.40	−211.85
Industrial Built-Up	0.28	8.56	1.26	35.23	1.54	31.88
Mangrove	9.91	5.05	60.82	29.49	70.74	26.48
Marshy Land	−12.62	−10.77	19.87	19.00	7.26	5.83
Mix-Vegetation	−50.37	−19.23	−67.68	−32.00	−118.06	−82.07
Mudflat	−21.11	−33.66	37.36	89.79	16.25	20.58
Ocean	−12.78	−1.04	61.73	5.10	48.94	3.84
Open Vegetation	12.11	5.33	20.86	8.71	32.97	12.67
Port/Harbor	2.43	53.73	3.46	49.78	5.90	56.57
Rural Settlement	79.36	48.49	47.64	19.60	127.00	43.69
Scrub Land	−5.03	−6.52	−3.20	−4.43	−8.23	−11.93
Urban Settlement	3.35	23.46	9.03	51.16	12.39	46.42
Water Bodies (Streams/Canals/Ponds/Lakes)	−9.84	−4.41	−19.95	−9.36	−29.79	−15.41

Negative value indicates degradation of area, and positive value indicates gain of area in an LULC class.
K = magnitude of change (%).

industrial built-up area shows a progressively increasing trend. In 1990, the estimated area is classified as 3.30 km² (0.05%) which increases to 4.84 km² (0.08%) in 2017 (Table 5.5). Moreover, the results also show that 0.28 km² (8.56%) areal changes have been found during 1990 and 2009, whereas from 2009 to 2017, the change in area is estimated as 1.26 km² (33.23%). The overall change of industrial built-up area is calculated as 1.54 km² with a change rate of 31.88% during the entire study period (Table 5.6). Most of the industrial built-up area is observed in Kujang block and Remuna block.

5.5.1.7 Mudflat

In and across the estuary, backwater is usually scattered throughout the coastal region, resulting in mudflats. Due to the flood inflow from multiple rivulets, the deeper regions in the hinterland are becoming inundated and the margins turned into mudflats. The intertidal, open mudflat region is very rich for the growth of algae. The surface area of mudflat is calculated as 62.72 km² (1.01%) in 1990, 41.61 km² (0.67%) in 2009, and 78.97 km² (1.28%) in 2017. The mudflat extensively decreased, losing an areal extent of 21.11 km², from 1990 to 2009, and during 2009–2017, the areal extent of mudflat increased by 37.36 km² in the study area at an increasing rate of 89.79% (Table 5.5). The maximum area of mudflat is observed in Bahanaga, Baleshwar, and Remuna blocks in 1990. In 2017, there is a dramatic increase of mudflat in Rajnagar and Chandabali blocks in the study site. There is a progressive decrease of mudflat observed in Baliapal and Basudevpur blocks, whereas a progressive increase is observed in Balikda, Erasma, Kujang, and Mahakalpada blocks.

5.5.1.8 Mangrove

The mangrove forest is increasing gradually along the coast, and there are natural and human-made explanations for the variations. The accreted soil due to the sediment at the river's mouth is allowing the mangrove to regenerate naturally, while mangrove habitat is being lost due to aquaculture growth. The dynamic spatio-temporal changes of mangrove forests in the study area between 1990 and 2017 are analyzed in Table 5.5. Mangrove in the study area is progressively increasing, mainly concentrated along riverbanks and coastlines of inhabited lands. In 1990, the estimated area of mangrove land is 196.34 km² (3.18%), followed by 206.26 km² (3.34%) in 2009 and 267.08 km² (4.32%) in 2017. During the period between 1990 and 2009, the mangrove area is increased by 9.91 km²; however, from 2009 to 2017, the areal change is calculated as 60.82 km² in the study area (Table 5.6). The highest percentage of mangrove area is recorded from Rajnagar block, followed by Mahakalpada and Chandabali blocks, located in the central part of the study area. The lowest mangrove area is documented in Bahanga block, Balipal block, Baleshwar block, and Remuna block, located mostly in the central part of the study site.

5.5.1.9 Marshy Land

Marshes are known to be inundated on a daily or intermittent basis with water that is marked by developing soft-stained vegetation and is adapted to too saturated soil conditions. The intertidal marshes are normally covered and exposed daily by the tide. In 1990, the estimated marshy land was recorded as 117.19km² (1.90%). In 2009, the estimated marshy land was less (104.57km²) and again it has been increased by 124.45km² (2.01%) in 2017. During the period between 1990 and 2009, there was a negative growth of marshy land, with an estimated area of 12.62 km² with a decreasing rate of 10.77%. From 2009 to 2017, positive growth of marshy land was estimated as 19.87km² with an increasing rate of 19.00%. Overall, positive growth of marshy land was an estimated area of 7.26km². The highest estimated area of marshy land was recorded from Ersama, Kujang, Rajnagar, Basudevpur, and Baleshwar blocks. Moreover, the marshy land was decreased in Baliapal, Balikuda, Bhograi, and Mahakalpada blocks in the study site.

5.5.1.10 Open Vegetation

Open vegetation along the coastal region comprises farmyard and other natural trees, including eucalyptus, casuarinas, and bamboos. The plantations along the coastal region play a vital role in mitigating destruction and preserving the coastal environment by protecting human lives during severe natural disasters (Misra and Balaji, 2015). The geographical extent of the land use is estimated at 227.25 km² (3.68%) in 1990, 239.36 km² (3.87%) in 2009, and 260.22 km² (4.21%) in 2017. The results showed a progressive increase of open vegetation cover in the study area: 12.11 km² positive growth of open vegetation cover during 1990–2009 at an increasing rate of 5.33%, whereas during 2009–2017, the estimated growth rate was 8.71% with an areal change of 20.86 km². However, the highest amount of open vegetation cover is recorded from Bhograi block, followed by Ersama and Kujang blocks. The lowest areal coverage is seen in Bahanaga, Chandabali, and Rajnagar blocks.

5.5.1.11 Rural and Urban Settlement

The land areas for human residences contained in artificial systems, transit networks, and facilities and supplementary services in connection with subsistence activities have been established and developed (Adeel, 2010). Development areas in the region are directly proportional to population increase, which is increasing demand of food and conversion of land use pattern (Ohri and Poonam, 2012). The geographical extent of rural settlements in the study area has progressively increased. In 1990, the total settlement area is calculated as 163.66 km² (2.65%), increasing to 243.02 km² (3.93%) in 2009 and 290.67 km² (4.70%) in 2017. The overall increase rate is calculated as 43.69% during the entire study period. From 1990 to 2009, the positive spatial extent is estimated as 79.36 km² with an increase rate of 48.49%, and during 2009–2017, the estimated spatial extent is 19.60 km² with an increase rate of 127%. The maximum rural settlement area is documented in Basudevpur, Bhograi, and Chandabali blocks. The geographical extent of rural settlement is lowest in Ersama, Kujang, and Balikuda blocks of the study area.

Moreover, the urban settlement in the study area has increased progressively. However, in the study area, urban population is mainly recorded in the Baleshwar, Kujang, and Remuna blocks. The estimated area of urban population is recorded as 14.30 km² (0.23%), followed by 17.65 km² in 2009 and 26.69 km² in 2017 (Table 5.5). During the period between 1990 and 2009, there was an increase of 3.35 km² in area at a rate of 23.46%. From 2009 to 2017, the estimated increase rate of the urban population is calculated as 51.16% (Table 5.6).

5.5.1.12 Scrub Land

In the area of research, scrub land, in conjunction with fallow and open land, was either naturally occurring or a result of human activities. Scrub, or broom and shrub, consists of stunted trees with a canopy cover of less than 20%. There is a particular pattern of plant development and zonation; for example, seaweed and microalgae mostly grow in and cover sub-tidal areas. The results of the analysis showed a decreasing trend for scrub land during the period between 1990 and 2017. In 1990, the estimated area of scrub land was recorded as 77.20 km² (1.25%), decreasing to 72.17 km² (1.17%) in 2009 and 68.97 km² (1.12%) in 2017 (Table 5.5). From 1990 to 2009, the estimated area of change is recorded as 5.03 km² with a change rate of 6.52%. In 2009–2017, the estimated change in scrub land is recorded as 3.20 km² with a decrease rate of 4.43%.

5.5.1.13 Water Bodies (Streams/Canals/Ponds/Lakes)

Surface water bodies constitute streams, canals, ponds, and lakes in different parts of the study area. This land usually occupies topographically low-lying areas. There are seven major rivers, namely Devi, Mahanadi, Brahmni, Baitarani, Budha Balaganga, Bardaia, and Subarnarekha, along with tributaries of the rivers, flowing easterly from the eastern Ghats. The geographical coverage of the rivers and streams is denoted as 223.10 km² (3.61%) in 1990, 213.26 km² (3.45%) in 2009, and 193.31 km² (3.13%) in 2017. Moreover, the water bodies throughout the study area decreased by 9.84 km² at a decrease rate of 4.41% from 1990 to 2009, and by 19.95 km² at a decrease rate of 9.36% during 2009–2017. The decrease of water bodies reduces water storage capacity and groundwater level, which turns cultivable land into fallow land, and this leads to increased barren land cover in the study area (Kaliraj et al., 2017). The maximum area of water bodies is recorded in Chandabali,

Mahakalpada, and Rajnagar blocks, and the minimum area is estimated in Bahanaga, Basudevpur, and Remuna blocks in the study site (Figure 5.3). In addition, due to human violations, water sources have been reduced to such an extent that the cost of sea water taxes on coastal aquifers has risen due to over-exploitation of the ground water in summer (Kaliraj et al., 2017).

5.5.1.14 Ocean

The subsidence and downturn of ocean water is easily observed from the satellite data. Ocean water is defined as an area covered by water in any tidal situation. In 1990, the ocean water is estimated at 1224.05 km^2 (19.80%), followed by 1211.27 km^2 (19.59%) in 2009 and 1272.99 km^2 (20.59%) in 2017. The areal changes of ocean are calculated as 12.78km^2 with a decrease rate of 10.77%, whereas from 2009 to 2017, there is positive growth, recorded as 61.73 km^2 with an increase rate of 5.10%. Sea water fluctuations have been related to the vagaries of equilibrium between water and transgression of land, which varies according to the different cycles of sea level and the effect of climate change seasonally, inter-annually, and globally. The maximum ocean area is recorded from Rajnagar block, followed by Mahakalpada and Chandabali blocks. On the other hand, the minimum ocean area is recorded from Balikuda block, followed by Bahanaga block and Remuna block.

5.5.2 ESTIMATION OF LAND DEGRADATION CHARACTERISTICS

Based on the magnitude and dynamism of LULC change, the study area has been divided into low, moderate, high, and very high degraded regions (Figure 5.3). This takes into account the magnitude (changes are either positive or negative) and the dynamism of the area (how fast the changes are occurring). Table 5.7 shows the block-wise resource degradation analysis of the study area. Maximum degradation of resources is estimated from the Ersama block (47.44%), followed by Balikuda block (39.04%) and Kujang block (35.72%) which are located in the southern side of the study area, while the maximum area is recorded in Chandabali block (599.33 km^2), followed by Rajnagar block (513.43 km^2). In the north of the study area, the maximum degradation of resources is estimated in the Baliapal block (30.95%), followed by Bhograi block (27.02%). The minimum degradation of resources is calculated for the Bahanaga block (16.57%), followed by Remuna block (19.95%), located in the central north of the study area (Table 5.7). The results also show that the maximum area of "very high" resource degraded region is estimated in Chandabali block (72.35 km^2), followed by Rajnagar block (63.64 km^2) and Mahakalpada block (62.69 km^2). However, the highest percentage resource degraded block is Balikuda block (17.22%), followed by Mahakalpada block (13.23%) and Rajnagar block (12.40%). The lowest resource degraded area is recorded in Bahanga block (5.39km^2), followed by Remuna block (12.34km^2) and Bhograi block (16.04km^2). Moreover, the lowest percentage of resource degraded area is recorded in Bahanga block (2.42%), followed by Remuna block (3.97%) and Bhograi block (4.63%). The areas of "high resource degraded" blocks are demarcated as Chandabali block (37.26 km^2), followed by Basudebpur block (33.30 km^2) and Baleshwar block (32.86 km^2). The maximum percentages of "high resource degraded" blocks are calculated for Ersama block (9.02%), followed by Baliapal block (8.05%) and Kujang block (7.79%). The maximum area of "moderate resource degraded" blocks is demarcated as Bhograi block (33.26 km^2), followed by Baliapal block (24.13 km^2) and Balikuda block (16.88 km^2). Subsequently, the maximum percentage of "moderate resource degraded" block is estimated in Balipal block (9.74%), followed by Bhograi block (9.60%) and Balikuda block (6.07%). The minimum areas of "medium resource degraded" blocks are Bahanaga block (3.35 km^2), followed by Ersama block (7.1 km^2) and Basudebpur block (8.65 km^2). The minimum percentages of "medium resource degraded" blocks are Bahanga block (1.50%), followed by Basudebpur block (1.87%) and Ersama block (2.15%). The maximum area of "low resource degraded" block is estimated in Ersama block (85.66 km^2), followed by Mahakalpada block (55.99 km^2) and Rajnagar block (35.89 km^2). The highest area of "low resource degraded" block is calculated from Ersama block (25.90%), followed by Kujang block (16.19%) and Mahakalpada block (11.82%). Moreover, the lowest area of "low resource degraded" block is

LULC Characteristics of Odisha Coastal Zone

TABLE 5.7
Block-Wise Risk Areas in Terms of Resource Degradation of Odisha Coast during the Period between 1990 and 2017

Name of the Blocks	Low Area (km²)	%	Moderate Area (km²)	%	High Area (km²)	%	Very High Area (km²)	%	Overall Risk Area (km²)	%
Bhograi	22.96	6.62	33.26	9.60	21.39	6.17	16.04	4.63	346.63	27.02
Baliapal	14.69	5.93	24.13	9.74	19.95	8.05	17.92	7.23	247.76	30.95
Baleshwar	31.22	7.26	9.85	2.29	32.86	7.64	26.48	6.16	430.02	23.35
Remuna	19.09	6.14	9.35	3.01	21.27	6.84	12.34	3.97	311.02	19.95
Bahanaga	14.72	6.60	3.35	1.50	13.52	6.06	5.39	2.42	223.12	16.57
Basudebpur	26.17	5.65	8.65	1.87	33.3	7.19	31.37	6.78	462.85	21.50
Chandabali	34.86	5.82	16.56	2.76	37.26	6.22	72.35	12.07	599.33	26.87
Rajnagar	35.89	6.99	13.15	2.56	17.00	3.31	63.64	12.40	513.43	25.26
Mahakalpada	55.99	11.82	15.03	3.17	20.64	4.36	62.69	13.23	473.74	32.58
Kujang	47.21	16.19	11.58	3.97	22.7	7.79	22.67	7.77	291.58	35.72
Ersama	85.66	25.90	7.1	2.15	29.84	9.02	34.3	10.37	330.73	47.44
Balikuda	29.5	10.61	16.88	6.07	14.29	5.14	47.88	17.22	278.08	39.04

recorded from Baliapal block (14.69 km²), followed by Bahanaga block (14.72 km²) and Bhograi block (22.96 km² and 5.65%), followed by Chandabali block (5.82%) and Remuna block (6.14%); the minimum percentage of "low resource degraded" block is estimated in Basudebpur block.

In the present analysis, the accuracy measurement is rendered using three different measures based on the error-matrix principle and the omission factor, for example, accuracy of the user, consistency of the producer, and total accuracy (Carlotto, 2009). The statistical information of accuracy assessment is shown in Tables 5.8, 5.9, and 5.10. Hence, the result describes the various LULC feature classes based on the Level II category with overall classification accuracies of 84.00%, 89.00%, and 93.00% in 1990, 2009, and 2017, respectively. Moreover, the overall Kappa coefficient value is calculated as 0.82, 0.87, and 0.90 for 1990, 2009, and 2017, respectively. The post-classified field verification suggests that the location and distribution of agricultural fallow, croplands, beach/sand dune (with vegetation), sandy beaches, aquaculture, mangrove, ocean, open vegetation, scrub land, and water bodies (streams/canals/ponds/lakes) are demarcated with high accuracy relative to other classes. This is mainly due to the mixture of spectral signatures of the digital image (Figure 5.4). Moreover, the change of rural and urban settlement shows significant changes in the relative accuracy of K^ values for the entire study period. During the period between 1990 and 2017, the ocean, mudflat, and beach areas are noted as having lower K^ values due to accretion, erosion, and regular activities of beach mining and coastal flooding that change the landform dynamically. Overall, the observation of LULC accuracy and reliability of the classification outputs have been calculated in terms of geographic location, surface area, and spatial dissemination of features in the study area.

TABLE 5.8
Accuracy Assessment Report (LULC for the Year 2017)

	Accuracy Assessment of 2017				
Class Name	Reference Totals	Classified Totals	Number Correct	Producers Accuracy (%)	Users Accuracy (%)
Water Bodies	6	5	5	83.33	100.00
Sea	10	10	10	100.00	100.00
Aquaculture	5	5	5	100.00	100.00
Scrub Land	3	4	3	100.00	75.00
Mudflat	0	0	0	—	—
Dense Vegetation	6	6	6	100.00	100.00
Fallow Land	1	0	0	—	—
Mangrove	4	4	4	100.00	100.00
Marshy Land	1	1	1	100.00	100.00
Agricultural Land	44	44	41	93.18	93.18
Open Vegetation	4	4	3	75.00	75.00
Agricultural Fallow	3	4	3	100.00	75.00
Urban Settlement	2	2	2	100.00	100.00
Rural Settlement	11	11	10	90.91	90.91
Built-Up/Industrial	0	0	0	—	—
Port/Harbor	0	0	0	—	—
Sand	0	0	0	—	
Totals	100	100	93		

Overall Classification Accuracy = 93.00%

End of Accuracy Totals

KAPPA (K^) STATISTICS

Overall Kappa Statistics = 0.8999

TABLE 5.9
Accuracy Assessment Report (LULC for the Year 2009)

Accuracy Assessment of 2009

Class Name	Reference Totals	Classified Totals	Number Correct	Producers Accuracy (%)	Users Accuracy (%)
Water Bodies	5	5	4	80.00	80.00
Sea	11	11	11	100.00	100.00
Aquaculture	4	4	4	100.00	100.00
Scrub Land	3	4	3	100.00	75.00
Mudflat	0	0	0	—	—
Dense Vegetation	5	5	4	80.00	80.00
Fallow Land	2	2	2	100.00	100.00
Mangrove	4	4	4	100.00	100.00
Marshy Land	1	1	1	100.00	100.00
Agricultural Land	46	45	40	86.96	88.89
Open Vegetation	5	5	4	80.00	100.00
Agricultural Fallow	3	3	2	66.67	66.67
Urban Settlement	2	2	2	100.00	100.00
Rural Settlement	9	9	8	88.89	88.89
Built-Up/Industrial	0	0	0	—	—
Port/Harbor	0	0	0	—	—
Sand	0	0	0	—	—
Totals	100	100	89		

Overall Classification Accuracy = 89.00%

End of Accuracy Totals

KAPPA (K^) STATISTICS

Overall Kappa Statistics = 0.8648

5.6 DISCUSSION

The coastal land uses and their deceptive variations are insufficient to guard the massive coastline and coastal resources of Odisha. Information on LULC is the key element for planning the management policy of the coastal zone. The present research work has also evaluated the land suitability classification for land use zoning and the socio-economic ailment of the coastal community. Landsat images are commonly used for land cover classifications due to the availability and feasibility of their temporal (16 days) and spatial resolution (30 m), spatial cover (185 × 185 km), and consistency. LULC is an indication of historical livelihood pattern for the local community. The structure of the eastern coast of India is generally affected by the shoaling, refraction, and diffraction of wave progressions, causing deposition on the up-drift side and erosion on the down-drift side of the coast (Misra and Balaji, 2015; Misra et al., 2013). The decreasing dune vegetation cover in several parts of the coastal land causes coastal instability and vulnerability along the Chandabali, Rajnagar, Ersama, and Baleshwar regions. Kaliraj et al. (2013) reported that severe erosion on the down-drift side of the coast was changing the morphological structure of the landform. Subsequently, due to the cumulative effects of waves and tides, the beach and sand dunes have noticeably increased in Mahakalpada, Chandabali, Bahanaga, Basudevpur, Baliapal, and Bhograi. In the study area, a larger area of fallow lands is observed in Chandabali, Mahakalpada, and Rajnagar blocks; this is rapidly being transformed into built-up areas, aqua farming, and other infrastructure. The mudflats along the estuaries and backwater are encroached on for aquaculture expansion and settlement. However, the increase in mudflat in the study area may be attributed to transgression of the Bay of

TABLE 5.10
Accuracy Assessment Report (LULC for the Year 1990)

	Accuracy Assessment of 1990				
Class Name	Reference Totals	Classified Totals	Number Correct	Producers Accuracy (%)	Users Accuracy (%)
Water Bodies	6	6	5	83.33	83.33
Sea	11	11	11	100.00	100.00
Aquaculture	2	2	2	100.00	100.00
Scrub Land	4	4	3	75.00	75.00
Mudflat	0	0	0	—	—
Dense Vegetation	6	5	4	66.67	80.00
Fallow Land	3	3	2	66.67	66.67
Mangrove	3	3	3	100.00	100.00
Marshy Land	1	1	1	100.00	100.00
Agricultural Land	49	48	41	83.67	85.42
Open Vegetation	5	6	4	80.00	66.67
Agricultural Fallow	3	3	2	66.67	66.67
Urban Settlement	1	1	1	100.00	100.00
Rural Settlement	6	7	5	83.33	71.43
Built-Up/Industrial	0	0	0	—	—
Port/Harbor	0	0	0	—	—
Sand	0	0	0	—	
Totals	100	100	84		

Overall Classification Accuracy = 84.00%

End of Accuracy Totals

KAPPA (K^) STATISTICS

Overall Kappa Statistics = 0.8238

Bengal with deposition of coastal mud in and around the backwater and estuaries in the coastal zone of the study area. As such, on the Odisha coastal track, several new pits with saltwater and saltpans have been noticed because of high tide waves and storm surges that deposited sea water on the profuse placer mining excavations (Kaliraj et al., 2014). In the study area, mangrove coverage shows an increasing trend. The mangroves are usually found in the estuaries of the coastal areas. where sweet river water meets the saline water of the sea. Some of the uninhabited areas are entirely bounded by mangroves. In the study area, mostly *Rhizophora* and *Avicennia* species are observed near estuaries and the confluence of deltas in coastal areas. The cultivators have changed their livelihood from farming to aqua farming because of salt water interference and recurrent cyclones. The natural aquaculture farming is increasing corrosion in the study area. Moreover, the reduction in sediment deposits observed after the creation of embankments has instigated serious issues in these areas. Additionally, the preservation of mixed plantation lands is another major concern in the study area. New, illegal Bangladeshi settlements are springing up adjacent to the sea and inside the plantation areas. Mudflats and sand sheets are also increasing trend showing aggradation nature of the coast. The results also suggested that the highest degradation of resources is observed in the Chandabali and Rajnagar blocks, located in the northern part of the study area. Moreover, Ersama and Balikuda blocks have recorded the highest percentages of resource degradation in the study area.

On the Odisha coast, it is observed that areas in aquaculture, beach/sand dune (with vegetation), industrial built-up area, mangrove, marshy land, mudflat, ocean, open vegetation, port/harbor, rural settlement, scrub land, and urban settlement have increased during the period between 1990 and 2017. Moreover, regular attacks by cyclones in the study area have damaged crops and caused huge

LULC Characteristics of Odisha Coastal Zone 67

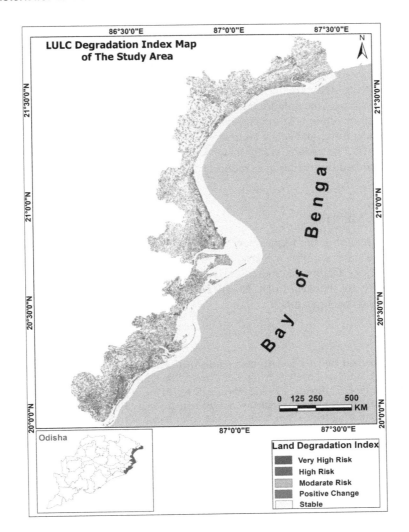

FIGURE 5.4 LULC degradation map of the study area.

losses of economic property in the study area. The results suggest that to meet the requirements of local people, the LULC features are being vigorously transformed to additional land uses. For instance, cultivable lands are converted to built-up areas and aqua farming due to human infringement activities, while natural progressions such as land dilapidation and surface overflow are changing cultivable land into fallow land, and fallow and mudflat are resulting in damage to soil fertility and agriculture production.

5.7 CONCLUSION

In the present study, it has been found that the coastal area is subjected to intensive changes of wetland category, which may be due to the restoration and destruction of land by natural and anthropogenic activities. During the period between 1990 and 2000, major changes have been found in marshy land and mudflats; at the same time, mangrove and aquaculture have increased at the expense of other wetland categories. According to the present study, it may be concluded that the coastal zone is rich in ecological and morphological features, which need to be considered for restoration, protection, and proper use. Large land alterations have been found in the study area, such as wetlands, mudflats, and reclaimed mudflats being converted into aquaculture land. Moreover, agricultural fallow, cultivated land, mixed vegetation, and surface water body have decreased due to anthropogenic encroachment and artificial construction. The decreasing area of these land covers affects the sediment load, which has the same effect as shoreline erosion in various parts of the study area. There are possibilities to maintain proper management of the coastal zone; for example, the paleo sand dune area can be used for development activities, where reclaimed land can be altered. Moreover, most LULC features are altered to built-up, industrial development, aquaculture, and other land uses without considering their adverse influences on the coastal systems. The present research work identifies key evidence for sustainable coastal management and scheduling relating to reducing the risk and susceptibility in the Odisha coastal region.

ACKNOWLEDGMENTS

The authors are grateful to the Department of Remote Sensing & GIS, Vidyasagar University for giving the lab and other support to carry out the study. We are also thankful to Dr. Jatisankar Bandyopadhyay, HoD, Dept. of Remote Sensing & GIS, Vidyasagar University for his continued encouragement. We are thankful to Dr. Manik Mahapatra, Scientist-B, NCSCM, MoEF, Chennai for his guidance and encouragement.

REFERENCES

Adeel, M. (2010). Methodology for identifying urban growth potential using land use and population data: a case study of Islamabad Zone IV. *Procedia Environ. Sci.* 1, 32–41.

Anderson, J.R., Hardy, E.E., Roach, J.T. and Witmer, R.E. (1976). A land use and land cover classification system for use with remote sensor data. *U.S. Geological Survey Professional Papers 964*, 28 p.

Baby, S. (2015). Monitoring the coastal land use land cover changes (LULCC) of Kuwait from spaceborne Landsat sensors. *Indian J. Geo-Mar. Sci.* 44 (6), 1–7.

Carlotto, M. J. (2009). Effect of errors in ground truth on classification accuracy. *Int. J. Remote Sens.* 30 (18), 4831–4849.

Chakraborty, A., Sachdeva, K. and Joshi, P.K. (2016). Mapping long-term land use and land cover change in the central Himalayan region using a tree-based ensemble classification approach. *Appl. Geogr.* 74, 136–150.

Chen, J., Gong, P., He, C., Pu, R. and Shi, P. (2003). Land-use/land-cover change detection using improved change-vector analysis. *Photogramm. Eng. Rem. S.* 69, 369–379.

Geological Survey of India (GSI) (1974). *Geology and mineral resources of the states of India*. Geological Society of India, Miscellaneous Publication 30(IV), pp. 1–78.

Haque, S., Bhatta, G.D., Hoque, N., Rony, M.H. and Rahman, M. (2008). Environmental impacts and their socioeconomic consequences of shrimp farming in Bangladesh. Paper presented at the Competition for resources in a changing world: new drive for rural development. Stuttgart, Germany.

El Asmar, H.M. and Hereher, M.E. (2011). Change detection of the coastal zone east of the Nile Delta using remote sensing. *Environ. Earth Sci.* 62, 769–777.

El Gammal, E.A., Salem, S.M. and El Gammal, A.E.A. (2010). Change detection studies on the world's biggest artificial lake (Lake Nasser, Egypt). *Egypt. J. Rem. Sens. Space Sci.* 13, 89–99.

Foody, M.G. and Mathur, A. (2004). A relative evaluation of multiclass image classification by support vector machines. *IEEE Trans. Geosci. Remote Sens.* 42, 1335–1343.

Jayanth, J., Koliwad, S. and Ashok Kumar, T. (2015). Classification of remote sensed data using Artificial Bee Colony algorithm. *Egypt. J. Rem. Sens. Space Sci.* 18, 119–126.

Joshi, R.R., Warthe, M., Dwivedi, S., Vijay, R. and Chakrabarti, T. (2011). Monitoring changes in land use land cover of Yamuna riverbed in Delhi: a multitemporal analysis. *Int. J. Remote Sens.* 32 (24), 9547–9558.

Kaliraj, S., Chandrasekar, N. and Magesh, N.S. (2013). Evaluation of coastal erosion and accretion processes along the southwest coast of Kanyakumari, Tamil Nadu using geospatial techniques. *Arab J Geosci.* 8, 239–253. doi: 10.1007/s12517-013-1216-7

Kaliraj, S., Chandrasekar, N., Ramachandran, K.K., Srinivas, Y. and Saravanan, S. (2017). Coastal landuse and land cover change and transformations of Kanyakumari coast, India using remote sensing and GIS. *Egypt. J. Rem. Sens. Space Sci.* 20, 169–185.

Kaliraj, S., Chandrasekar, N., Simon Peter, T., Selvakumar, S. and Magesh, N. S. (2014). Mapping of coastal aquifer vulnerable zone in the south west coast of Kanyakumari, South India, using GIS-based drastic model. *Monit. Assess. Environ.* 10.1007/s10661-014-4073-2.

Kawakubo, F.S., Morato, R.G., Nader, R.S. and Luchiari, A. (2011). Mapping changes in coastline geomorphic features using Landsat TM and ETM imagery: examples in southeastern Brazil. *Int. J. Remote Sens.* 32 (9), 2547–2562.

Macarof, P. and Statescu, F. (2017). Comparison of NDBI and NDVI as indicators of surface urban heat island effect in landsat 8 imagery: a case study of Iasi. *Present Environ. Sustain. Dev.* 11 (2). 10.1515/pesd-2017-0032

Misra, A. and Balaji, R. (2015). Decadal changes in the land use/land cover and shoreline along the coastal districts of southern Gujarat, India. *Environ. Monit. Assess.* 10.1007/s 10661-015-4684-2.

Misra, A., Murali, R.M. and Vethamony, P. (2013). Assessment of the land use/land cover (LU/LC) and mangrove changes along the Mandovi-Zuari estuarine complex of Goa. India. *Arab. J. Geosci.* 8 (1), 267–279.

Mujabar, P.S. and Chandrasekar, N. (2013). Shoreline change analysis along the coast between Kanyakumari and Tuticorin of India using remote sensing and GIS. *Arab. J. Geosci.* 6, 647–664

Mukhopadhyay, A., Hornby, D.D., Hutton, C.W., Lázár, A.N., Amoako Johnson, F. and Ghosh, T. (2018). Land cover and land use analysis in coastal Bangladesh. In: Nicholls R., Hutton C., Adger W., Hanson S., Rahman M., Salehin M. (eds) *Ecosystem Services for Well-Being in Deltas*. Palgrave Macmillan, Cham.

Mukhopadhyay, A., Mukherjee, S., Hazra, S. and Mitra, D. (2011). Sea level rise and shoreline changes: a geoinformatic appraisal of Chandipur Coast, Orissa. *Int. J. Geol. Earth Environ. Sci.* 1 (1), 9–17.

NASA (Ed.) (2011). Landsat 7 science data users handbook landsat project science office at NASA's Goddard Space Flight Center in Greenbelt, 186. http://landsathandbook.gsfc.nasa.gov/pdfs/Landsat7_Handbook.pdf.

Nayak, S. (2002). Use of satellite data in coastal mapping. *Indian Cartographer* 22, 147–156.

Nayak, S. (2004). Application of remote sensing for implementation of coastal zone regulation: a case study of India. Global Spatial Data Infrastructure – 7th Conf held during 2–6 February, 2004 in Bangalore, India.

Neumann, B., Vafeidis, A.T., Zimmermann, J. and Nicholls, R.J. (2015). *PLoS One* 10 (3), e0118571. Published online 2015 March 11. 10.1371/journal.pone.0118571.

Ohri, A. and Poonam, Y. (2012). Urban sprawl mapping and land use change detection using remote sensing and GIS. *Int. J. Remote Sens.* 1 (1): 12–25.

Rawat, J.S., Biswas, V. and Kumar, M. (2013). Changes in land use/cover using geospatial techniques: a case study of Ramnagar town area, district Nainital, Uttarakhand, India. *Egypt. J. Rem. Sens. Space Sci.* 16, 111–117.

Richards, J.A. and Jia, X. (2006). *Remote Sensing Digital Image Analysis: An Introduction.* Springer, Heidelberg, New York, pp. 247–268.

Rouse, J.W, Haas, R.H., Scheel, J.A. and Deering, D.W. (1974). Monitoring vegetation systems in the great plains with ERTS. Proceedings 3rd Earth Resource Technology Satellite (ERTS) Symposium, vol. 1, p. 48–62.

SAC (2012). 'Coastal Zones of India', Space Application Centre (SAC), ISRO, Ahmadabad, Sponsored by Ministry of Environment & Forests (MoEF), Govt. of India. www.sac.gov.in.

Tian, B., Wu, W., Yang, Z. and Zhou, Y. (2016). Drivers, trends, and potential impacts of long-term coastal reclamation in China from 1985 to 2010. *Estuar. Coast. Shelf Sci.* 170, 83–90.

UNEP (United Nations Environment Programme) (2007). Physical alteration and destruction of habitats. www.unep.org (accessed February 27, 2008).

Yuan, H., Cynthia, F., Wiele, V.D. and Khorram, S., (2009). An automated artificial neural network system for land use/land cover classification from landsat TM imagery. *Remote Sens.* 1, 243–265.

6 Evaluating Landscape Dynamics in Jamunia Watershed, Jharkhand (India) Using Earth Observation Datasets

Mukesh Kumar, Piyush Gourav, Lakhan Lal Mahato,
Arnab Kundu, Alex Frederick, K. M. Sharon,
Akash Pal, and Deepak Lal

CONTENTS

6.1 Introduction	71
6.2 Materials and Methods	72
6.2.1 Study Area	72
6.2.2 Data Processing	72
6.2.3 LULC Classification	74
6.2.4 Landscape Metrics Analysis	74
6.3 Results and Discussion	76
6.3.1 Classification and Accuracy Assessment of LULC Images	76
6.3.2 Fragmentation Analyses	79
6.4 Summary and Conclusions	79
References	79

6.1 INTRODUCTION

Change in composition of land use/land cover (LULC) is a dynamic, extensive, and accelerating procedure driven by natural phenomena and anthropogenic activities, which are regarded as important factors of global change affecting landscapes (Southworth et al., 2004; Kamusoko and Aniya, 2007; Günlü et al., 2009). These changes are ultimately affecting humans (Paudel and Yuan, 2012). Human activity on the land cover has become a serious problem in the last few decades, ultimately changing the structure and pattern of the landscape (Naveh and Lieberman, 1990; Petropoulos et al., 2015; Srivastava et al., 2012, Singh et al., 2016). In this way, a huge volume of the earth's terrestrial ecosystem has been converted into either agriculture or settlement, which also catalyzes forest fragmentation in the region (Botequilha Leitáo et al., 2006). To understand and quantify changes in landscape structure, pattern, and dynamics, it is important to have a clear understanding of landscape indices. Hundreds of landscape indices have been developed in the past few decades to provide useful information about the composition and configuration of landscapes (Li et al., 2000; Olsen et al., 2006). However, researchers select a limited number of landscape indices according to their research demands. To explain landscape dynamics, several approaches can be used in the collection, investigation, and demonstration of data. However, obtaining accurate, faithful, timely, and lower-cost information has been a challenge for scientists, local communities, and policy decision-makers. The practice of remote sensing (RS) and geographical information system (GIS) technologies shows excellent promise for data collection because of its faithful, repetitive coverage of the whole of an area at various spatial and temporal scales; even datasets for inaccessible locations are available at low cost (Sanchez-Hernandez et al., 2007; Gupta and Srivastava, 2010; Hansen and Loveland, 2012). Otherwise, limited spatial statistics programs like FRAGSTAT and Patch Analyst can be effectively used to compute landscape metrics. FRAGSTATS is a computer-based software program intended to compute an extensive variety of landscape metrics for categorical map designs in the Windows operating environment (McGarigal et al., 2002; Paudel and Yuan, 2012). However, Patch Analyst works as an extension in ArcView and ArcGIS and contains the latest software code updates (Rempel, 2008). Therefore, FRAGSTATS is the most frequently used software rather than Patch Analyst because of its user-friendly Windows operating environment. Earlier works have demonstrated the applications of these tools (RS data with FRAGSTATS) to demonstrate the spatio-temporal dynamics of LULC and forest cover change (Nagashima et al., 2002; Turner et al., 2003; Southworth et al., 2004; Wakeel et al., 2007; Singh et al., 2016; Kumar et al., 2018; Gabril et al., 2019). On the other hand, QGIS delivers a free and open-source desktop and server environment and ships with all the functionalities of a recent GIS system (QGIS Development Team, 2013). Jung (2012) described in detail automated landscape ecology analysis through a QGIS plugin called LecoS (Landscape ecology Statistics). In their study, Rajendran and Mani (2015) quantified the dynamics of landscape configuration to assess the impact of urban sprawl effects on spatial heterogeneity using QGIS.

This research aims to address LULC changes and landscape fragmentation for 1989 and 2016 in the Jamunia watershed of Jharkhand, India. Landsat imageries have been used in this study, since data are continuously available from 1972 with moderate resolution (Gupta and Srivastava, 2010; Hansen and Loveland, 2012; Singh et al., 2016). Jamunia watershed is selected for study because of high percentage of forest area and also because in the report of Jharkhand Space Application Centre (JSAC) (2010), it is one of the most vulnerable watersheds.

6.2 MATERIALS AND METHODS

6.2.1 STUDY AREA

The study area, Jamunia watershed, comprises part of the Dhanbad, Bokaro, and Hazaribag districts of Jharkhand, India, from 86°13'44.601"E to 85°40'44.923"E longitude and from 23°44'48.432"N to 24°1'46.2"N latitude (Figure 6.1). The study area comes under the northeastern and north central plateau region, which is rich in forest dominated by *Shorea robusta*, *Madhuca longifolia*, and bamboo trees. The season of the study area is mainly comprised of summer, winter, and rainy seasons. May is the hottest month, when the temperature goes up to 45 °C, and December is the coolest month of the year. Jamunia watershed elevation varies from 210 to 390 m above mean sea level. The study area and its surroundings are mainly dominated by coalfields. The Dhanbad coal mine is one of the examples of the study area. Giridih and Gandi blocks of Giridih district contain one of the best qualities of metallurgical coal in India. Hazaribagh district is famous for limestone, fireclay, quartz, and mica. The soil of the study area is generally sandy. The study area covers Bishnugarh block, part of Hazaribagh, Bagodar, Dumri, Pirtanr block of Giridih, Baghmara, Topchanchi block of Dhanbad, and Bermo and Nawadih blocks of Bokaro district.

6.2.2 DATA PROCESSING

Two scenes of Landsat Thematic Mapper (TM)/Operational Land Imager (OLI) data covering the study area for the years 1989 and 2016 were obtained from the United States Geological Survey

Evaluating Landscape Dynamics in Jamunia

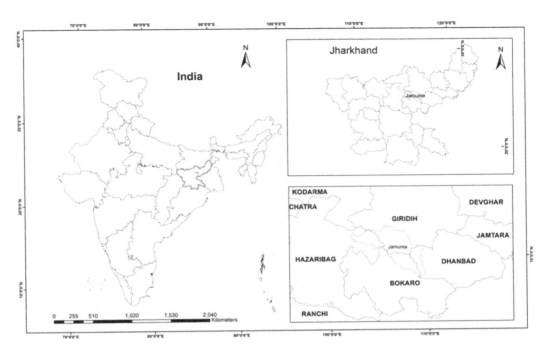

FIGURE 6.1 Location map of Jamunia watershed.

TABLE 6.1
Details of Satellite Images

Satellite/ Sensor	Year/Date	Path/Row	Bands Used	Spatial Resolution (m)
Landsat 5/TM	November 9, 1989	140/43	1,2,3,4,5,7	30
Landsat 5/TM	November 25, 1989	140/44	1,2,3,4,5,7,	30
Landsat 8/ OLI	November 19, 2016	140/43	2,3,4,5,7	30
Landsat 8/ OLI	November 19, 2016	140/44	2,3,4,5,7	30

(USGS) Viewer (www.usgs.gov.in); details are given in Table 6.1. Then, digital numbers of all the band information were computed with top of atmosphere (TOA) reflectance values. The solar zenith angle and sun elevation angle effect derived from satellite image was used to determine the digital number of apparent reflectance. Atmospheric correction of TOA is done using Equations 6.1 and 6.2 (Given et al., 2014).

Conversion to TOA Reflectance:

$$\rho\lambda' = M\rho Qcal + A\rho \tag{6.1}$$

Where
  $M\rho$ = Band-specific multiplicative rescaling factor from the metadata (REFLECTANCE_MULT_BAND_x, where x is the band number)
  $A\rho$ = Band-specific additive rescaling factor from the metadata (REFLECTANCE_ADD_BAND x, where x is the band number)
  Qcal = Quantized and calibrated standard product pixel values (DN)

TOA reflectance with a correction for the sun angle is given by

$$\rho\lambda' = \frac{\rho\lambda'}{\cos(\theta sz)} + \frac{\rho\lambda'}{\sin(\theta se)} \tag{6.2}$$

Where

$\rho\lambda$ = TOA planetary reflectance

Θ_{SE} = Local sun elevation angle. The scene center sun elevation angle in degrees is provided in the metadata (SUN_ELEVATION)

Θ_{SZ} = Local solar zenith angle; $\Theta_{SZ} = 90° - \Theta_{SE}$

The accurate boundary of the Usri watershed was obtained from the JSAC, Jharkhand, India. Both the scenes were mosaicked, and by using this boundary, a subset of the study area was prepared and extracted from satellite scenes.

6.2.3 LULC Classification

This study used the supervised classification technique with the maximum likelihood algorithm on extensive ground data collected using global positioning system (GPS) and Toposheet. The Forest Survey of India (FSI) report was used to produce LULC maps for the two time periods. Supervised classification was carried out with ERDAS IMAGINE to find the LULC classes. On the basis of the Food and Agriculture Organization (FAO) classification system (FAO, 2010; 5.3), four LULC classes (Forest, Agriculture, Barren Land, and Mining) have been prepared. A detailed description of LULC classes is given in Table 6.2.

Furthermore, the accuracy of the maps was checked. For each classified image, 105 random ground control points were selected, and each point was separately evaluated to assess the accuracy of the image.

The information on classes for all images was determined by Survey of India topographic maps, Google Earth (GE) data, published literature and interviews with forest officers, a classified map published by the National Remote Sensing Centre (NRSC), and ground truth information.

6.2.4 Landscape Metrics Analysis

FRAGSTATS (McGarigal and Marks, 1994) was used in this study to quantify fragmentation. The advantage of using FRAGSTATS is that it is a Windows operating environment where classified satellite images can be used directly (Rempel et al., 1999; Singh et al., 2016; McGarigal et al., 2002). This study carries out landscape metrics at the patch and class level. The landscape metrics used in this study are described in Table 6.3.

The results of different matrices, namely, NP, PD, LPI, TE, ED, LSI, IJI, and MESH (Number of patches, Patch density, Largest patch index, Total edge, Edge density, Landscape shape index, Interspersion-juxtaposition index, and MESH – MESH equals the sum of patch area squared,

TABLE 6.2
Land Use/Cover Class Descriptions for the Study Area

Land Use/Land Cover Classes	Description
Agriculture	Agricultural land
Barren Land	Contains open soil, rock, and sand
Forest	Refers to land with a tree canopy cover of more than 10 percent and area of more than 0.5 ha
Mining	Areas encompassed under surface mining operations, mine/quarry, land fill area

TABLE 6.3
Metrics Used at Class Level to Quantify Fragmentation

Metric (Unit)	Description	Formula
PLAND (%)	Percentage of like adjacencies as proportion of a given class type related to the total area	$Pi = \left(\dfrac{\sum_{j=1}^{n} a_{ij}}{A}\right)100$ Pi = proportion of the landscape occupied by patch type (class) i a_{ij} = area (m^2) of patch ij A = total landscape area (m^2)
NP	Total number of patches in this class	NP = ni ni = number of patches in the landscape of patch type (class) i
PD (per unit ha)	Number of patches of the corresponding patch type divided by total landscape area	$\dfrac{ni}{A}(10,000)(100)$ A = total landscape area (m^2) ni = number of patches in the landscape of patch type (class) i
LPI (%)	Ratio of largest patch area to region of interest	$\dfrac{\max_{j=1}^{n} A_{ji}}{A}(100)$ a_{ij} = area (m^2) of patch ij A = total landscape area (m^2)
TE (m)	Sum of length of all edge segments for the corresponding patch type class	$TE = \sum_{k=1}^{m} e_{ik}$ e_{ik} = total length (m) of edge in landscape involving patch type (class) i; includes landscape boundary and background segments involving patch type i
ED (m/ha)	Total length of edge involving the corresponding patch type class divided by total area	$ED = \dfrac{\sum_{k=1}^{m} e_{ik}}{A}(10,000)$ e_{ik} = total length (m) of edge in landscape involving patch type (class) i; includes landscape boundary and background segments involving patch type i A = Total landscape area (m^2)
LSI	Average complexity of the landscape as a whole	$LSI = \dfrac{0.25\sum_{k=1}^{m} e_{ik}^{*}}{\sqrt{A}}(10,000)$ e^{*}_{ik} = total length (m) of edge in landscape between patch types (classes) i and k; includes the entire landscape boundary and some or all background edge segments involve class I A = Total Landscape Area (m^2)
IJI	Degree of interspersion of patches of this class with all other classes	
MESH (ha)	Landscape Division Index expresses the probability that two randomly placed landscapes are in the same patch, and the MESH corresponds to area defined by the division index; division is a probability, while MESH is an area	$\dfrac{\sum_{k=1}^{n} a_{ij}^{2}}{A}(10,000)$ a_{ij} = area (m^2) of patch ij A = total landscape area (m^2)

Source: McGarigal, K. and Marks, B.J., *Spatial pattern analysis program for quantifying landscape structure*, General Technical Report PNW-GTR-351, U.S. Department of Agriculture, Forest Service, Pacific Northwest Research Station, 1995; Jaeger, J.A.G., *Landscape Ecology*, 15, 115–130, 2000.

TABLE 6.4
Classification Accuracy of Satellite Images

Years	Overall Accuracy (%)	Kappa Coefficient
1989	85.20	0.8177
2016	85.83	0.8636

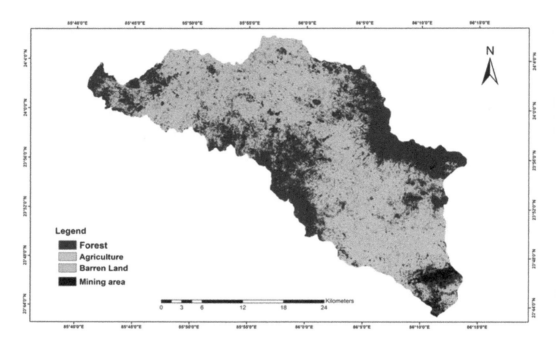

FIGURE 6.2 Classified map of study area (1989).

summed across all patches of the corresponding patch type, divided by the total landscape area [m²], divided by 10,000 [to convert to hectares]) for the classified maps were compared for the different years.

6.3 RESULTS AND DISCUSSION

6.3.1 CLASSIFICATION AND ACCURACY ASSESSMENT OF LULC IMAGES

The accuracy of the classified satellite images in this study is higher than 85 percent (Table 6.4). Much previous research has adopted this level of accuracy for further analysis (Anderson et al., 1976; Kamusoko and Aniya, 2007; Singh et al., 2016).

The classified map for the year 1989 (Figure 6.2) showed that the southern part of the study area was covered with dense forest, which has been gradually reduced by the year 2016 (Figure 6.3). Also, the western portion of the forested area has been changed into agricultural land. However, some of the barren land has also been converted into agricultural land. In the classified image from 1989, the central part was dominated by barren land and forest, which has mostly been converted into agriculture in the year 2016. On the other hand, the extreme south of the study area was a mining area, which has become agricultural land in the studied period.

Evaluating Landscape Dynamics in Jamunia

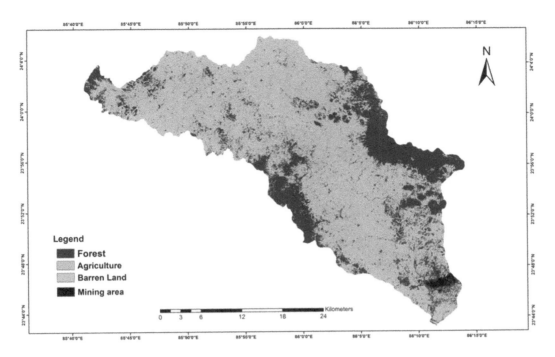

FIGURE 6.3 Classified map of study area (2016).

TABLE 6.5
Area Statistics of Different Land Use Classes of Different Years

LULC Class	1989		2016	
	Area (ha)	Area (%)	Area (ha)	Area (%)
Agriculture	40,854.9	42.42	46,432.20	48.21
Barren land	21,020	21.82	25,066.40	26.03
Dense forest	33,169	34.44	24,292.70	25.22
Settlement/mining	1,270.21	1.32	522.81	0.54
	96,314.11		96,314.11	

According to the analysis of the 1989 satellite image, agricultural land and dense forest were the dominant LULC classes, however, in the classified image from 2016, the dominance of the agriculture class is followed by barren land (Table 6.5). The reported change in agricultural land area from 1989 to 2016 is from 40854.9 to 46922.2 ha. Dense forests exhibited a decreasing trend in the studied period. The forested area, which had been 33169 ha in 1989, shrank to 24292.7 ha by 2016. An increase in agricultural practices in the forested landscapes has been identified as the main reason for forest degradation in the study area. As per the report of the Department of Forest and Environment, Government of Jharkhand (2014), the demand for agricultural land is one of the main factors driving deforestation. Many researchers reported that the demand for agricultural crops is the main cause of forest degradation (Lepers et al., 2005; Geist and Lambin, 2002; Southworth et al., 2004; Abdullah and Nakagoshi, 2007). The barren land area increased from 21020 to 25060.4 ha between the studied years. It has been confirmed from ground truth that agricultural practices are gradually increasing in the study area.

TABLE 6.6
Class-Level Landscape Metrics

Year	LULC		PLAND	NP	PD	LPI	TE	ED	LST	IJI	MESH
1989	Forest	F	13.9951	13,356	5.6353	4.2965	7,335,000	30.94388	100.6198	59.8492	906.7746
1989	Agriculture	AG	17.238	7,775	3.2805	13.0858	9,062,820	38.2391	112.0534	57.3632	4,065.976
1989	Barren Land	BL	8.869	7,181	3.0299	3.7837	6,013,230	25.3718	103.6432	48.768	370.5295
1989	Mining Area	M	0.7469	764	0.3224	0.5787	468,150	1.9753	27.7687	53.0266	7.9555
2016	Forest	F	10.2499	11,278	4.7586	3.2759	5,997,030	25.3035	96.1096	34.6104	334.4845
2016	Agriculture	AG	19.7981	6,109	2.5776	16.5532	11,991,870	50.6977	138.3163	45.3995	6,518.81
2016	Barren Land	BL	10.5738	9,198	3.8809	2.1705	7,690,020	32.4468	121.3712	22.6215	197.3276
2016	Mining Area	M	0.2206	154	0.065	0.1885	107,580	0.4539	11.719	30.3563	0.8432

6.3.2 Fragmentation Analyses

Statistical analysis of the results showed that the forest in the study area changed gradually from 1989 to 2016. However, to understand more about the pattern of forest changes, a fragmentation study is vital. FRAGSTATS analyses of the LULC classifications in the two years delivered evidence about how patterns of forest cover were transformed over time. The pattern of forest patches for different year dates is shown in Table 6.6. The number of forest patches in forest classes decreased from 13,356 to 11,278 between 1989 and 2016. This decrease in forest patches is because of the decrease in forest cover, mainly in the southern part as well as the northwestern part of the study area. In our case, the decreased area of the forests causes a decrease in forest patches; thus, the fragmentation was also suppressed (Table 6.6). The patch density of the forest also decreased, and accordingly, LPI decreased, i.e., the lower the number of patches, the lower the patch density became, and the largest patches became smaller (Table 6.6). In the case of forest classes, MESH decreased to 334.4845 ha in 2016 from 906.7746 ha in 1989. The decrease in ED and PLAND in the studied period indicates that the forest in this area has been degraded. IJI for the forest decreased from 59.8492 in 1989 to 34.6104 in 2016. A continuous increase in agricultural area has been reported in this study. It has been observed that NP was 7775 in the year 1989 and 6109 in 1916, whereas LPI increased from 13.0858 percent in 1989 to 16.5532 in 2016.

6.4 SUMMARY AND CONCLUSIONS

Spatial and temporal LULC changes to landscape fragmentation characteristics in the Jamunia watershed over a period of 27 years were analyzed. Also, open source software, Integrated Land and Water Information System (ILWIS) and QGIS, successfully assisted in fulfilling our objective. Supervised classification using maximum likelihood classification provided a satisfactory result for further analysis. Analysis of this study explains that in Jamunia watershed, the forest is declining, whereas agricultural land has increased during the studied period. FRAGSTATS analysis showed that much of the forested land had been converted into agricultural land, causing a decrease in the number of forest patches. As the forest patches continue to decrease, various medicinal and economically significant plants may be lost from this forested watershed. The decrease in forest density of the area is also alarming for animals present in the region. These types of change in LULC can also promote soil loss in the region.

REFERENCES

Abdullah, S.A. and Nobukazu Nakagoshi, N. (2007). Forest fragmentation and its correlation to human land use change in the state of Selangor, peninsular Malaysia. *Forest Ecology and Management* 241: 39–48. doi:10.1016/j.foreco.2006.12.016

Alkan Günlü, Ali Ihsan Kadıoğulları, Sedat Keleş and Emin Zeki Başkent (2009). Spatiotemporal changes of landscape pattern in response to deforestation in Northeastern Turkey: a case study in Rize. *Environmental Monitoring and Assessment* 148: 127–137.

Anderson, J.R., Hardy, E.E., Roach, J.T. and Witmer, R.E. (1976). A land use and land cover classification system for use with remote sensor data. *Geological Survey Professional Paper 964*. U.S. Govt. Printing Office. Washington.

Botequilha Leitáo, L., Miller, J., Ahern, J. and McGarigal, K. (2006). *Measuring Landscapes. A Planner's Handbook*. Washington: Island Press.

FAO (2010). *The State of World Fisheries and Aquaculture*. Rome: Food and Agriculture Organization of the United Nations, 2010. ISBN 978-2-5-106675-1.

Gabril, E.M.A., Denis, D.M., Nath, S., Paul, A. and Kumar, M. (2019). Quantifying LULC change and landscape fragmentation in Prayagraj district, India using geospatial techniques. *The Pharma Innovation Journal* 8(5), 670–675.

Geist, H.J. and Lambin, E.F. (2002). Proximate causes and underlying driving forces of tropical deforestation. *Bioscience* 52: 143–150.

Given, D.D., Cochran, E.S., Heaton, T., Hauksson, E., Allen, R., Hellweg, P., Vidale, J. and Bodin, P. (2014). Technical implementation plan for the ShakeAlert production system – An Earthquake Early Warning system for the West Coast of the United States: U.S. Geological Survey Open-File Report 2014–1097, 25 p., https://dx.doi.org/10.3133/ofr20141097.

Gupta, M. and Srivastava, P.K. (2010). Integrating GIS and remote sensing for identification of groundwater potential zones in the hilly terrain of Pavagarh, Gujarat, India. *Water International* 35: 233–245.

Hansen, M.C. and Loveland, T.R. (2012). A review of large area monitoring of land cover change using Landsat data. *Remote Sensing of Environment* 122: 66–74.

Jaeger, J.A.G. (2000). Landscape division, splitting index, and effective mesh size: new measures of landscape fragmentation. *Landscape Ecology* 15: 115–130.

Jharkhand Space Application Centre (JSAC) (2010). Status of Watershed Projects In Jharkhand State By B. Nijalingappa CEO, JSWM-SLNA 20-21 May 2010, New Delhi.

Jung, M. (2012). LecoS – A QGIS plugin to conduct landscape ecology statistics. http://plugins.qgis.org/plugins/LecoS.

Kamusoko, C. and Aniya, M. (2007). Land use/cover change and landscape fragmentation analysis in the Bindura District, Zimbabwe. *Land Degradation and Development* 18(2): 221–233.

Kumar, M., Denis, D.M., Singh, S.K., Szabó, S. and Suryavanshi, S. (2018). Landscape metrics for assessment of land cover change and fragmentation of a heterogeneous watershed. *Remote Sensing Applications: Society and Environment* 10, 224–233.

Lepers, E., Lambin, E.F., Janetos, A.C., DeFries, R., Achard, F., Ramankutty, N. and Scholes, R.J. (2005). A synthesis of information on rapid land-cover change for the period 1981–2000. *BioScience* 55: 115–124.

Li, X., Lu, L., Cheng, G. and Xiao, H. (2000). Quantifying landscape structure of the Heihe River Basin, north-west China using FRAGSTAT. *Journal of Arid Environment* 48: 521–535.

McGarigal, K. and Cushman, S.A. (2002). The gradient concept of landscape structure: or, why are there so many patches. In Wiens J., Moss, M., (eds.) *Issues and Perspectives in Landscape Ecology.* Cambridge.

McGarigal, K. and Marks, B. (1994). *FRAGSTATS: Spatial pattern analysis program for quantifying landscape structure.* Reference manual. Forest Science Department, Oregon State University, Corvallis Oregon. 62 pp.#Appendix.

McGarigal, K. and Marks, B.J. (1995). *Spatial pattern analysis program for quantifying landscape structure.* General Technical Report PNW-GTR-351 US Department of Agriculture, Forest Service, Pacific Northwest Research Station.

Nagashima, K., Sands, R., Whyte, A.G.D., Bilek, E.M. and Nakagoshi, N. (2002). Regional landscape change as a consequence of plantation forestry expansion: an example in the Nelson region, New Zealand. *Forest Ecology and Management* 163: 245–261.

Naveh, Z. and Lieberman, A.S. (1990). *Landscape Ecology: Theory and Application.* 2nd ed. New York, NY: Springer-Verlag. ISBN: 9780387940595.

Olsen, L.M., Dale, V.H. and Foster, T. (2006). Landscape patterns as indicators of ecological change at Fort Benning, Georgia, USA. *Landscape and Urban Planning* 79: 137–149.

Paudel, S. and Yuan, F. (2012). Assessing landscape changes and dynamics using patch analysis and GIS Modelling. *International Journal of Applied Earth Observation and Geoinformation* 16: 66–76.

Petropoulos, G.P., Kalivas, D.P., Georgopoulou, I.A. and Srivastava, P.K. (2015). Urban vegetation cover extraction from hyperspectral imagery and geographic information system spatial analysis techniques: case of Athens, Greece. *Journal of Applied Remote Sensing* 9: 096088–096088.

QGIS Development Team (2013). QGIS Geographic Information System. Open Source Geospatial Foundation Project. Available at: http://qgis.osgeo.org

Rajendran, P. and Mani, K. (2015). Quantifying the dynamics of landscape patterns in Thiruvananthapuram corporation using open source GIS tools. *International Journal of Research in Engineering and Applied Sciences* 5: 77–87.

Rempel, R. (2008). *Patch analyst 4.* Centre for Northern Forest Ecosystem Research. Ontario Ministry of Natural Resources. http://flash.lakeheadu.ca/rrempel/patch/ (accessed on 28.11.15).

Rempel, R.S., Carr, A. and Elkie, P.C. (1999). *Patch analyst and patch analyst (grid) function reference.* Centre for Northern Forest Ecosystem Research, Ontario Ministry of Natural Resources, Lakehead University, Thunder Bay, ON.

Sanchez-Hernandez, C., Boyd, D.S. and Foody, G.M. (2007). Mapping specific habitats from remotely sensed imagery: support vector machine and support vector data description based classification of coastal saltmarsh habitats. *Ecological Informatics* 2: 83–88.

Singh, S., Pandey, A. and Singh, D. (2016). Land use fragmentation analysis using remote sensing and Fragstats. In Srivastava, P.K., Mukherjee, S., Gupta, M., Islam, T., editors. *Remote Sensing Applications in Environmental Research*. Society of Earth Scientists Series. Switzerland: Springer International Publishing. pp. 151–176. ISBN: 978-3-319-05905-1.

Southworth, J., Munroe, D. and Nagendra, H. (2004). Land cover change and landscape fragmentation-comparing the utility of continuous and discrete analyses for a western Honduras region. *Agriculture, Ecosystems and Environment* 101: 185–205. doi:10.1016/j.agee.2003.09.011

Srivastava, P.K., Han, D., Rico-Ramirez, M.R., Bray, M. and Islam, T. (2012). Selection of classification techniques for land use/ land cover change investigation. *Advances in Space Research* 50: 1250–1265.

Turner, M.G., Pearson, S.M., Bolstad, P. and Wear, D.N. (2003). Effects of land-cover change on spatial pattern of forest communities in the Southern Appalachian Mountains (USA). *Landscape Ecology* 13: 449–464.

Wakeel, A., Rao, K.S., Maikhuri, R.K. and Saxena, K.G. (2007). Forest management and land-use/cover changes in a typical micro watershed in the mid-elevation zone of Central Himalaya, India. *Forest Ecology and Management* 213: 229–242.

7 Drought Frequency and Soil Erosion Problems in Puruliya District of West Bengal, India
A Geo-Environmental Study

Abhisek Santra and Shreyashi S. Mitra

CONTENTS

7.1 Introduction 83
7.2 Role of Open Source Software Packages 85
    7.2.1 SAGA 85
    7.2.2 QGIS 85
    7.2.3 GRASS GIS 85
    7.2.4 Integrated Land and Water Information System (ILWIS) 86
    7.2.5 Others 86
7.3 Materials and Methods 86
    7.3.1 Study Area 86
    7.3.2 Methodology 88
        7.3.2.1 Assessment and Monitoring of Vegetative Drought 88
        7.3.2.2 Modeling Soil Erosion: 89
7.4 Results and Discussion 91
    7.4.1 Spatio-Temporal Severity of Drought 91
    7.4.2 Threat of Soil Erosion 91
7.5 Conclusion 94
Acknowledgment 97
References 97

7.1 INTRODUCTION

Drought is considered to be an intermittent natural hazard with the potential to damage the socio-economic and environmental scenario of a country. The international community has perceived and accepted the importance of studying drought in recent decades because of its frequency and capacity to cause damage (Zhou et al., 2012; Zhang et al., 2017). From the environmental perspective, in the long term, this hazard causes increases in atmospheric CO_2 by reducing the ecosystem's CO_2 uptake (Ciais et al., 2005; Jiao et al., 2019a). Furthermore, its immediate effects include soil erosion, desertification, sandstorms, forest fires, water scarcity, and the deterioration of plants' health or even their death (Guo et al., 2017). As a consequence, it indirectly affects crop growth, leading to food crises and political unrest (Wu et al., 2013; Zhang et al., 2016; Han et al., 2020). It is estimated that in comparison with other natural hazards, drought is the most devastating in monetary terms (Wilhite et al., 2000; Vicente-Serrano, 2007; Caparrini and Manzella, 2009; Gebrehiwot et al., 2011; Du et al., 2013; Ghoneim et al., 2017; Gidey et al., 2018; Kundu et al., 2020). The severity of drought is dependent upon the extent of the affected area and the associated effects on anthropogenic activities, agriculture, and the surrounding environment (Caparrini and Manzella, 2009). Therefore, it is essential to understand, classify, identify, and monitor drought and to assess the possible damage caused by it.

Droughts can be broadly classified into the four following categories – meteorological drought resulting from abnormally low precipitation; agricultural drought due to shortage of soil moisture for plant growth; hydrological drought entailing scarcity of surface and underground water supply; and socio-economic drought, referring to an insufficient supply of water to meet the demand for certain commodities (Orville, 1990; Quiring and Ganesh, 2010; Du et al., 2013; Hao et al., 2015; Zhang et al., 2017). This classification is based on several factors, including relief, vegetation type, land use/cover, antecedent moisture content, etc. However, considering the interrelated nature of these factors, it is very difficult to classify a drought into a single category (Jiao et al., 2019b).

Traditionally, the classification and the monitoring of drought were dependent upon temperature and rainfall data from meteorological station–based point observations (AghaKouchak et al., 2015; Zhang and Jia, 2013). However, problems arise in areas where the number of meteorological stations is not sufficient to assess the phenomenon. In this situation, remote sensing technology offers the only feasible solution to monitor soil moisture and vegetation conditions from the local to the global scale. In other words, it provides a valuable source of information for drought monitoring (Rhee et al., 2010; Wu et al., 2013). Satellite images are constantly available to monitor drought events comprehensively at regular intervals without even going into the field. A series of image-derived indices are quite helpful in this regard to capture the start, duration, cessation, frequency, and severity of drought (Gidey et al., 2018). However, the accuracy depends upon the resolution of the imageries.

Among the different natural resources, soil is considered as important yet highly susceptible (Lal, 2001; Yesuph and Dagnew, 2019). Soil erosion is a global, multifaceted, and predominant process that degrades agricultural land by elevating the nutrient loss and surface runoff and causing decreasing water availability for plants from the sub-soil (Ganasri and Ramesh, 2016; Pal and Chakrabortty, 2019). As a result, the moisture condition of the plants and soil decreases, causing agricultural and vegetative drought in arid and semi-arid regions. It is estimated that soil degradation is responsible for more than four-fifths of the total global land degradation and reduction of nearly one-fifth of the total crop productivity of the world (Wijesundara et al., 2018). Moving rainwater is the primary cause of soil erosion on harvested and bare lands (Yahiaoui et al., 2015; Karamage et al., 2017). The slope of the ground and land use/cover are also very important factors in potential soil loss. However, in India, soil erosion is dominated by traditional agricultural practices and unplanned land utilization. As a consequence, the agricultural sector is suffering, and there is an increasing trend toward siltation problems in rivers and reservoirs (Prasannakumar et al., 2012; Ghosh et al., 2013). Therefore, the estimation of soil erosion through the use of proper models is essential. Over the years, various models have been developed and used globally. These models may be classified as physical, empirical, and conceptual (Brady and Weil, 2012; Farhan and Nawaiseh, 2015; Phinzi and Ngetar, 2019). Empirical models are preferred and used around the world because of their computation-friendly nature (Eisazadeh et al., 2012). In Puruliya district, the area chosen for this research is suffering from both drought and soil erosion problems due to its semi-arid climate. Crop productivity is greatly affected by these problems (Santra and Mitra, 2016). Studies have been conducted to address these two problems separately in this district (Hazra et al., 2017; Sarkar and Mishra, 2018). However, no studies have yet reported any relation between these two slowly damaging phenomena.

In this chapter, a brief account of the different satellite image–derived spectral indices that address the drought problem will be discussed. Similarly, light will be shed on the Revised Universal Soil Loss Equation (RUSLE), an empirical soil erosion model. Several commercial software packages and datasets are available and utilized; however, considering the importance of free and open source software packages and datasets, a case study of the geo-environmental problems of drought and soil erosion in Puruliya district of West Bengal, India is presented using the free and open source geographical information system (GIS) software QGIS.

7.2 ROLE OF OPEN SOURCE SOFTWARE PACKAGES

The adoption of open source software packages and open data has increased worldwide since the beginning of the twenty-first century. To the user community, the terms "open source" and "free" are synonymous. However, there is a basic difference between the two. According to the Free Software Foundation (FSF), free software possesses the freedom of running for any purpose; studying and adapting based on user requirements; redistributing the copies to help others; and improving the software for the benefit of the community (Steiniger and Bocher, 2009). However, the users of open source software packages have access to the source codes without any authority to modify and redistribute them. The Open Source Geospatial Foundation (OSGeo), representing the open source GIS community since 2006, follows a specific software development and licensing approach that ensures transparency through a series of rights that safeguard the source code copyright. These software packages have a wide range of libraries, tools, applications, and bases developed and released under different Open Source Initiative (OSI) licenses (Coetzee et al., 2020). Some examples of these libraries and application programming interfaces (APIs) are System for Automated Geoscientific Analyses (SAGA) GIS Tool library, QGIS API, GDAL/OGR API, PyQt4 API, Numpy library, etc. (Duarte et al., 2016). The capabilities and pros and cons of some of the widely used open source GIS software packages are described in the following subsections.

7.2.1 SAGA

SAGA is free open source software developed by researchers of the Department of Physical Geography, Gottingen in 2007. The latest SAGA v7.6.3 has a wide range of GIS tools, specifically for weather and climate data analysis, erosion, and several other grid- and image-based analyses. However, except for SAGA-API, most of the source codes are licensed under GNU General Public License (GPL) (SAGA, 2020).

7.2.2 QGIS

The most widely used open source GIS software package licensed under GNU-GPL is Quantum GIS (QGIS). It operates in almost every operating system with numerous raster- and vector-based GIS operations (QGIS, 2020). QGIS was originally developed in 2002 by a group of volunteers for fast geographic data visualization under a Linux-based system (Hugentobler, 2008). The software package supports more than 1200 Python-based plug-ins that support several GIS operations, including VERSAO Vega Monitor, which extracts phonological metrics and computes different vegetative drought indices (QGIS, 2020). However, due to its fast computation capacity, different sensor data handling capability, and large open source plug-in support for different applications, QGIS is preferred to the other open source and free software packages. Considering the demands of the present research, QGIS was chosen for processing and analyzing the satellite datasets.

7.2.3 GRASS GIS

Geographic Resources Analysis Support System (GRASS) is free and open source GIS software for geospatial data management, analysis, and modeling (GRASS GIS, 2020). It offers more than 350 Windows- and Linux-based OSGeo add-ons through the github web portal. However, the command-based old user interface demands expertise to operate the software.

7.2.4 Integrated Land and Water Information System (ILWIS)

ILWIS is PC-based remote sensing and GIS software developed by ITC, University of Twente in 2005. However, the first official DOS version was released in 1998. Since 2007, ILWIS has been regarded as free and open source software under the GPL License of 52° North, and the support by ITC has stopped (ITC, 2020). The latest ILWIS 4 Alpha licensed under GNUGPL V3 offers various GIS analytical tools and Python-supported computations. The Insitu and Online Data (ISOD) toolbox is capable of retrieving and processing time-series data from various online data archives (52North, 2020). Such datasets, related to elevation information from SRTM, ASTER, and GMTED; potential evapotranspiration; satellite and gauge rainfall, etc., may be considered a valuable resource for soil erosion and drought studies at both global and regional scales. However, the software has less community support compared with most of the previously mentioned open source software packages.

7.2.5 Others

Nowadays, considering the importance of free and open source software technology and free data, several institutes and agencies are developing such software products. These are user-friendly and capable of various GIS analyses. Examples include μDig, gvSIG, OpenJump, GeoDa, Kosmo, MapWindow, etc. However, this list is not exclusive. Among these, the Generalitat Valenciana Sistema d'Información Geográfica (gvSIG) is reported as the largest open source GIS software project in terms of financial and development resources (Steiniger and Bocher, 2009). Many of these software packages are quite efficient in handling specific tasks, e.g., GeoDa offers an efficient solution to exploratory spatial data analysis (Anselin et al., 2006). The gvSIG provides various gridded visualization techniques for temporal components (gvSIG, 2020). It offers significant three-dimensional terrain visualization similar to that of ESRI's ArcScene. This is quite useful in terrain monitoring for soil erosion. The JUMP, OpenJump, and Kosmo fall into the family of Open Java Unified Mapping Platform, which provides various facility management tools, including precision farming (Steiniger and Bocher, 2009).

7.3 MATERIALS AND METHODS

7.3.1 Study Area

The area chosen for this case study is Puruliya district of West Bengal, India. The latitudinal and longitudinal extents of the district are 22°40′ N to 23°15′ N and 85°45′ E to 86°45′ E (Figure 7.1). The westernmost district of West Bengal lies at the eastern fringe of Chotanagpur Plateau, covering an area of approximately 6259 km^2, with a general slope toward the east. The average elevation of the district is about 115 m above the mean sea level (Santra and Mitra, 2016). The residual plateau land projects from the Ranchi Plateau to the west. It functions as the watershed of the Subarnarekha–Kasai–Damodar group of rivers. These elevated upland areas are dominated by residual hills and isolated peaks. The broad erosional plain is marked by various interfluves and river valleys, e.g., the Damodar valley, the Subarnarekha valley, the Damodar–Dwarakeshwar interfluves, the Damodar–Kasai interfluves, etc. (Bhattacharya et al., 1985). The general slope of the river valley is from northwest to southeast. However, the Subarnarekha valley does not follow this trend. The Kangsabati or Kasai River is the main river, draining more than 60 percent of the district (Mukhopadhyay, 1989). All the above-mentioned rivers form the drainage system of the district. These non-perennial rivers are prone to flash floods that occur almost every year after a continued long spell of rainstorms in August over the Chotanagpur Plateau. The climate of the district is monsoonal in nature. The average annual rainfall decreases diagonally from the southwest flank to the northeast corner. The presence of crystalline basement hinders the groundwater development. Still, groundwater can be traced

Drought and Soil Erosion in Puruliya

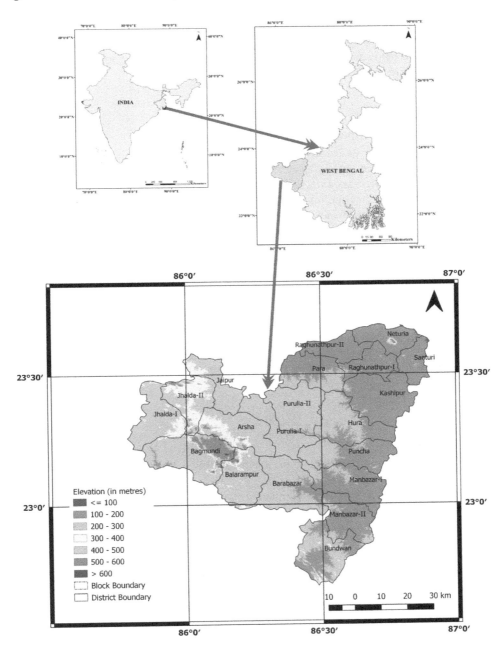

FIGURE 7.1 Study area.

in the surrounding areas of fractures, joints, and lineaments found along hard crystalline rocks and weathered sedimentary rock beds near major rivers. Groundwater adjacent to the major streams and within the unconsolidated sand, silt, and clay occurs as discontinuous patches (Mukhopadhyay, 1989). Red-brown Aridisol is the dominant soil group of this district. Latosol patches are observed in the upland areas, whereas alluvial deposits are found along the river courses (Santra and Mitra, 2016). The pH of the soil ranges from neutral to slightly acidic and gradually increases toward the east. The western and southern plateau lands are covered with dense forest. Otherwise, the occurrence of forest is patchy in nature. The residual hills, composed of intrusive rocks, are bare surfaces devoid of vegetation cover. The agricultural sector acts as a major source of local employment. However, low productivity in comparison with the neighboring districts is the major drawback of this sector. Agrarian fields are still mainly dependent upon erratic monsoonal rainfall and suffer from drought in the pre-monsoon season.

7.3.2 Methodology

7.3.2.1 Assessment and Monitoring of Vegetative Drought

Vegetative droughts can be assessed and monitored at regular intervals in terms of vegetation health from remotely sensed data (Tucker and Choudhury, 1987; Nemani et al., 2009). However, in this regard, the derived index images, which capture the inherent spectral information, offer better results than the raw satellite images. The Normalized Difference Vegetation Index (NDVI) (Rouse et al., 1974) is the most widely applied vegetation index to estimate vegetative and agricultural drought (Thenkabail et al., 2004; Karnieli et al., 2010; Gebrehiwot et al., 2011; Dutta et al., 2015; Park et al., 2016; Zhang et al., 2017; Jiao et al., 2019a; Kundu et al., 2020; Santra and Santra Mitra, 2020). To assess the vegetative drought of Puruliya district, the time series MODIS NDVI (MOD13C2) for the period from 2000 to 2013 was used. Since drought is a dry seasonal phenomenon, only the months from January to May were considered in this study. From MODIS NDVI data, the Vegetation Condition Index (VCI) was computed. The VCI acts as an indicator of vegetation cover as a function of maximum and minimum NDVI for a given area. The index normalizes the NDVI data and separates the short-term climate signal from the long-term ecological signal. It has proved to be a better indicator of water stress conditions than NDVI (Jain et al., 2009; Park et al., 2016; Zhang et al., 2017; Jiao et al., 2019b; Kundu et al., 2020; Santra and Santra Mitra, 2020). The VCI can be computed using Equation 7.1:

$$\text{VCI} = \frac{NDVIi - NDVImin}{NDVImax - NDVImin} \times 100 \tag{7.1}$$

Where *NDVIi*, *NDVImin*, and *NDVImax* are mean monthly, long-term minimum, and long-term maximum NDVI values, respectively.

The VCI ranges from 0 to 100, showing relative changes in vegetation conditions from worst to ideal (Kogan, 1995). In extremely dry months, the VCI value lies close to or equal to zero, while in optimal conditions, the index value rises close to 100.

Similarly, the Temperature Condition Index (TCI), which denotes the temperature-related vegetation stress, was computed for the same months from the long-term mean monthly MODIS Land Surface Temperature (LST) (MOD11C3). It has been proved that the subtle change in vegetation health due to thermal stress can be addressed through the analysis of TCI data (Kogan, 1995, 1997). The TCI can be computed using the following equation:

$$\text{TCI} = \frac{Tempmax - Tempi}{Tempmax - Tempmin} \times 100 \tag{7.2}$$

Where *Tempi, Tempmin, and Tempmax* are mean monthly, long-term minimum, and long-term maximum LST values, respectively.

After that, the Vegetation Health Index (VHI) was computed by taking the mean values of VCI and TCI for each cell. This VHI represents the overall vegetation health condition of the study area (Hazra et al., 2017; Santra and Santra Mitra, 2020).

$$\text{VHI} = \frac{VCI + TCI}{2} \tag{7.3}$$

Drought and Soil Erosion in Puruliya

Finally, the drought frequency was computed to show the total number of drought-affected months for the considered temporal period. The delineation of the drought-affected months from the VHI threshold values was done after collecting relevant information from the Block Development Offices of the district.

7.3.2.2 Modeling Soil Erosion:

Soil erosion is a combination of three complex processes: detachment of soil particles from the main soil mass, transportation of detached soil particles, and deposition of transported soil particles to the soil mass as sediment (USDA, 2011). Together, these processes cause both onsite and offsite effects. The onsite effects include loss of organic matter, nutrients, soil depth, fertility, and productivity of soil. The offsite effects encompass reduction of the carrying capacity of rivers due to sediment deposition, clogging up irrigation canals, water inundation due to flooding, etc. (Morgan, 2005; Ketema and Dwarakish, 2019). Therefore, it is essential to model the soil erosion and identify the possible erosion risk zones in an area. Several models were developed based on the type of erosion processes, the spatio-temporal scales, and the implementation criteria of remote sensing and GIS (Mitasova et al., 2013; Karan et al., 2019). Statistical models were developed depending on the controlling factors that are derived from the ground measurements. These are fast but demand long-term data (Beach, 1987). The Universal Soil Loss Equation (USLE) was developed in the mid-1960s and enhanced in the late 1970s (Wischmeier and Smith, 1965, 1978). It is the most widely applied and recognized empirical erosion model to predict the long-term average soil loss in relation to rainfall, soil, slope gradient and length, vegetation cover, land use and land cover (LULC), and land management practices (Wischmeier and Smith, 1978). The USLE model was improved, and the Revised USLE (RUSLE) model was developed. The model estimates the long-term annual soil degradation rate from raindrop impact and runoff (Renard et al., 1997). It retains all the six factors of USLE, but the mechanism of determining these factors has been revised (Ketema and Dwarakish, 2019). This revised version of USLE has been widely applied for the precise determination of soil erosion (Arekhi et al., 2012). RUSLE successfully integrates the remote sensing and GIS platforms to identify the spatial variation of erosion potential zones in a cost-effective and time-efficient manner (Lufafa et al., 2003; İrvem et al., 2007; Dabral et al., 2008; Pal and Chakrabortty, 2019). The USLE model was also modified into Modified USLE (MUSLE) to estimate the sediment yield for a certain storm event (Williams, 1975). The factors of MUSLE include peak discharge and runoff variables, replacing the rainfall erosivity factor of USLE (Ketema and Dwarakish, 2019). In this case study, the potential soil erosion of the Puruliya district was estimated using RUSLE, as this model yields acceptable results in many countries under different bio-climatic conditions (Lu et al., 2004; Alkharabsheh et al., 2013; Kumar and Kushwaha, 2013; Napoli et al., 2016; Tadesse et al., 2017).

The soil erosion of the district was estimated using a RUSLE model that integrates five factors by Equation 7.4.

$$A = R \times K \times LS \times C \times P \tag{7.4}$$

where A = average annual soil loss per unit area (t ha^{-1}yr^{-1})
R = rainfall–runoff erosivity factor (MJ mm ha^{-1}h^{-1}yr^{-1})
K = soil erodibility factor (t ha h MJ^{-1}mm^{-1})
LS = slope length and steepness factor (dimensionless)
C = land cover and management factor (dimensionless)
P = conservation support practice factor (dimensionless)

These factors were resampled to 30 m spatial resolution under a common datum of WGS84 and a uniform projection system of Universal Transverse Mercator's (UTM) projection with zone 45. The layers were multiplied at the cell level to obtain the potential loss of soil for the entire district.

The rainfall erosivity (R) is the most significant controlling force and the dynamic indicator of soil disintegration and transportation (Antoneli et al., 2018; Rodrigo Comino et al., 2018). Several algorithms were developed to estimate this factor based on hourly, daily, monthly, and yearly rainfall data (Mondal et al., 2016; Amanambu et al., 2019; Hu et al., 2019). However, this requires long-term information. In countries like India, it is difficult to obtain reliable long-term data for every region. Therefore, the R factor was calculated from the long-term average annual rainfall data (1910–2010) using Equation 7.5, suggested by Singh and Rambabu (1981).

$$R = 81.5 + 0.375 \times P \tag{7.5}$$

(Singh and Rambabu, 1981)
where R = rainfall–runoff erosivity factor (MJ mm ha^{-1}h^{-1}yr^{-1})
P = average annual rainfall in mm (340 mm < P < 3500 mm)]

Soil erodibility (K) describes the resistance property of soil material to the raindrop impact on the soil surface and reflects the vulnerability of soil to erosion (Pal and Shit, 2017; Yesuph and Dagnew, 2019). Therefore, the K factor may be considered as an empirical index of soil vulnerability to erosion. Different physico-chemical soil property point data were collected from the Department of Agriculture, Government of West Bengal, India and National Bureau of Soil Survey & Land Use Planning (NBSS & LUP), Government of India to prepare the K factor map using Equation 7.6, from Wischmeier and Smith (1978).

$$K = 2.8 \times 10^{-7} \times (12 - OM) \times M^{1.14} + 4.3 \times 10^{-3}(s-2) + 3.3 \times 10^{-3}(p-3) \tag{7.6}$$

(Wischmeier and Smith, 1978)
Where K = soil erodibility factor (t ha h MJ^{-1}mm^{-1})
M = (% of sand + % of silt) × (100 – % of clay)
OM = % of organic matter
s = soil structure code (4 for massive, 3 for coarse granular, 2 for fine granular, and 1 for very fine granular)
p = soil permeability code (6 for very slow, 5 for slow, 4 for moderately fast to slow, 3 for moderately fast, 2 for fast to moderately fast, and 1 for fast)]

The slope LS factor expresses the joint effects of length (L) and steepness (S) of the area's slope on the soil erosion rate (Nyesheja et al., 2019). In Puruliya district, the LS factor was estimated from the ASTER GDEM data using Equation 7.7 (Wischmeier and Smith, 1978; Thomas et al., 2018).

$$LS = \left(\frac{\lambda}{22.13}\right)^m \times (65.41 \times \sin^2\theta + 4.56 \times \sin\theta + 0.065) \tag{7.7}$$

(Wischmeier and Smith, 1978)
Where λ = slope length in m
θ = slope angle
m = 0.5 (if % of slope ≥5)
0.4 (if 3.6 ≤ % of slope ≤ 4.9)
0.3 (if 1 ≤ % of slope ≤ 3.5)
0.2 (if % of slope <1)

The dimensionless cover and management (C) factor highlights the influence and mechanisms by which different land use and cover types affect soil erosion rates (Renard et al., 1997). Originally, this factor used the information related to the roughness, moisture, and management condition of the soil and the role of plant canopy and agricultural dregs as soil cover (Yesuph and Dagnew, 2019).

However, it is very difficult to evaluate and combine all the necessary information in data-sparse regions (Farhan and Nawaiseh, 2015). Therefore, the C factor map of the district was prepared from the LULC map. The LULC map was generated from the Landsat 8 OLI image. However, before preparing the LULC map, the Landsat 8 OLI image was corrected radiometrically, as the corrected image provides earth surface reflectance after omitting possible atmospheric attenuation effects (Santra et al., 2019).

The conservation support practice (P) factor represents the ratio of soil erosion with a specific support practice to the equivalent loss with up- and down-slope tillage (Wischmeier and Smith, 1978; Renard et al., 1997). It was estimated from the classified LULC map. However, due to constraints on identifying the actual field characteristics, the P factor is considered the least reliable among all the RUSLE factors (Renard et al., 1991). Finally, the soil erosion potential was estimated by multiplying the above-mentioned factors.

7.4 RESULTS AND DISCUSSION

7.4.1 Spatio-Temporal Severity of Drought

Investigation through the application of meteorological indices is not efficient enough to capture vegetation health. According to Caparrini and Manzella (2009), even in extremely scanty rainfall conditions, certain species of crop and vegetation will not die. Instead, a small amount of stored water is beneficial for their critical growth phase. Therefore, the simple NDVI-based drought index VHI yields better results in successful monitoring of vegetational drought (Ghoneim et al., 2017).

The MODIS NDVI and LST products were used to estimate the drought risk of the study area. The NDVI values range from −1 to 1. Healthy vegetation is located near NDVI value 1, while low positive values near 0 indicate a bare and degraded land surface. Negative NDVI values show water bodies and built-up areas. The VCI and TCI images range from 0 to 85% and 0 to 100%, respectively, for the considered months of January to May from 2000 to 2013. The VHI values range from 0 to 70% in this temporal span. From the values, it is observed that dry conditions are dominant in the central, northern, eastern, and extreme western parts of the Puruliya district. This dry condition is more pronounced from February to April. In general, the drought severity has temporally increased from 2000 to 2013, but stronger effects were observed in 2008 and 2009. The mean VHI values for each cell show the cell-wise average VHI value for the entire time series considered in this study. The value ranges from 6 to 58. The average mean VHI value is 31 for the whole study area. High values of VHI are found in the adjacent areas of Damodar River and other perennial channels. A temporal variation of VHI values over the agricultural lands is observed due to the seasonal characteristics of rainfall and cropping patterns.

However, the mean VHI values do not efficiently reflect the drought condition of the district. Therefore, the drought frequency map (Figure 7.2) was developed to address the cell-wise prevailing drought conditions within this period (Santra and Santra Mitra, 2020). The supportive information for the VHI threshold values was taken from the literature and the Block Development offices of the district. The drought frequency ranges from 0 to 29. It is observed that the western, middle, and lower parts of the district show frequent occurrence of drought, whereas the northern part and some patches around the central part of the district are less affected by drought severity. Out of the 70 months under consideration, the mean drought frequency was estimated to be 7 months. Table 7.1 shows five drought frequency classes with respective coverage areas. Overall, the district is under low to moderate frequency of drought.

7.4.2 Threat of Soil Erosion

Research on soil erosion has highlighted that rainfall is one of the most sensitive eroding agents (Dabral et al., 2008; Fayas et al., 2019; Ganasri and Ramesh, 2016). The R factor depicts the rainfall–runoff erosivity. The long-term (1910–2010) rainfall data from India Meteorological

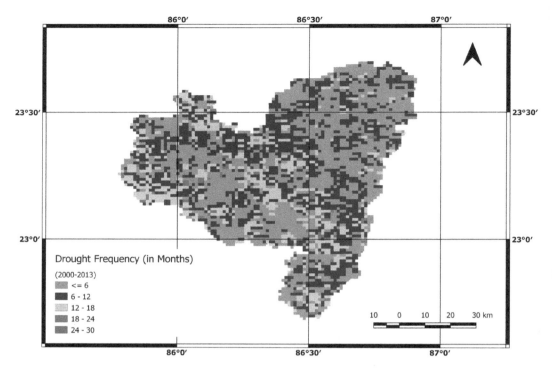

FIGURE 7.2 Frequency of drought.

TABLE 7.1
Severity of Drought

Categories	Very Low Frequency	Low Frequency	Moderate Frequency	High Frequency	Very High Frequency
Frequency of Drought (in months)	0–6	6–12	12–18	18–24	24–30
Percentage of Area	45.6	35.0	16.9	02.4	00.1

Department (IMD) weather station data were used to generate the R factor raster layer. It is observed that the R factor increases proportionately with the increase of intensity and amount of average annual rainfall in the study area. The average amount of precipitation was estimated to be 1330 mm. The R factor varies from 537 to 584 MJ mm ha^{-1} h^{-1} yr^{-1} with a mean value of 563 MJ mm ha^{-1} h^{-1} yr^{-1} (Figure 7.3).

The K factor quantifies the soil's cohesive properties based on its physical and chemical characteristics (Gupta and Kumar, 2017). Therefore, soil structure, texture, organic matter, and permeability area are considered to be important factors in deciding erodibility (Shinde et al., 2011). The K factor value ranges from 0.1 to 0.5 t ha h MJ^{-1} mm^{-1}. Higher K values are observed in the fine loamy soil region of the southern part, whereas lower K values are reported from the fine to gravelly loamy soil-covered areas (Figure 7.4).

The combined LS factor portrays the effect of topography on soil erosion. The higher the LS factor value, the more susceptible the soil is to erosion. The LS factor map was derived from the ASTER GDEM. The elevation in the area ranges from 0 to 697 m, and the LS factor value ranges from 0 to 140. It is observed that harvested agricultural fields and bare ground are the places showing high LS factor values. However, each of the height information sources (e.g., digital elevation

Drought and Soil Erosion in Puruliya

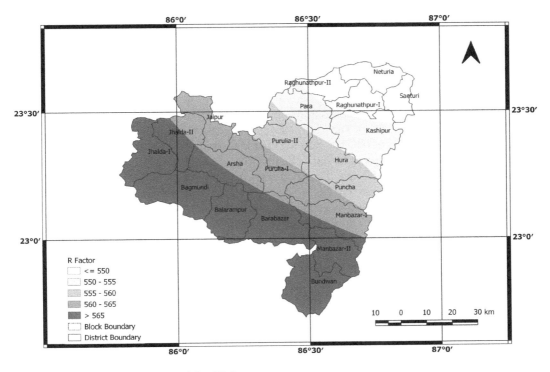

FIGURE 7.3 Rainfall–runoff erosivity (R) factor.

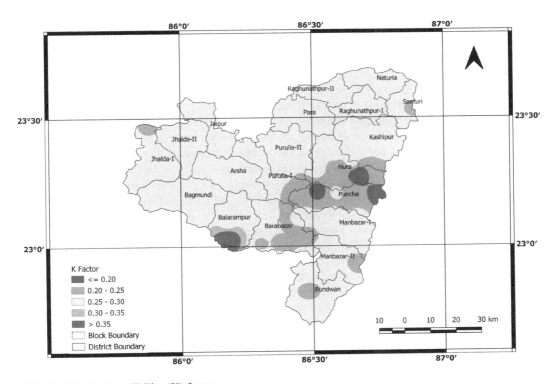

FIGURE 7.4 Soil erodibility (K) factor.

model [DEM], Survey of India [SOI] toposheet, etc.) has its own limitations in creating the LS factor map (Figure 7.5).

The C factor map shows the role of LULC in managing erosion. Areas covered with good vegetation cover show low C factor values in comparison with ground having less or no vegetation growth (Figure 7.6). In agricultural fields, the C factor values are dependent upon the type of crop and the management practices.

The P factor ranges from 0 to 1, representing excellent to deprived conservation practices, respectively (Ganasri and Ramesh, 2016). The natural vegetated areas in the lower and western parts of the district show higher P values of 0.8. The rain-fed crop lands where irrigation facilities are available show P factor values ranging from 0.2 to 0.4 (Figure 7.7).

Combining all the RUSLE factors, the soil erosion map was generated. The soil erosion of the Puruliya district ranges from 0 to 150 t ha^{-1} yr^{-1} with a mean erosion value of 76 t ha^{-1} yr^{-1}. The study area was finally classified into five potential erosion severity zones (Figure 7.8). It is noteworthy that due to their lesser areal contribution, some classes (more erosion-prone areas) lack visual clarity. However, magnification of the image would enhance their visibility. To support this, an inset of a specific area is provided in Figure 7.8.

It is estimated that around half of the area is in the low-erosion zone, whereas one-fourth of the area, covering mainly the rills and gullies, comes within the very high to severe soil erosion–prone zones (Table 7.2). The lack of moisture in the soil, the non-perennial condition of the tributaries, and the proneness of the district to drought are probably the causes of the high risk of soil erosion. Also, agricultural fields where traditional cultivation practices are mainly followed fall under low to medium risk of soil erosion. However, modern farming practices and irrigation facilities may convert such areas to low–erosion risk zones in the future. It is also observed that among the five factors, the LS factor contributed most to the erosion of the area.

7.5 CONCLUSION

Remote sensing and GIS provide improved technology to study, assess, and monitor different natural as well as man-made hazards like drought and soil erosion. The uniqueness of the technology is the offering of solutions at the user-required scale, from global to local. This characteristic is substantially beneficial from the decision- and policy-making points of view. To achieve this goal, since the origin of the subject, several datasets and software packages have been developed, utilized, and modified for better analysis. Now, in the era of open source and free software and data, the community is wide enough. New findings are coming out from different corners of the world. As a result, the precision and accuracy of the analysis and prediction of such hazards are increasing day by day.

In this chapter, light has been shed upon the utility and applicability of free data and open source software packages like QGIS in the geo-environmental study of drought and soil erosion of Puruliya district of West Bengal, India. Vegetative drought for the entire district was assessed and analyzed using the freely available MODIS global NDVI and LST products. VHI was found to be a reliable indicator of vegetative drought assessment for the period from 2000 to 2013. The spatiotemporal analysis of VHI helped to identify the critical areas where proper land management is required. The average frequency of drought-affected months was estimated to be 7 months in this period, though the range extends from 0 to 29 months. On the other hand, the severity of the creeping soil erosion hazard was assessed using a RUSLE model. This model is widely accepted because of its limited data requirements and is free to acquire. The average annual soil erosion of the district was estimated to be 76 tonnes per hectare per year. The LS factor was found to be the most important factor controlling soil erosion. However, this is subject to the sensitivity of the LS factor to different freely available DEM and other height information sources. Overall, it is observed that in this district, these two hazards are somehow positively interrelated. It has been observed that the risk of soil erosion is more pronounced in areas of higher drought frequency. Similarly, a high risk

Drought and Soil Erosion in Puruliya

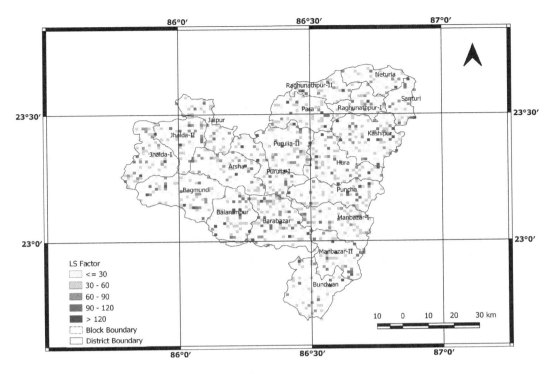

FIGURE 7.5 Slope length and steepness (LS) factor.

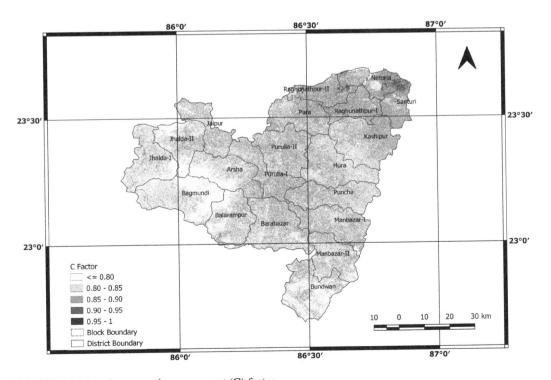

FIGURE 7.6 Land cover and management (C) factor.

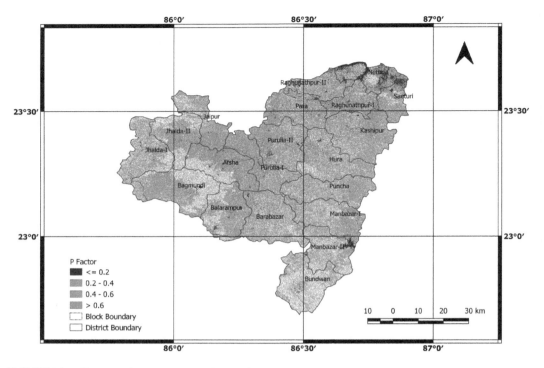

FIGURE 7.7 Conservation support practice (P) factor.

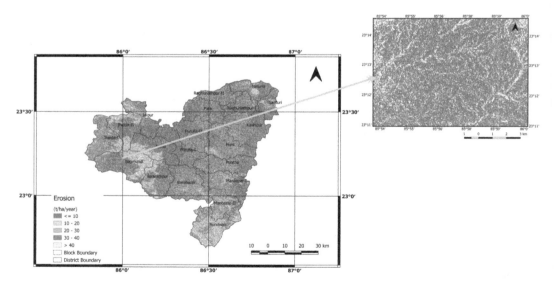

FIGURE 7.8 Average annual soil loss.

of vegetative drought is observed where the risk of soil erosion is quite high. A plausible explanation could be found in the status of the antecedent soil moisture conditions. This could open a new direction of research. In general, this chapter will surely help to grasp the ideas and methodologies for assessing these two hazards. For planners and policy-makers, this chapter provides a case study in which the critical areas are earmarked for necessary planning and management.

TABLE 7.2
Severity of Different Soil Erosion Zones

Categories	Soil Erosion				
	Low	Moderate	High	Very High	Severe
Average annual soil loss in t ha^{-1}yr^{-1}	≤10	10–20	20–30	30–40	≥40
Percentage of area	65	10	6	4	15
Average frequency of drought in months (2000–2013)	7	9	10	10	12

ACKNOWLEDGMENT

The authors are sincerely grateful to the University Grants Commission (UGC), Government of India, for providing financial support to carry out the research work under the BSR project scheme (No. F. 20-7 (19) / 2012 (BSR)).

REFERENCES

52North. 2020. *Innovative Open Source Software* [cited 10th May, 2020]. Available from https://52north.org/software/software-projects/.

AghaKouchak, A., A. Farahmand, F. S. Melton, J. Teixeira, M. C. Anderson, B. D. Wardlow, and C. R. Hain. 2015. Remote sensing of drought: progress, challenges and opportunities. *Reviews of Geophysics* 53 (2):452–480.

Alkharabsheh, M. Minwer, T. K. Alexandridis, G. Bilas, N. Misopolinos, and N. Silleos. 2013. Impact of land cover change on soil erosion hazard in northern Jordan using remote sensing and GIS. *Procedia Environmental Sciences* 19:912–921.

Amanambu, Amobichukwu C., Lanhai Li, Christiana N. Egbinola, Omon A. Obarein, Christophe Mupenzi, and Ditao Chen. 2019. Spatio-temporal variation in rainfall-runoff erosivity due to climate change in the Lower Niger Basin, West Africa. *CATENA* 172:324–334.

Anselin, Luc, Ibnu Syabri, and Youngihn Kho. 2006. GeoDa: an introduction to spatial data analysis. *Geographical Analysis* 38 (1):5–22.

Antoneli, V., E. A. Rebinski, J. A. Bednarz, J. Rodrigo-Comino, S. D. Keesstra, A. Cerdà, M. Pulido Fernández. 2018. Soil erosion induced by the introduction of new pasture species in a Faxinal Farm of Southern Brazil. *Geosciences* 8 (5):166, https://doi.org/10.3390/geosciences8050166.

Arekhi, Saleh, Yaghoub Niazi, and Aman Mohammad Kalteh. 2012. Soil erosion and sediment yield modeling using RS and GIS techniques: a case study, Iran. *Arabian Journal of Geosciences* 5 (2):285–296.

Beach, T. 1987. *A Review of Soil Erosion Erosion Modelling. Design for a Water-Resources GIS*. Water Resource Research Centre, University of Minnesota.

Bhattacharya, B. K., B. R. Chakraborti, N. N. Sen, et al. 1985. West Bengal District *Gazetters: Puruliya*. Kolkata, India: Govt. of West Bengal.

Brady, N. C., and R. C. Weil. 2012. *The Nature and Properties of Soils*. New Delhi: Pearson Education.

Caparrini, F., and F. Manzella. 2009. Hydrometeorological and vegetation indices for the drought monitoring system in Tuscany Region, Italy. *Advances in Geosciences* 17:105–110.

Ciais, Ph, M. Reichstein, N. Viovy, et al. 2005. Europe-wide reduction in primary productivity caused by the heat and drought in 2003. *Nature* 437 (7058):529–533.

Coetzee, Serena, Ivana Ivánová, Helena Mitasova, and Maria A. Brovelli. 2020. Open geospatial software and data: a review of the current state and a perspective into the future. *ISPRS International Journal of Geo-Information* 9 (2):90, https://doi.org/10.3390/ijgi9020090.

Dabral, P. P., Neelakshi Baithuri, and Ashish Pandey. 2008. Soil erosion assessment in a hilly catchment of North Eastern India using USLE, GIS and remote sensing. *Water Resources Management* 22 (12):1783–1798.

Du, Lingtong, Qingjiu Tian, Tao Yu, et al. 2013. A comprehensive drought monitoring method integrating MODIS and TRMM data. *International Journal of Applied Earth Observation and Geoinformation* 23:245–253.

Duarte, L., A. C. Teodoro, J. A. Gonçalves, D. Soares, and M. Cunha. 2016. Assessing soil erosion risk using RUSLE through a GIS open source desktop and web application. *Environmental Monitoring and Assessment* 188 (6):351.

Dutta, D., A. Kundu, N. R. Patel, S. K. Saha, and A. R. Siddiqui. 2015. Assessment of agricultural drought in Rajasthan (India) using remote sensing derived Vegetation Condition Index (VCI) and Standardized Precipitation Index (SPI). *The Egyptian Journal of Remote Sensing and Space Science* 18 (1):53–63.

Eisazadeh, L., R. Sokouti, M. Homaee, and E. Pazira. 2012. Comparison of empirical models to estimate soil erosion and sediment yield in micro catchments. *Eurasian Journal of Soil Science* 1:28–33.

Farhan, Yahya, and Samer Nawaiseh. 2015. Spatial assessment of soil erosion risk using RUSLE and GIS techniques. *Environmental Earth Sciences* 74 (6):4649–4669.

Fayas, Cassim Mohamed, Nimal Shantha Abeysingha, Korotta Gamage Shyamala Nirmanee, Dinithi Samaratunga, and Ananda Mallawatantri. 2019. Soil loss estimation using RUSLE model to prioritize erosion control in KELANI river basin in Sri Lanka. *International Soil and Water Conservation Research* 7 (2):130–137.

Ganasri, B. P., and H. Ramesh. 2016. Assessment of soil erosion by RUSLE model using remote sensing and GIS – a case study of Nethravathi Basin. *Geoscience Frontiers* 7 (6):953–961.

Gebrehiwot, Tagel, Anne van der Veen, and Ben Maathuis. 2011. Spatial and temporal assessment of drought in the Northern highlands of Ethiopia. *International Journal of Applied Earth Observation and Geoinformation* 13 (3):309–321.

Ghoneim, E., A. Dorofeeva, M. Benedetti, D. Gamble, L. Leonard, and M. AbuBakr. 2017. Vegetation drought analysis in Tunisia: a geospatial investigation. *Journal of Atmospheric & Earth Sciences* 1:002.

Ghosh, K., S. K. De, S. Bandyopadhyay, and S. Saha. 2013. Assessment of soil loss of the Dhalai River Basin, Tripura, India using USLE. *International Journal of Geosciences* 4 (1):11–23.

Gidey, Eskinder, Oagile Dikinya, Reuben Sebego, Eagilwe Segosebe, and Amanuel Zenebe. 2018. Analysis of the long-term agricultural drought onset, cessation, duration, frequency, severity and spatial extent using Vegetation Health Index (VHI) in Raya and its environs, Northern Ethiopia. *Environmental Systems Research* 7 (1):13.

GRASS GIS. 2020. *GRASS GIS – Bringing Advanced Geospatial Technologies to the World* [cited 9th May, 2020]. Available from https://grass.osgeo.org/.

Guo, Enliang, Xingpeng Liu, Jiquan Zhang, et al. 2017. Assessing spatiotemporal variation of drought and its impact on maize yield in Northeast China. *Journal of Hydrology* 553:231–247.

Gupta, Surya, and Suresh Kumar. 2017. Simulating climate change impact on soil erosion using RUSLE model—a case study in a watershed of mid-Himalayan landscape. *Journal of Earth System Science* 126 (3):43.

gvSIG. 2020. *Conoce gvSIG Desktop, el Sistema de Información Geográfica libre* [cited 9th May, 2020]. Available from www.gvsig.com/es/productos/gvsig-desktop.

Han, Yang, Ziying Li, Chang Huang, et al. 2020. Monitoring droughts in the greater Changbai Mountains using multiple remote sensing-based drought indices. *Remote Sensing* 12 (3):530, https://doi.org/10.3390/rs12030530.

Hao, Cui, Jiahua Zhang, and Fengmei Yao. 2015. Combination of multi-sensor remote sensing data for drought monitoring over Southwest China. *International Journal of Applied Earth Observation and Geoinformation* 35:270–283.

Hazra, S., S. Roy, and S. Mitra. 2017. *Enhancing Adaptive Capacity and Increasing Resilience of Small and Marginal Farmers of Purulia and Bankura Districts, West Bengal to Climate Change.* Kolkata: School of Oceanographic Studies, Jadavpur University.

Hu, Sai, Long Li, Longqian Chen, et al. 2019. Estimation of soil erosion in the Chaohu Lake Basin through modified soil erodibility combined with gravel content in the RUSLE model. *Water* 11 (9):1806, https://doi.org/10.3390/w11091806.

Hugentobler, M. 2008. Quantum GIS. In *Encyclopedia of GIS*, edited by S. Shekhar, and H. Xiong. Boston, MA: Springer-Verlag, 935–939.

İrvem, Ahmet, Fatih Topaloğlu, and Veli Uygur. 2007. Estimating spatial distribution of soil loss over Seyhan River Basin in Turkey. *Journal of Hydrology* 336 (1):30–37.

ITC. 2020. *Integrated Land and Water Information System (ILWIS)* [cited 10th May, 2020]. Available from www.itc.nl/ilwis/.

Jain, S. K., R. Keshri, A. Goswami, A. Sarkar, and A. Chaudhry. 2009. Identification of drought-vulnerable areas using NOAA AVHRR data. *International Journal of Remote Sensing* 30 (10):2653–2668.

Jiao, Wenzhe, Chao Tian, Qing Chang, Kimberly A. Novick, and Lixin Wang. 2019a. A new multi-sensor integrated index for drought monitoring. *Agricultural and Forest Meteorology* 268:74–85.

Jiao, Wenzhe, Lixin Wang, Kimberly A. Novick, and Qing Chang. 2019b. A new station-enabled multi-sensor integrated index for drought monitoring. *Journal of Hydrology* 574:169–180.

Karamage, Fidele, Chi Zhang, Xia Fang, L. Nahayo, A. Kayiranga, and J. B. Nsengiyumva. 2017. Modeling rainfall-runoff response to land use and land cover change in Rwanda (1990–2016). *Water* 9 (2):1–24.

Karan, Shivesh Kishore, Somaparna Ghosh, and Sukha Ranjan Samadder. 2019. Identification of spatially distributed hotspots for soil loss and erosion potential in mining areas of Upper Damodar Basin – India. *CATENA* 182:104144.

Karnieli, Arnon, Nurit Agam, Rachel T. Pinker, et al. 2010. Use of NDVI and land surface temperature for drought assessment: merits and limitations. *Journal of Climate* 23 (3):618–633.

Ketema, A., and G. S. Dwarakish. 2019. Water erosion assessment methods: a review. *ISH Journal of Hydraulic Engineering*:1–8, DOI: 10.1080/09715010.2019.1567398.

Kogan, F. N. 1995. Application of vegetation index and brightness temperature for drought detection. *Advances in Space Research* 15 (11):91–100.

Kogan, Felix N. 1997. Global drought watch from space. *Bulletin of the American Meteorological Society* 78 (4):621–636.

Kumar, Suresh, and S. P. S. Kushwaha. 2013. Modelling soil erosion risk based on RUSLE-3D using GIS in a Shivalik sub-watershed. *Journal of Earth System Science* 122 (2):389–398.

Kundu, A., N. R. Patel, D. M. Denis, and D. Dutta. 2020. An estimation of hydrometeorological drought stress over the central part of India using geo-information technology. *Journal of the Indian Society of Remote Sensing* 48 (1):1–9.

Lal, R. 2001. Soil degradation by erosion. *Land Degradation & Development* 12 (6):519–539.

Lu, D., G. Li, G. S. Valladares, and M. Batistella. 2004. Mapping soil erosion risk in Rondônia, Brazilian Amazonia: using RUSLE, remote sensing and GIS. *Land Degradation & Development* 15 (5):499–512.

Lufafa, A., M. M. Tenywa, M. Isabirye, M. J. G. Majaliwa, and P. L. Woomer. 2003. Prediction of soil erosion in a Lake Victoria basin catchment using a GIS-based Universal Soil Loss model. *Agricultural Systems* 76 (3):883–894.

Mitasova, H., M. Barton, I. Ullah, J. Hofierka, and R. S. Harmon. 2013. 3.9 GIS-based soil erosion modeling. In *Treatise on Geomorphology*, edited by J. F. Shroder. San Diego: Academic Press, pp. 228–258.

Mondal, Arun, Deepak Khare, and Sananda Kundu. 2016. Change in rainfall erosivity in the past and future due to climate change in the central part of India. *International Soil and Water Conservation Research* 4 (3):186–194.

Morgan, R. P. C. 2005. *Soil Erosion and Conservation*. 3rd ed. Blackwell Science Ltd. National Soil Resources Institute, Cranfield University, pp. 1–316, ISBN: 978-1-405-11781-4.

Mukhopadhyay, M. 1989. *Resource Planning and its Impact on Rural Development (A Case Study of Purulia District)*. Allahabad, India: Vohra Publishers & Distributors.

Napoli, Marco, Stefano Cecchi, Simone Orlandini, Gabriele Mugnai, and Camillo A. Zanchi. 2016. Simulation of field-measured soil loss in Mediterranean hilly areas (Chianti, Italy) with RUSLE. *CATENA* 145:246–256.

Nemani, Ramakrishna, Hirofumi Hashimoto, Petr Votava, et al. 2009. Monitoring and forecasting ecosystem dynamics using the Terrestrial Observation and Prediction System (TOPS). *Remote Sensing of Environment* 113 (7):1497–1509.

Nyesheja, Enan M., Xi Chen, Attia M. El-Tantawi, Fidele Karamage, Christophe Mupenzi, and Jean Baptiste Nsengiyumva. 2019. Soil erosion assessment using RUSLE model in the Congo Nile Ridge region of Rwanda. *Physical Geography* 40 (4):339–360.

Orville, Harold D. 1990. AMS Statement on Meteorological Drought*. *Bulletin of the American Meteorological Society* 71 (7):1021–1025.

Pal, Subodh Chandra, and Rabin Chakrabortty. 2019. Simulating the impact of climate change on soil erosion in sub-tropical monsoon dominated watershed based on RUSLE, SCS runoff and MIROC5 climatic model. *Advances in Space Research* 64 (2):352–377.

Pal, Subodh Chandra, and Manisa Shit. 2017. Application of RUSLE model for soil loss estimation of Jaipanda watershed, West Bengal. *Spatial Information Research* 25 (3):399–409.

Park, Seonyoung, Jungho Im, Eunna Jang, and Jinyoung Rhee. 2016. Drought assessment and monitoring through blending of multi-sensor indices using machine learning approaches for different climate regions. *Agricultural and Forest Meteorology* 216:157–169.

Phinzi, Kwanele, and Njoya Silas Ngetar. 2019. The assessment of water-borne erosion at catchment level using GIS-based RUSLE and remote sensing: a review. *International Soil and Water Conservation Research* 7 (1):27–46.

Prasannakumar, V., H. Vijith, S. Abinod, and N. Geetha. 2012. Estimation of soil erosion risk within a small mountainous sub-watershed in Kerala, India, using Revised Universal Soil Loss Equation (RUSLE) and geo-information technology. *Geoscience Frontiers* 3 (2):209–215.

QGIS. 2020. *QGIS—A Free and Open Source Geographic Information System* [cited 10th May, 2020]. Available from www.qgis.org/en/site/.

Quiring, Steven M., and Srinivasan Ganesh. 2010. Evaluating the utility of the Vegetation Condition Index (VCI) for monitoring meteorological drought in Texas. *Agricultural and Forest Meteorology* 150 (3):330–339.

Renard, K. G., G. R. Foster, G. A. Weesies, D. K. McCool, and D. C. Yoder. 1997. Predicting soil erosion by water: a guide to conservation planning with the Revised Universal Soil Loss Equation (RUSLE). In *Agricultural Handbook No: 703*. Washington, DC: United States Department of Agriculture.

Renard, Kenneth G., George R. Foster, Glenn A. Weesies, and Jeffrey P. Porter. 1991. RUSLE: revised universal soil loss equation. *Journal of Soil and Water Conservation* 46 (1):30.

Rhee, Jinyoung, Jungho Im, and Gregory J. Carbone. 2010. Monitoring agricultural drought for arid and humid regions using multi-sensor remote sensing data. *Remote Sensing of Environment* 114 (12):2875–2887.

Rodrigo Comino, Jesús, Saskia D. Keesstra, and Artemi Cerdà. 2018. Connectivity assessment in Mediterranean vineyards using improved stock unearthing method, LiDAR and soil erosion field surveys. *Earth Surface Processes and Landforms* 43 (10):2193–2206.

Rouse, J. W., R. W. Haas, J. A. Schell, D. W. Deering, and J. C. Harlan. 1974. *Monitoring the Vernal Advancement and Retrogradation (Greenhouse Effect) of Natural Vegetation*. NASA / GSFCT Type III Final Report, NASA.

SAGA. 2020. *System for Automated Geoscientific Analyses* [cited 9th May, 2020]. Available from www.saga-gis.org/en/index.html.

Santra, A., and S. S. Mitra. 2016. Multi criteria decision analysis for assessing crop suitability in drought prone Puruliya district, West Bengal, India. *Journal of Environment* 5:7–12.

Santra, Abhisek, and Shreyashi Santra Mitra. 2020. Space-time drought dynamics and soil erosion in Puruliya District of West Bengal, India: a conceptual design. *Journal of the Indian Society of Remote Sensing*. doi: https://doi.org/10.1007/s12524-020-01147-y.

Santra, Abhisek, Shreyashi Santra Mitra, Debashis Mitra, and Ashis Sarkar. 2019. Relative radiometric normalisation – performance testing of selected techniques and impact analysis on vegetation and water bodies. *Geocarto International* 34 (1):98–113.

Sarkar, Tanmoy, and Mukunda Mishra. 2018. Soil erosion susceptibility mapping with the application of logistic regression and artificial neural network. *Journal of Geovisualization and Spatial Analysis* 2 (1):8.

Shinde, Vipul, Arabinda Sharma, Kamlesh N. Tiwari, and Manjushree Singh. 2011. Quantitative determination of soil erosion and prioritization of micro-watersheds using remote sensing and GIS. *Journal of the Indian Society of Remote Sensing* 39 (2):181–192.

Singh, G., and S. Rambabu. 1981. Soil loss prediction research in India. Dehradun: CSWCR & TI.

Steiniger, Stefan, and Erwan Bocher. 2009. An overview on current free and open source desktop GIS developments. *International Journal of Geographical Information Science* 23 (10):1345–1370.

Tadesse, Lemlem, K. V. Suryabhagavan, G. Sridhar, and Gizachew Legesse. 2017. Land use and land cover changes and soil erosion in Yezat Watershed, North Western Ethiopia. *International Soil and Water Conservation Research* 5 (2):85–94.

Thenkabail, P. S., M. S. D. N. Gamage, and V. U. Smakhtin. 2004. *Use of Remote Sensing Data for Drought Assessment and Monitoring in South West Asia*. Research Report 85. Colombo, Sri Lanka: International Water Management Institute.

Thomas, Jobin, Sabu Joseph, and K. P. Thrivikramji. 2018. Estimation of soil erosion in a rain shadow river basin in the southern Western Ghats, India using RUSLE and transport limited sediment delivery function. *International Soil and Water Conservation Research* 6 (2):111–122.

Tucker, Compton J., and Bhaskar J. Choudhury. 1987. Satellite remote sensing of drought conditions. *Remote Sensing of Environment* 23 (2):243–251.

USDA. 2011. *National Agronomy Manual*. 4th ed. Vol. 190-V-NAM, part 501, Natural Resource Conservation Service. Washington, DC: United States Department of Agriculture.

Vicente-Serrano, Sergio M. 2007. Evaluating the impact of drought using remote sensing in a mediterranean, semi-arid region. *Natural Hazards* 40 (1):173–208.

Wijesundara, N. C., N. S. Abeysingha, and D. M. S. L. B. Dissanayake. 2018. GIS-based soil loss estimation using RUSLE model: a case of Kirindi Oya river basin, Sri Lanka. *Modeling Earth Systems and Environment* 4 (1):251–262.

Wilhite, D. A., M. J. Hayes, and M. D. Svoboda. 2000. Drought monitoring and assessment: status and trends in the United States. In *Drought and Drought Mitigation in Europe. Advances in Natural and Technological Hazards Research*, edited by J. V. Vogt, and F. Somma. Dordrecht: Springer, pp. 1–316, ISBN: 978-94-015-9472-1.

Williams, J. R. 1975. Sediment-yield prediction with universal equation using runoff energy factor. In *Present and Prospective Technology for Predicting Sediment Yield and Sources*. Washington, DC: US Department of Agriculture, Agriculture Research Service, pp. 244–252.

Wischmeier, W. H., and D. D. Smith. 1965. Predicting rainfall erosion losses from Cropland East of the Rocky Mountains. In *Agricultural Handbook No: 282*. Washington, DC: United States Department of Agriculture, p. 47.

Wischmeier, W. H., and D. D. Smith. 1978. Predicting rainfall erosion losses: a guide to conservation planning. In *Agriculture Handbook No: 537*. Washington, DC: United States Department of Agriculture, pp. 1–67.

Wu, Jianjun, Lei Zhou, Ming Liu, Jie Zhang, Song Leng, and Chunyuan Diao. 2013. Establishing and assessing the Integrated Surface Drought Index (ISDI) for agricultural drought monitoring in mid-eastern China. *International Journal of Applied Earth Observation and Geoinformation* 23:397–410.

Yahiaoui, Ibrahim, Abdelkader Douaoui, Qiang Zhang, and Ahmed Ziane. 2015. Soil salinity prediction in the Lower Cheliff plain (Algeria) based on remote sensing and topographic feature analysis. *Journal of Arid Land* 7 (6):794–805.

Yesuph, Asnake Yimam, and Amare Bantider Dagnew. 2019. Soil erosion mapping and severity analysis based on RUSLE model and local perception in the Beshillo Catchment of the Blue Nile Basin, Ethiopia. *Environmental Systems Research* 8 (1):17.

Zhang, Anzhi, and Gensuo Jia. 2013. Monitoring meteorological drought in semiarid regions using multi-sensor microwave remote sensing data. *Remote Sensing of Environment* 134:12–23.

Zhang, Jie, Qiaozhen Mu, and Jianxi Huang. 2016. Assessing the remotely sensed Drought Severity Index for agricultural drought monitoring and impact analysis in North China. *Ecological Indicators* 63:296–309.

Zhang, Lifu, Wenzhe Jiao, Hongming Zhang, Changping Huang, and Qingxi Tong. 2017. Studying drought phenomena in the Continental United States in 2011 and 2012 using various drought indices. *Remote Sensing of Environment* 190:96–106.

Zhou, Lei, Jie Zhang, Jianjun Wu, et al. 2012. Comparison of remotely sensed and meteorological data-derived drought indices in mid-eastern China. *International Journal of Remote Sensing* 33 (6):1755–1779.

8 Effects of Cyclone Fani on the Urban Landscape Characteristics of Puri Town, India
A Geospatial Study Using Free and Open Source Software

Anindya Ray Ahmed, Asit Kumar Roy, Suman Mitra, and Debajit Datta

CONTENTS

8.1 Introduction 103
8.2 Materials and Methods 105
    8.2.1 Study Area 105
    8.2.2 Data Source 105
    8.2.3 Image Processing 106
    8.2.4 LULC Classification 107
    8.2.5 Normalized Difference Vegetation Index (NDVI) based assessment 108
    8.2.6 Community Perception–Based Vegetation Appraisal 108
    8.2.7 Change Detection Analysis 109
8.3 Results and Discussion 109
    8.3.1 Land Use and Land Cover Pattern of Puri Town 109
    8.3.2 Pre-Fani Vegetation Scenario 111
    8.3.3 Post-Fani Vegetation Scenario 112
    8.3.4 Cyclone-Induced Changes in Vegetation Cover 113
    8.3.5 Community-Formulated Plantation Guidelines 116
8.4 Conclusions 117
Acknowledgments 118
References 118

8.1 INTRODUCTION

Unbridled alteration of natural coastal landscapes has become one of the most alarming causes for concern in the last five decades, as almost 44% of the global population is presently living within 150 km of the coastline. This offers situational advantages of diversified terrestrial and marine ecosystem services, both tangible and intangible, such as supply of euryhaline species, rich source of minerals, benefits of water transportation, better environmental quality, recreational opportunities, etc. (Elliott et al., 2007; UN Ocean Atlas, 2010). In contrast to these, coastal tracts are exposed to several environmental issues, such as cyclonic storms, periodic tidal surges, coastal erosion, recurrent tsunami waves, aggravated salinity in aquifers, and perceptible rise of sea levels (Nicholls and Small, 2002). Among these, tropical cyclones, originating over the warm ocean surface of the tropics (4°S to 22°S and 4°N to 35°N), are such commonly occurring phenomena, which take 1 to 2 weeks from their genesis to their dissipation and turn into hazardous events owing to their high rotational wind speed (>119 km h^{-1}) and the potential of bringing heavy rainfall along the coastal region (William, 1975; Emanuel and Nolan, 2004). Throughout the world, densely populated coastal regions, dotted with several massive urban agglomerations, are often affected by these cyclonic events causing immense damage to urban infrastructures, communication networks, urban vegetation, and human lives, primarily due to the accompanying trio of gusty winds, heavy rainfall, and storm surges (Resio and Irish, 2015). Vegetation cover spread along the coastal areas acts as a natural buffer between sea and human habitations, but this is being noticeably reduced as a consequence of the multifaceted natural (e.g., cyclonic winds, storm surges, coastal erosion, etc.) and anthropogenic disturbances (e.g., illegal small-scale logging, wood pilferage, extending urban infrastructures, etc.) in recent years (Cortés-Ramos et al., 2020; Das and DSouza, 2020; Nandi et al., 2020). Sometimes, the coupled effects of tropical cyclones and anthropogenic disturbances have the potential to eradicate the urban greenery and thereby create ecological havoc in the coastal urban landscape (Guha et al., 2020; Nandi et al., 2020). To elaborate, not much importance has been given to key landscape characteristics such as selection of cyclone-resilient plants for beautification of coastal cities, maintaining an appropriate depth of plantation, and species selection based on the site-specific status of ecological succession. This has cast some doubts on land use management strategies in terms of urban greening, causing it to collapse frequently without fulfilling its prime objectives.

The 7516.6 km long coastline of India, inhabited by 40% of the country's total population, experiences 10% of the annual global cyclones (NCRMP, 2019). Generally, the eastern coast of the Indian peninsula is more prone to cyclonic events than the western coast, since 58% of the total cyclones are generated in the Bay of Bengal. These mostly affect the eastern coastal states of Tamil Nadu, Andhra Pradesh, Odisha, and West Bengal. Between 1891 and 2000, 103 out of 308 cyclones generated in the Bay of Bengal struck the eastern coast (NCRMP, 2019). Among these states, Odisha was the worst affected by the 98 recurrent cyclonic events from 1891 to 2002. Each year, Odisha experiences a series of such devastations, which immensely damage the existing vegetation cover as well as the associated livelihoods pursued by the local marginal coastal populace (Sahoo and Bhaskaran, 2018). In recent times, the densely populated municipal town of Puri, Odisha, experienced massive devastation due to the "extremely severe cyclonic storm" Fani, which made landfall there on May 3, 2019 (GoO, 2019). With an inconceivable wind speed of 170–205 km h^{-1}, Fani took approximately 21 lives in the Puri district alone and cumulatively 89 lives across India and Bangladesh. About 1.4 million trees were uprooted, among which 0.5 million were outside forested land. More than 36% of the total population of Odisha (16 million approximately) experienced the direct consequences of Fani through the loss of 508,467 residential houses, 123,128 km electric lines, 6791 school buildings, 1031 health centers, and 980 km riverine and marine embankments (IRCS, 2019). Fani also snatched away the livelihoods of the coastal populace to a greater extent by damaging 152,985.40 ha of agricultural lands, 6390 traditional boats used for marine fishing, and uncountable accessories needed for daily wage earning (IRCS, 2019).

Puri or "Jagannatha-Dhama," the district headquarter of Puri district of Odisha, is an ancient town mainly known as a famous pilgrimage site for Hindus. Although this town has been populated since ancient times, it became a popular place for coastal tourism in the last century, attracting millions (Surjan and Shaw, 2008). In spite of having a long history of experiencing dreadful hazards, the population of Puri increased largely due to its prospective economic activities (Surjan and Shaw, 2008; Zhuang et al., 2019). Cyclone Fani once again evoked the history of Puri's desolation after the "super-cyclone" of 1999 and ravaged the entire municipality area, causing heavy damage to the urban infrastructure, significantly the urban vegetal cover. Since this town was noticeably deprived of urban greenery from earlier times, and additionally, the recurrent cyclonic devastations had annihilated the remaining vegetation cover, Puri municipality requires serious attention from researchers. The focus should be on restoring it into an ecologically viable and thriving town without negating its environmental aspect. In this context, it has been observed that the multidimensional effects of tropical cyclones on the coastal urban landscapes, with special reference to the urban vegetation cover, have remained unexplored and are ignored in peer-reviewed literature (Das and DSouza, 2020; Cortés-Ramos et al., 2020).

Considering the substantial research gap and topical relevance, this study was mainly designed to assess the urban vegetation dynamics of the Puri municipality area after the occurrence of cyclone Fani, using a hybrid methodological framework integrating geospatial techniques with community-based appraisal. Furthermore, this study also tried to put forward certain site-specific implementable strategies for regenerating the urban vegetation following a sustainable standpoint of environmental management. The entire study was carried out using freely available open source software systems. In developing nations, the adoption of openly sourced software systems has gained huge importance in recent times due to their easy access and cost effectiveness (Camara and Onsrud, 2004; Moreno, 2015). Eventually, this cost effectiveness completely eliminates the inappropriate use of licensed, proprietary-based software (Maurya et al., 2015).

8.2 MATERIALS AND METHODS

8.2.1 STUDY AREA

Puri town, being a place of religious importance with a thriving coastal tourism industry, is densely populated (Surjan and Shaw, 2008). Located along the margin of the Bay of Bengal, this town enjoys a maritime climate with only a slight difference between summer (29.7°C) and winter (22.1°C) temperatures. It receives an annual average rainfall of 1517 mm with the maximum and minimum amounts received in the months of August and December, respectively (Vijay et al., 2011). Puri is prone to tropical cyclones of varying size on a regular basis and experiences irreparable destruction to its socio-economic and ecological setup (Table 8.1).

Topographically, Puri town has developed on the Mahanadi delta and is typically characterized by low-lying coastal plain comprised of recent alluvium, gentle seaward (south or southeast) slope, and marshy littoral tract dissected by small distributaries (e.g., Nuanai, Dhaudia, etc.) (Jana et al., 2018; Sahu et al., 2018). However, the morphology of this area has been modified to a great extent through excessive human intervention. Coexisting with the major soil types of this region (e.g., Alfisols, Eridisols, and Ultisols), semi-evergreen forest, peninsular coastal and littoral forest, and moist peninsular Sal (*Shorea robusta*) forest also grow naturally in this region (Lal, 1990; GoI, 2013). However, certain areas are gradually becoming dominated by *Casuarina* species and *Anacardium occidentale*, which have been planted by the Odisha Forest Development Corporation (OFDC) since the 1960s primarily as bio-shields against the fury of cyclones (Panda and Patnaik, 1993; Das and Sandhu, 2014). The Puri municipality was formed in 1881 to administer the sixth largest urban area of Odisha (Figure 8.1). This municipality is comprised of 32 wards with a spatial extent of 16.84 km² and approximately 0.2 million inhabitants at present (Census of India, 2011).

8.2.2 DATA SOURCE

Two multi-spectral satellite datasets of Sentinel 2A with 10 m spatial resolution were downloaded from the open source data archive of the United States Geological Services (USGS) Glovis (http://glovis.usgs.gov) website for April 28, 2019 and May 18, 2019 as representatives of pre-Fani and post-Fani scenarios, respectively, to analyze the cyclone-induced vegetation loss in Puri municipality (Malmgren-Hansen et al., 2020; Nandi et al., 2020). A ward boundary map of Puri municipality was downloaded from its official website (https://purimunicipality.nic.in) and further georeferenced

TABLE 8.1
Historical Cyclonic Events Experienced by the Coastal Districts of Odisha and Its Neighboring States

Cyclonic Event	Date of Landfall	Magnitude (km h^{-1})	Area of Landfall	Affected District
Amphan	May 20, 2020	155–260	Sagar Island, West Bengal	East Midnapur, North 24 Parganas, South 24 Parganas, Kolkata, Hooghly, Howrah, Balasore, Bhadrak, Kendrapada, Jagatsinghpur, Mayurbhanj, Cuttack, Jajpur, Keonjhar, Khordha, and Puri
Bulbul	November 9, 2019	140–155	Sagar Island, West Bengal	Jagatsinghpur, Kendrapara, Bhadrak, Balasore, and Mayurbhanj
Fani	May 3, 2019	170–205	Puri, Odisha	Puri, Khurdha, Cuttack, Nayagarh, and Jagatsinghpur
Titli	October 11, 2018	150–180	Palasa, Andhra Pradesh	Srikakulam, Vishakhapatnam, Vijayanagram, Ganjam, Gajapati, Puri, Kendrapada, Nayagarh, Bhadrak, Jagatsingpur, Balasore, Jajpur, Bhuweneshwar, and Sambhalpur
Hudhud	October 12, 2014	185–215	Visakhapatnam, Andhra Pradesh	Visakhapatnam, Vijayanagaram, Srikakulam, Jagatsinghpur, Puri, Ganjam, Mayurbhanj, Jajpur, Cuttack, Khurdha, Nayagarh, and Gajapati
Phailin	October 12, 2013	215–260	Gopalpur, Odisha	Ganjam, Kendrapada, Jagatsinghpur, Puri, Khurda, Balasore, and Srikakulam
Super-cyclone	October 29, 1999	260	Puri, Odisha	Puri, Kendrapara, Balasore, Dhenkanal, Cuttack, Jagatsinghpur, and Mayurbhanj

for a detailed ward-wise investigation. A community-based questionnaire survey was conducted from August 2 to 4, 2019 chiefly to collect information regarding pre-existing vegetation patterns, species-specific damage caused by the cyclone, and people's perception of vegetation management in the locality. A total of 114 respondents were surveyed during the field investigation. Additionally, 60 ground control points (GCPs) were collected from prevailing land use/land covers (LULCs) from 10 selected sites of Puri municipality using a handheld Garmin E-Trex 20 global positioning system (GPS) device (Hansen and Loveland, 2012).

8.2.3 Image Processing

The downloaded Level-1C satellite datasets of Sentinel-2A were pre-processed using the freely available open source geospatial platform of QGIS 2.18.9 software. Primarily, these multi-temporal datasets, orthorectified with Universal Transverse Mercator (UTM) projection and WGS84 datum, were geometrically corrected using the field reference points collected as GCPs during the field investigation (Roy and Datta, 2018). Since the dates of image acquisition were in close proximity to the date of the cyclonic event, the downloaded images, having 61.11% and 57.98% cloud coverage during the pre- and post-Fani period, respectively, required radiometric calibration. This radiometric calibration was performed using the Semi-Automatic Classification Plugin (SCP) of QGIS, whereby the top of atmosphere (TOA) datasets of Sentinel-2A were transformed into the bottom of atmosphere (BOA) spectral reflectance product (Congedo, 2016; Gemusse et al., 2018). Specifically, SCP, a plugin of QGIS, is a one-stop solution for the entire image processing technique: from downloading, to pre-processing, to classification as well as post-classification analysis for

Effects of Cyclone Fani on Urban Landscape

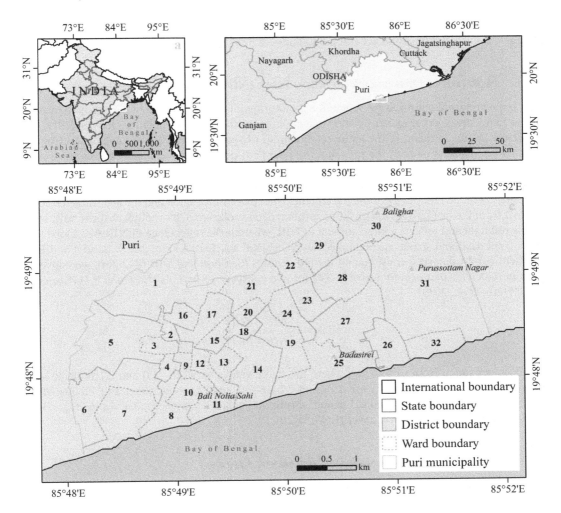

FIGURE 8.1 Location map of Puri municipality.

images of LANDSAT, Sentinel-2 satellite systems, etc. Hence, it reduces time and the need for other dedicated proprietary software systems (Mourya et al., 2015).

8.2.4 LULC Classification

Using the geospatial application platform of QGIS software, LULC classification maps were prepared from the Sentinel-2A datasets for the pre-Fani period (April 28, 2019). Supervised classification technique was used to prepare the LULC map following the widely accepted maximum likelihood algorithm (Shalaby and Tateishi, 2007; Mitra et al., 2020). Based on the investigator's a priori knowledge and information gathered during the field survey, it was attempted to identify seven dominant LULC classes from the satellite dataset. These were beach, built-up, greenery, marsh land, road, vacant lot, and water body. The sandy coastal areas were designated as beach, and all kinds of impervious or metallic surfaces were pronounced as built-up. Similarly, land with perceptible vegetal cover was classified as greenery. Water bodies covered with hygrophytes was designated as marsh land, and the inland water-filled depressions were separately termed as water body. Areas covered with metaled pathways were designated as roads, and the areas free from any of the above-mentioned characteristics were classified as vacant lot. The classification accuracy had also been measured for the pre-Fani LULC map by determining the producer's accuracy, user's accuracy, overall classification accuracy, and kappa statistics.

8.2.5 Normalized Difference Vegetation Index (NDVI) based assessment

NDVI, the preferred approach for vegetation monitoring globally, was used in this study to measure the cyclonic stress experienced by the vegetation cover of Puri municipality (Erener, 2011; Roy and Datta, 2018). To quantify the temporal dynamics of vegetation, NDVI was calculated from the Sentinel-2A datasets of both the pre- and the post-Fani period. This calculation was done using the raster calculator in QGIS. In principle, NDVI is the measure of plant health with respect to chlorophyll content present in the living plant. This is determined by assessing the ratio of the dual combination of the reflectivity of the red channel from the visible region and the near infrared (NIR) channel from the infrared region of the electromagnetic spectrum. Chlorophyll absorbs most of the visible red light and reflects a large amount of NIR, whereas the reflectance of NIR was found to be lower from dead and sparse vegetation cover (Sahebjalal and Dashtekian, 2013). Band 4 and Band 8 of Sentinel-2A datasets, consisting of the red channel (665 μm) and NIR (842 μm), were given as input in the QGIS platform to derive NDVI for two consecutive periods (Equation 8.1) (Lee et al., 2009; Dutta et al., 2013).

$$\text{NDVI} = \frac{(NIR - Red)}{(NIR + Red)} \tag{8.1}$$

NDVI values range from −1.0 to 1.0, where the lower values between −1.0 and 0.1 indicate areas without vegetation such as empty areas of rocks, bare soil, sand or snow, water and, clouds. While moderate values between 0.2 and 0.3 denote shrubs and meadows, higher values range from 0.6 to 0.8, indicating dense vegetation cover (EOS, 2020).

8.2.6 Community Perception–Based Vegetation Appraisal

A community-based questionnaire survey had also been conducted during the field investigation for assessing the impacts of Fani on vegetation dynamics using the freely accessible Google form, which provided ready-to-use data after the survey was over. Based on the initial observations regarding the damaged status of vegetation caused by cyclone Fani, the total 32 wards of Puri municipality were categorized into three cyclone impact zones: areas of high impact or category 1 (C1), areas of medium impact or category 2 (C2), and areas of low impact or category 3 (C3). Considering the varying degree of cyclonic impacts on urban greenery, Wards No. 7, 11, 12, 14, 19, 23, and 31 came under C1; Wards No. 5, 8, 9, 15, 17, 18, 20, 21, 24, 25, and 32 fell under C2; and Wards No. 1–4, 6, 10, 13, 16, 22, 26–29, and 30 were found to be in C3. The sampling procedure was designed in such a way that six, four, and two respondents would be surveyed from each ward under C1, C2, and C3 respectively.

The respondents were asked specific structured and semi-structured questions asked during the questionnaire survey in order to collect data regarding changes that had occurred in the number of tree species in their own garden and overall vegetation changes witnessed by them in their own ward. Respondents were also asked to mention the most prevalent plant species found in their vicinity in order to make an appraisal of the species-specific resilience capability and also to infer the changing status of those species in post-Fani conditions. In addition, considering the responses to initiatives taken by the local governmental bodies regarding replenishing the vegetation losses, suitable and implementable recommendations for ward-wise plantation of cyclone-resilient plants were formulated.

8.2.7 Change Detection Analysis

The spatio-temporal changes in vegetation health had been assessed from derived NDVI maps of pre- and post-Fani periods by using the change detection analysis technique in QGIS. The image difference data acquired from this analysis were mapped further to identify the wards with varying degrees of NDVI change (Datta and Deb, 2012; Roy and Datta, 2018; Nandi et al., 2020). It had been hypothesized that cyclone Fani was primarily responsible for the NDVI dynamics in the pre- and post-Fani periods in Puri municipality, which needed to be tested for statistical significance. A paired-sample one-tailed t-test was conducted between two sets of pre- and post-Fani NDVI values extracted from 166 randomly chosen pixels representing 0.001% of the entire dataset.

A hybrid scoring technique was adopted using both the NDVI percentage change values and the questionnaire survey–based perception scores to determine the degree of impact of this cyclonic hazard on urban vegetation. Initially, ward-wise mean NDVI values were calculated for both the pre- and post-Fani periods, and subsequently, difference values were estimated. It was attempted to standardize these values by transforming them into percentages of change, which were further scored from 1 to 5 based on their mean ($\bar{x}$) and standard deviation (σ) values. In the five-point scores, 1 was assigned to higher percentages of NDVI change, and vice versa.

Ward-wise perception-based scores were also determined from responses acquired through the questionnaire-based survey. Initially, the responses on species diversity given by individuals were assessed to identify the predominant vegetation species, along with their status, in the pre- and post-Fani periods. These perception-based vegetation statuses were scored subsequently by assigning some specific values, such as 0 for not reported in that ward, 1 for reported as decreased in number due to the cyclonic event, and 2 for reported as unchanged. The species-wise weightages were calculated using Equation 8.2):

$$\text{Weightage}(W_i) = \frac{\text{Number of times the species is mentioned by the respondents}}{\text{Total number of wards}} \qquad (8.2)$$

After these values and weightages were obtained, ward-wise perception-based scores were calculated using Equation 8.3:

$$\text{Perception Score} = \sum_{i=1}^{i=15} S_i \times W_i \qquad (8.3)$$

where S_i = changed status of the ith tree and W_i = weightage assigned to the ith tree.

After combining the derived ward-wise NDVI difference score and perception score, a composite score (CS_i) on tree resilience and diversity was estimated using Equation 8.4):

$$CS_i = \sqrt{(\text{NDVI Difference Percentage Score}_i) \times (\text{Perception Score}_i)} \qquad (8.4)$$

Here, the square root was used to keep the composite score in the same score domain as its components (i.e., NDVI difference score and perception score). In this regard, the wards with higher composite scores were more resilient to cyclones as well as having higher species diversity, and vice versa. This entire calculation was done using Google sheets available online.

8.3 RESULTS AND DISCUSSION

8.3.1 Land Use and Land Cover Pattern of Puri Town

The classified LULC map of the pre-Fani period showed the prevailing urbanization pattern of Puri municipality (Figure 8.2). The overall classification accuracy and kappa coefficient measured

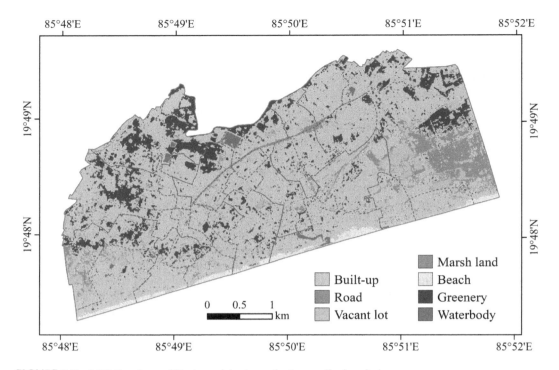

FIGURE 8.2 LULC pattern of Puri municipal area in the pre-Fani period.

TABLE 8.2
Pre-Fani Land Use/Land Cover Statistics of Puri Municipality

Land Use/Land Cover Class	Area of April, 2019	
	Hectare (ha)	Percentage (%)
Built-up	1185.64	71.37
Road	35.60	2.14
Vacant lot	56.70	3.42
Marsh land	121.09	7.29
Beach	28.89	1.74
Greenery	215.79	12.99
Waterbody	17.48	1.05
Total	1661.19	100.00

for this LULC map were 89.79% and 0.88, respectively. It is shown that 215.83 ha of land (comprising 12.99% of the municipal area) came under the class of greenery, though it was found to be notably (Indian standards indicate a minimum of 33%) low before the occurrence of cyclone Fani (Table 8.2). Large and medium patches of vegetation were mostly restricted to Wards No. 1, 5, 6, 17, 21, 30, and 31. The other areas were characterized by sparse vegetation cover, carrying the signs of landscape fragmentation induced by dominant alternative land uses, such as built-ups, road network, and vacant impervious surfaces that comprise almost 76.92% of the entire study area.

In spite of local governing bodies' claims to have planted *Tamarix diocia* along the shoreline, it was observed that the beach front areas were almost devoid of such vegetation, causing high exposure to tropical cyclones and other coastal hazards to hit the urban built-ups (Das and Sandhu,

Effects of Cyclone Fani on Urban Landscape

2014). The wards (e.g., 2, 11, 14, 15, 18–20, 23–25, and 27) situated in close proximity to functional nodes such as pilgrimage sites, tourist beaches, and the railway station were almost devoid of prominent vegetation patches. Eventually, in those wards, a negligible amount of greenery was observed appearing in the gaps between two buildings or beside the narrow roads in a linear pattern. A noteworthy share of land (121.09 ha) was also found to be under marsh land, which was 7.29% of the total area. Although this marsh land was restricted to Wards No. 6 and 31, the fact that it was covered with a perceptible amount of greenery might be a consequence of higher presence of soil moisture content. Other LULC classes such as water bodies (17.48 ha) and beaches (28.89 ha) were completely deprived of any such signs of greenery.

It is evident that only the wards located far away from the functional center and tourist hub were capable of possessing greenery within the municipal limits. The quantity of vegetation was found to be alarmingly sparse specifically in the wards located near the coast. Since coastal vegetation acts as a major buffer against cyclonic storms and resists the initial force of the gusty wind, the lack of urban greenery in Puri made it more vulnerable to catastrophic cyclonic events and other coastal hazards. Additionally, the increased rate of landscape modifications in recent decades had degraded the ecological resilience of this area to a greater extent, as reflected clearly in the LULC scenario (Ramachandra et al., 2012).

8.3.2 Pre-Fani Vegetation Scenario

The NDVI map derived for the pre-Fani period showed patterns that largely matched the vegetation scenario found from the LULC classification. Here, the values ranging between –0.06 and 0.62 were represented with graduated gray shades (Figure 8.3). The pre-Fani NDVI map showed that Wards No. 1, 5, 6, 10, 17, 21, and 30 were characterized by high average NDVI (+ve) values (0.21–0.30), indicating the presence of moderately healthy vegetation cover. Wards No. 12–14, 16, 22, 28, 29, and 31 also displayed high maximum NDVI (+ve) values, ranging between 0.50 and 0.62. In contrast,

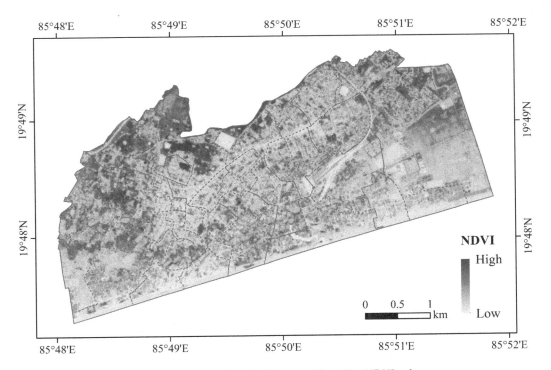

FIGURE 8.3 Pre-Fani condition of vegetation health as manifested by NDVI values.

TABLE 8.3
Dominant Vegetation Species Identified within the Boundary of Puri Municipality

Common Name	Scientific Name	Family
Mango	*Mangifera indica*	Anacardiaceae
Coconut	*Cocos nucifera*	Arecaceae
Banyan	*Ficus benghalensis*	Moraceae
Neem	*Azadirachta indica*	Meliaceae
Jackfruit	*Artocarpus heterophyllus*	Moraceae
Jhau	*Tamarix dioica*	Tamaricaceae
Cashew	*Anacardium occidentale*	Anacardiaceae
Bel	*Aegle marmelos*	Rutaceae
Debdaru	*Polyalthia longifolia*	Annonaceae
Guava	*Psidium guajava*	Myrtaceae
Banana	*Musa acuminata*	Musaceae
Chinese pistache	*Pistacia chinensis*	Anacardiaceae
Pineapple	*Ananas comosus*	Bromeliaceae
Akashi	*Acacia auriculiformis*	Leguminosae
Tulsi	*Ocimum tenuiflorum*	Lamiaceae

Wards No. 8, 11, 25–27, and 32 showed a lower range of average NDVI (+ve) values (0.10–0.14), which confirmed the presence of poor or unhealthy vegetation cover. Notably, the maximum NDVI (+ve) values were found to be extremely low in range (0.39 to 0.47) in Wards No. 7, 8, 11, 25, 26, and 32 located along the beach front areas, where urban greenery was rarely witnessed. Wards No. 2, 3, 13, 18, 19, 20, and 23 showed a moderate range of average NDVI (+ve) values (0.15–0.19). These parts serve as functional nodes of the Puri municipality area, primarily near the railway station and the administrative localities, and are reported as sites of urban beautification through social forestry.

Although it has already been shown from the NDVI and LULC statistics that the quality and quantity of vegetation were not satisfactory in the pre-Fani period within the municipal limits, a similar pattern was also inferred from the community-based appraisals. Only 15 plant species were identified as prevalent within the town in these appraisals (Table 8.3). However, these 15 species were not evenly distributed in the municipal area. Only *Cocos nucifera* was found in all the wards. Regionally common plant species such as *Azadirachta indica*, *Aegle marmelos*, *Musa acuminata*, *Psidium guajava*, and *Ocimum tenuiflorum* were found to be limited in number throughout Puri.

8.3.3 Post-Fani Vegetation Scenario

The post-Fani NDVI values, ranging between −0.065 and 0.593, were mapped further for the entire municipal area in a similar manner to those of the pre-Fani period (Figure 8.4). This demonstrated that a few areas located far away from the shoreline, viz. Wards No. 1, 5, and 17, were characterized by a higher range of average NDVI (+ve) values (0.17–0.25) indicating the presence of remnant patches of vegetation even after the cyclone. *Ficus benghalensis*, *Azadirachta indica*, *Acacia auriculiformis*, and *Polyalthia longifolia* were sturdy tree species that were still found to be existing in these wards. On the contrary, a large number of wards (e.g., 4, 7, 8, 11, 15, 25–27, and 32) situated nearer the shoreline were characterized by a lower range of average NDVI (+ve) values (0.07–0.10). This confirmed the minimum presence of vegetation cover consequent to the devastation caused by cyclone Fani along the beachfront. However, the maximum and minimum NDVI values observed in these wards ranged from −0.06 to 0.45. Uprooting and snapping of tree species such as *Cocos nucifera*, *Acacia auriculiformis*, *Anacardium occidentale*, and *Tamarix dioica*, found occasionally

Effects of Cyclone Fani on Urban Landscape 113

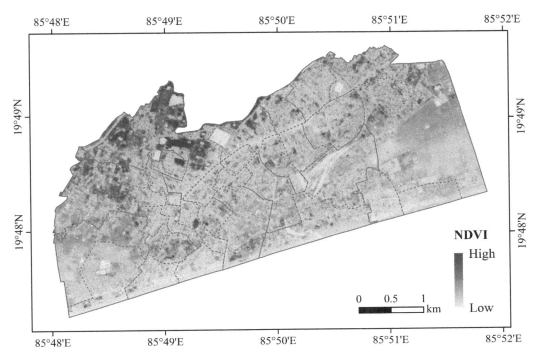

FIGURE 8.4 Post-Fani condition of vegetation health as manifested by NDVI values.

near the coast, were mainly responsible for these widespread lower NDVI values. All other wards, where higher and moderate ranges of average NDVI (+ve) values (0.14–0.25) were witnessed in the pre-Fani period, were transformed into areas having a lower range of average NDVI (+ve) values (0.11–0.16) due to the falling, uprooting, and shattering of large numbers of trees and shrubs. It was clearly demonstrated from the post-Fani NDVI statistics that only a few vegetation patches of inland areas somehow managed to safeguard themselves from the fatal effects of Fani. Otherwise, all vestiges of urban greenery were ravaged in such a way that they would be hard to restore in future years.

8.3.4 Cyclone-Induced Changes in Vegetation Cover

Change detection analysis of the NDVI status of the pre- and post-Fani periods helped in portraying the overall changing pattern of vegetation health (Figure 8.5). The changes in NDVI values of the pre- and post-Fani scenarios were found to be statistically significant by the paired-sample one-tailed t-test ($p < 0.05$). A massive loss of vegetation health had been detected in all 32 municipal wards in terms of change in mean NDVI values from 16.5% to 38.9%. Wards No. 1, 16, and 17 showed minimum changes in mean NDVI (16.5–22.3%), indicating negligible damage to urban vegetation in terms of its quality and quantity. In contrast, Wards No. 7, 11, 14, 25, 26, and 32 (located along the near shore area) as well as Wards No. 19, 23, 27, and 30 (associated with urban functional nodes) showed the maximum change of mean NDVI values (33.1–38.9%), representing a massive loss of urban canopy coverage. The remaining wards (e.g., 2–6, 8–10, 12, 13, 15, 18, 20–22, 24, 28, 29, and 31) portrayed a moderate to high change of mean NDVI values (23.9–31.9%), which depicted a considerable loss of vegetation health due to the cyclone. It had already been established from the pre-Fani NDVI map that the beachfront areas were deprived of any substantial vegetation cover prior to the cyclone. After this disastrous event, even the remaining sparse vegetation patches had also been obliterated. Usually, the vegetal cover of these areas helps to build a natural shield or

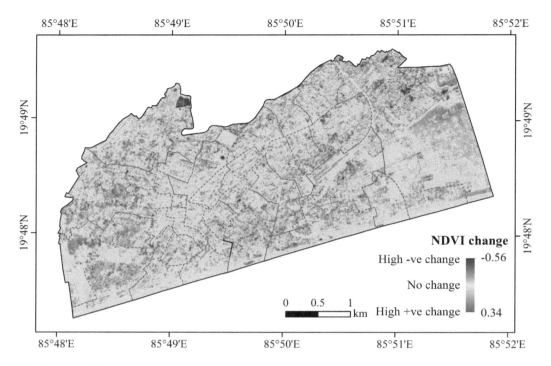

FIGURE 8.5 NDVI-based vegetation health dynamics between pre-Fani and post-Fani periods.

rather, a cyclone resilience system by reducing the speed of initial gusty winds and protecting the inland areas from sea surges. However, this system of resilience was mostly missing in the coastal wards in Puri municipality even before the cyclone. Consequently, it failed to protect the inland areas from unquantifiable damage to urban infrastructure as well as vegetation cover.

The status of 15 individual plant species in the post-Fani period was also revealed from the feedback of local respondents (Figure 8.6). Commonly found *Mangifera indica*, *Tamarix dioica*, and *Cocos nucifera* were the worst affected and reported as having decreased in most of the wards. Notably, *Acacia auriculiformis* and *Tamarix dioica* were also found to be in an alarming condition in Wards No. 14, 19, 23–28, 31, 32, etc., mostly located near the shoreline. *Azadirachta indica* was found to be lower in number in more inland wards. Several major species such as *Aegle marmelos*, *Anacardium occidentale*, *Ananas comosus*, *Artocarpus heterophyllus*, *Musa acuminata*, *Ocimum tenuiflorum*, *Pistacia chinensis*, *Psidium guajava*, and *Polyalthia longifolia* were inferred to be decreased in most of the municipal wards. Furthermore, species such as *Ficus benghalensis*, an evergreen tree species with large canopy cover, were found uprooted and smashed in a few wards (e.g., 10, 11, and 26). However, this was the only tree species that had shown moderate resilience against this cyclonic fury, as its numbers had remained unchanged in Wards No. 1, 5, 23, 24, 27, 28, and 29 from the pre-Fani to the post-Fani period. It was physically verified in the field that except for the wards located more inland (Wards No. 1, 17, 20, 21, 30, and 31), all other wards reported the wiping out of the majority of their roadside vegetation.

The cumulative composite scores based on ward-wise tree diversity and resilience patterns derived from community perceptions, and the status of urban greenery derived from the NDVI change statistics, revealed the zonal character of environmental vulnerability (or resilience) in Puri. Wards No. 3, 7, 23, 25, 26, 30, and 32, having lower composite scores ranging from 1.85 to 2.77, portrayed immense loss of vegetation cover as well as reduction of species diversity (Figure 8.7). Moderate values of composite scores ranging from 2.9 to 3.9 were found in Wards No. 2, 4, 6, 8, 10–22, 27–29, and 31. These wards were located further inland and comprised the major urban

Effects of Cyclone Fani on Urban Landscape 115

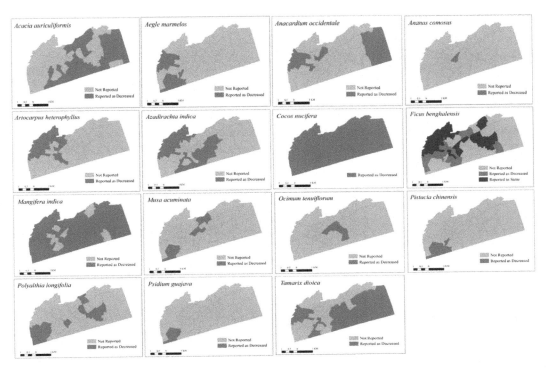

FIGURE 8.6 Post-Fani status of major plant species assessed from perception surveys (N = 114).

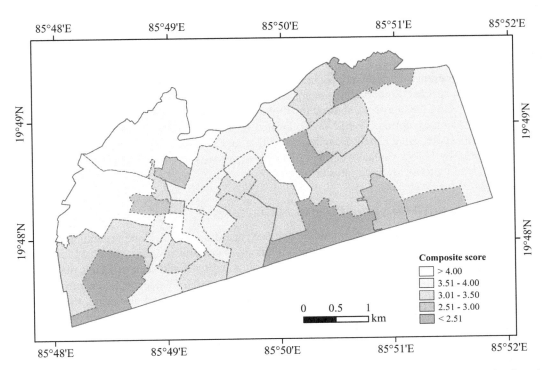

FIGURE 8.7 Cyclone resilience of the urban vegetation of Puri municipality derived from perception-based community appraisals and NDVI change values between pre- and post-Fani periods (N = 114).

FIGURE 8.8 Evidence of the devastated urban greenery in Puri municipal area.

functional nodes of Puri. Eventually, these wards experienced a considerable reduction in the diversity and health of vegetation after this hazardous event (Figure 8.8). A very limited number of remaining wards, such as 1, 5, 9, and 24, displayed higher composite score values (4.14–5.02), indicating the presence of relatively higher vegetation diversity and less damage experienced by each vegetation species. Remarkably, among all the municipal wards, Ward No. 1, located inland, and 7, located along the coastline, showed the highest (5.02) and the lowest (1.85) composite scores, respectively. As an exception, Ward No. 30, located far inland from the shoreline, showed a very low composite score value (2.40). Despite its location and coverage with large vegetation patches (as manifested by NDVI values), very low plant diversity was reported there during community appraisals, leading to its lower composite score.

8.3.5 Community-Formulated Plantation Guidelines

By integrating the responses of the coastal populace regarding post-cyclone vegetation scenarios, an implementable and realistic set of recommendations of site-specific plantation was formulated. As suggested by the respondents, all wards of Puri municipality should seek extensive plantation of *Cocos nucifera*, taking into account its noticeable obliteration after Fani, except in Wards No. 3, 25, 26, and 30. Although this particular species had experienced widespread uprooting, its economic, social, and ecological significance remained unchanged. The respondents mentioned that the problem was not with the species; rather, during the built-up expansion, most of the urban surface of Puri had been paved, thereby weakening the root structure of many tree species. Other sturdy vegetation species, such as *Mangifera indica*, *Azadirachta indica*, and *Ficus benghalensis*, which experienced a notable drop in number, also need to be given similar priority in certain wards during the regeneration process (Figure 8.9). Among all 15 major plant species identified, *Carica papaya*, *Musa*

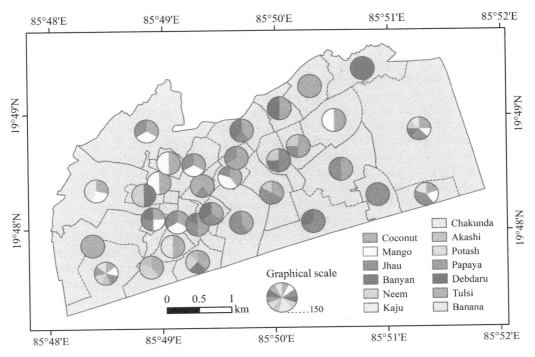

FIGURE 8.9 Recommended vegetation species for site-specific plantation in Puri municipality (N = 114).

acuminata, and *Ocimum tenuiflorum* were least often mentioned for plantation by the respondents, since these were commonly found and occur naturally in their opinion. However, the validity of this claim is still doubtful and needs to be further investigated.

Giving importance to the present situation, it could be recommended that the site-specific plantation of ecologically resilient and socio-economically relevant plant species should be initiated immediately, since this region experiences frequently occurring cyclones of varying severity. The narrowing lag time between two repetitive major cyclonic events eventually restricts the natural succession of plants (Nandi et al., 2020). Similarly, it was also shown that naturally grown vegetation was inadequate along the beachfront area, and accordingly, suitable species such *Tamarix diocia* should be planted here to overcome this deficit. Furthermore, a substantial amount of seedling growth was thought to be required within a thick span of area instead of planting in a narrow linear pattern to create a cyclone-resilient vegetation buffer. Otherwise, the plants would be unable to resist the force of the storm winds and fail to protect inland vegetation cover as well as urban structures. It was also recommended that hotels located near the seaside need to develop their own plantation strategies. Moreover, deep-rooted indigenous species must be prioritized for plantation to prevent the uprooting of roadside trees, serving the purpose of both increasing urban greenery and beautification. In general, to deal with these issues in the coming years, proper plantation and management regulations need to be adopted for choosing suitable plant species along with their planting sites, incorporating feedback from the inhabitants of Puri.

8.4 CONCLUSIONS

As it involved an assessment of cyclone-induced vegetation loss in an urban landscape using open source software systems, this study could be conceived as a pioneering effort in this part of India. Open source software systems have huge inherent potential, since they are provided absolutely free of cost. Furthermore, they are supported by worldwide software developers, eliminating any kind of discrimination (Maurya et al., 2015). During this study, it was realized that the use of open source geospatial software systems along with other freely accessible data handling tools saves time in data collection, image processing, and data analysis. Eventually, this study successfully applied a novel approach to identify cyclone-resilient wards considering the prevailing vegetation species of the municipal area. Furthermore, this study developed an implementable ward-wise regeneration plan of deep-rooted, sturdy, and fast-growing plant species to replenish the degradation of urban greenery after the occurrence of cyclones by identifying site-specific changes of vegetation cover and incorporating the views of local inhabitants. The prime causes of the huge vegetation loss within the Puri municipality were found to be mostly an outcome of three factors: first, the proximity of the municipal wards to the shoreline; second, the density of the vegetation patches; and last, plant species richness. However, since this research was conducted in a very short span of time just after the cyclone, in-depth investigation of the resilience characteristics of all tree species occurring in Puri could not be conducted. Moreover, detailed surveys incorporating more respondents' views and willingness to conserve urban green spaces could not be conducted. Despite these shortcomings, this study clearly explained the impacts of catastrophic hazards on an urban landscape triggered by unsustainable land use planning and long-term mismanagement of urban ecological structure. Taking inspiration from the geospatial methodology used in this study, dependent on open source datasets and software, local governmental and non-governmental organizations may take certain measures by investigating the vegetation loss as early as possible after the hazard, not only in forested lands but also in urban areas. Accordingly, urban administrative bodies should prepare a floral inventory specifically for their territory, which would help them to take immediate post-hazard mitigation measures and manage the landscape more efficiently.

ACKNOWLEDGMENTS

We sincerely acknowledge the assistance of the students of the Department of Geography, Jadavpur University during this investigation. We are thankful for the assistance of the residents of Puri Municipal area during the field surveys. We also express our gratitude towards those who provided logistical support during fieldwork.

REFERENCES

Câmara, G., and Onsrud, H. 2004. Open-source geographic information systems software: myths and realities. In *Open Access and the Public Domain in Digital Data and Information for Science: Proceedings of an International Symposium.* National Research Council, US National Committee for CODATA, Washington, DC.

Census of India. 2011. *District Census Handbook Puri*, Odisha. www.census2011.co.in/census/city/271-puri-town.html.

Congedo, L. 2016. *Semi-Automatic Classification Plugin Documentation (release 5.3.6.1).* http://dx.doi.org/10.13140/RG.2.2.29474.02242/1.

Cortés-Ramos, J., Farfán, L. M., and Herrera-Cervantes, H. 2020. Assessment of tropical cyclone damage on dry forests using multispectral remote sensing: the case of Baja California Sur, Mexico. *J. Arid Environ.* 178:1–19.

Das, S., and DSouza, N. M. 2020. Identifying the local factors of resilience during cyclone Hudhud and Phailin on the east coast of India. *Ambio* 49:950–961.

Das, S., and Sandhu, H. 2014. Role of exotic vegetation in coastal protection: an investigation into the ecosystem services of Casuarina in Odisha. *Econ. Political Wkly.* 49:42–50.

Datta, D., and Deb, S. 2012. Analysis of coastal land use/land cover changes in the Indian Sunderbans using remotely sensed data. *Geo Spat. Inf. Sci.* 15:241–250.

Dutta, D., Kundu, A., and Patel, N. R. 2013. Predicting agricultural drought in eastern Rajasthan of India using NDVI and standardized precipitation index. *Geocarto Int.* 28:192–209.

Elliott, M., Burdon, D., Hemingway, K. L., and Apitz, S. E. 2007. Estuarine, coastal and marine ecosystem restoration: confusing management and science – a revision of concepts. *Estuar. Coast. Shelf Sci.* 74:349–366.

Emanuel, K., and Nolan, D. S. 2004. Tropical cyclone activity and the global climate system. In *26th Conference on Hurricanes and Tropical Meteorology*. https://miami.pure.elsevier.com/en/publications/tropical-cyclone-activity-and-the-global-climate-system.

EOS. 2020. *NDVI.* https://eos.com/ndvi/ (accessed July 3, 2020).

Erener, A. 2011. Remote sensing of vegetation health for reclaimed areas of Seyitömer open cast coal mine. *Int. J. Coal Geol.* 86:20–26.

Gemusse, U., Lima, A., and Teodoro, A. 2018. Pegmatite spectral behavior considering ASTER and Landsat 8 OLI data in Naipa and Muiane mines (Alto Ligonha, Mozambique). In *Earth Resources and Environmental Remote Sensing/GIS Applications* IX. https://sigarra.up.pt/sasup/en/pub_geral.pub_view?pi_pub_base_id=327949.

GoI. 2013. *Ground Water Information Booklet, Puri District, Orissa.* Central Ground Water Board (CGWB). http://cgwb.gov.in/District_Profile/Orissa/Puri.pdf.

GoO. 2019. *Cyclone Fani: Damage, Loss, and Needs Assessment.* Government of Odisha. www.recoveryplatform.org/pdna/key_documents_on_country_pdnas (accessed January 13, 2020).

Guha, P., Aitch, P., and Bhandari, G. 2020. Effect of changing vegetation coverage and meteorological parameters on the hazard characteristics of Indian Sundarban region and its impact there on. In *An Interdisciplinary Approach for Disaster Resilience and Sustainability*, eds. I. Pal, J. V. Meding, S. Shrestha, I. Ahmed, and T. Gajendran, 451–475. Singapore: Springer.

Hansen, M. C., and Loveland, T. R. 2012. A review of large area monitoring of land cover change using Landsat data. *Remote Sens. Environ.* 122:66–74.

IRCS. 2019. *Odisha Fani Cyclone Assessment Report.* Indian Red Cross Society, India. https://reliefweb.int/report/india/odisha-fani-cyclone-assessment-report (accessed July 11, 2020).

Jana, S., Mohanty, W. K., Gupta, S., Rath, C. S., and Patnaik, P. 2018. Palaeo-channel bisecting Puri town, Odisha: vestige of the lost river "Saradha"? *Curr. Sci.* 115:300–309.

Lal, R. 1990. Tropical soils: distribution, properties and management. In *Tropical Resources: Ecology and Development*, eds. J. I. Furtado, W. B. Morgan, J. R. Pfafflin, and K. Ruddle, 39–52. USA: Harwood Academic Publishers.

Lee, T. M., and Yeh, H. C. 2009. Applying remote sensing techniques to monitor shifting wetland vegetation: a case study of Danshui River estuary mangrove communities, Taiwan. *Ecol. Eng.* 35:487–496.

Malmgren-Hansen, D., Sohnesen, T., Fisker, P., and Baez, J. 2020. Sentinel-1 change detection analysis for cyclone damage assessment in urban environments. *Remote Sens.* 12:1–16.

Maurya, S. P., Ohri, A., and Mishra, S. 2015. Open source GIS: a review. In *Proceedings of National Conference on Open Source GIS: Opportunities and Challenges*.

Mitra, S., Roy, A. K., and Tamang, L. 2020. Assessing the status of changing channel regimes of Balason and Mahananda river in the sub-Himalayan West Bengal, India. *Earth Syst. Environ.* 4:409–425.

Moreno, R. 2015. Free and Open Source Software for Geospatial Applications (FOSS4G). University of Colorado Denver, NCAR Geopatial Talks Series.

Nandi, G., Neogy, S., Roy, A. K., and Datta, D. 2020. Immediate disturbances induced by tropical cyclone Fani on the coastal forest landscape of eastern India: a geospatial analysis. *Remote Sens. Appl. Soc. Environ.* 20:1–13.

NCRMP. 2019. *Cyclones & Their Impact in India.* Government of India. https://ncrmp.gov.in/cyclones-their-impact-in-india/ (accessed June 27, 2020).

Nicholls, R. J., and Small, C. 2002. Improved estimates of coastal population and exposure to hazards released. *Eos Trans. AGU* 83:301–305.

Panda, P. C., and Patnaik, S. N. 1993. Flora of Puri district-I: forest vegetation. *Plant Sci. Res.* 15:7–17.

Ramachandra, T. V., Aithal, B. H., and Sanna, D. D. 2012. Insights to urban dynamics through landscape spatial pattern analysis. *Int. J. Appl. Earth. Obs. Geoinf.* 18:329–343.

Resio, D. T., and Irish, J. L. 2015. Tropical cyclone storm surge risk. *Curr. Clim. Chang. Rep.* 1:74–84.

Roy, A. K., and Datta, D. 2018. Analyzing the effects of afforestation on estuarine environment of river Subarnarekha, India using geospatial technologies and participatory appraisals. *Environ. Monit. Assess.* 190:1–16.

Sahebjalal, E., and Dashtekian, K. 2013. Analysis of land use-land covers changes using normalized difference vegetation index (NDVI) differencing and classification methods. *Afr. J. Agric. Res.* 8:4614–4622.

Sahoo, B., and Bhaskaran, P. K. 2018. Multi-hazard risk assessment of coastal vulnerability from tropical cyclones – a GIS based approach for the Odisha coast. *J. Environ. Manage* 206:1166–1178.

Sahu, S., Pati, D. P., Mohanty, C., Prasad, G., and Mandal, D. N. 2018. Aquifer salinization and ground water security in Puri urban area in the coastal tract of Odisha; understanding local geomorphology as a tool. In *International Ground Water Conference, 2017*.

Shalaby, A., and Tateishi, R. 2007. Remote sensing and GIS for mapping and monitoring land cover and land-use changes in the northwestern coastal zone of Egypt. *Appl. Geogr.* 27:28–41.

Surjan, A. K., and Shaw, R. 2008. "Eco-city" to "disaster-resilient eco-community": a concerted approach in the coastal city of Puri, India. *Sustain. Sci.* 3:249–265.

UN Ocean Atlas. 2010. *Human Settlements on the Coast.* www.oceansatlas.org/subtopic/en/c/114/ (accessed June 22, 2020).

Vijay, R., Khobragade, P., and Mohapatra, P. K. 2011. Assessment of groundwater quality in Puri City, India: an impact of anthropogenic activities. *Environ. Monit. Assess.* 177:409–418.

William, M. G. 1975. *Tropical Cyclone Genesis (Atmospheric Science* Paper No. 234). Colorado State University, USA. https://mountainscholar.org/bitstream/handle/10217/247/0234_Bluebook.pdf;sequence=1 (accessed May 12, 2020).

Zhuang, X., Lin, L., and Li, J. 2019. Puri vs. Varanasi destinations: local residents' perceptions, overall community satisfaction and support for tourism development. *J. Asia Pac. Econ.* 24:127–142.

9 Land Resource Mapping and Monitoring
Advances of Open Source Geospatial Data and Techniques

Gouri Sankar Bhunia and Pravat Kumar Shit

CONTENTS

9.1	Introduction	121
9.2	Combinatorial Innovation in Sustainable Land Management	122
9.3	Big (Geo) Data	123
	9.3.1 Coarse Resolution Satellite Data (>100 m Pixel Size)	125
	9.3.2 Medium Resolution Satellite Data (10–99 m Pixel Size)	125
	9.3.3 High Resolution Satellite Data (<10 m Pixel Size)	126
9.4	Web Search Engine for Free Access Remote Sensing Data	127
	9.4.1 Open Source Vector Data	127
9.5	Open Source Software for Land Resource Mapping and Monitoring	130
9.6	Crowdsource Platform	131
9.7	High-Quality Ground Truth and Land Use Management	135
9.8	Cloud Computing	136
9.9	Conclusion	141
References		141

9.1 INTRODUCTION

Land resources (economically referred to as property or raw materials) arise naturally within ecosystems that remain largely untouched by humans. The complexities of population development, growing demands from diverse players on scarce and already declining resources, soil depletion, loss of biodiversity, and climate change require a sound resource use strategy that promotes and improves sustainability and retains environmental resilience. Both natural and manmade land resources are a critical component for informed decision-making in policy, development, planning, and resource management. Land resource planning (also called land use planning) is a method for making sustainable and effective usage of energy, taking biophysical and socio-economic aspects into account. It is the comprehensive evaluation of land resources and land use possibilities for attaining optimal land usage and enhanced socio-economic conditions through a collaborative multi-sectoral, multi-stakeholder, and scale-dependent mechanism. Precise and timely land resource knowledge plays a critical role in a number of third world sectors, including food security, land use planning, hydrology modeling, and natural resource management planning. The tools and methods used for the land resource planning will help the varied and sometimes overlapping land resource users to choose land use and management choices that maximize production, encourage efficient farming and food systems, facilitate better land and water supply management, and satisfy the needs of the population. Technological, power, and institutional challenges, however, inhibit the development of clear and appropriate land resources for use in developing regions.

As part of an integrated land resource management (LRM) program, mapping for tracking and measuring natural resources can and should be undertaken by local people and provide local information. To improve the expertise of district and regional staff in land resource planning and tracking, one step is to educate students at regional and district universities (future government and nongovernmental organization [NGO] employees) in subjects including, but not limited to, forestry, agriculture, ecology, and mining. In recent years, the use of remotely sensed data in mapping land resources and as a source of input data for modeling environmental processes has been widespread. With the emergence of satellite data from different device sensors having a vast array of spatial, temporal, radiometric, and spectral resolutions, remote sensing has become, possibly, the best source of information for large-scale applications and studies. In terms of data challenges, the lack of consistent (spatially and temporally) pre-processed satellite data was expressed as a common obstacle that often led to inaccurate maps being generated. High spatial satellite data and proprietary software costs are barriers to their use for land resource mapping and wider community monitoring, particularly in developing countries. Due to budgetary restrictions, licensing problems, or language barriers, geospatial professionals frequently lacked exposure to emerging new technologies. Skilled geospatial growth depended on donor-related activity participation centered on free and open source solutions. Many of these participatory solutions to geographic information systems (GIS) have centered on low-tech methods to allow societies to spatially map and visualize geographical data. The capacity for unified use of new low-cost digital spatial technologies and data to enhance governance is also growing (Fisher and Myers, 2011). The development of free geospatial data and applications has been growing steadily over the last decade (Moreno-Sanchez, 2012). It has been argued that free GIS software, combined with free spatial data, creates new possibilities for improved openness and expanded citizen participation in environmental policy (Fuller, 2006; Fisher, 2012). Land information systems (LIS) provide a framework, in a given geographical context, for a wide range of thematic environmental applications. The main data, such as satellite imagery, cadastral records, climate measurements, and projections, usually define the natural, immediate characteristics of land and other resources. This record can be combined with other resources (Figure 9.1). Modeling applications may also investigate the possible effect of soil densitites and components or the influence of shifting climatic conditions or land use on the soil functions. The LIS results enable key stakeholders to make informed decisions and adopt successful strategies as well as to promote education and awareness about the environment (Hallett and Caird, 2017).

Geospatial tools are very useful for the mapping, tracking, and simulation of land resources include essential data overarching natural resources management planning (NRM). The new generation of free and open source (FOS) geospatial tools and open data has eliminated the need for licenses and data acquisition, which has been a major obstacle to greater participation of local government in land management planning, tracking, and modeling. In addition, the availability of free satellite imagery, elevation data, and raster-based spatial analysis tools offers city members new opportunities to merge quantitative interpretation of "strong" data with qualitative local information (Spiegel et al., 2012).

9.2 COMBINATORIAL INNOVATION IN SUSTAINABLE LAND MANAGEMENT

Combinatorial innovation pushes progress with unparalleled precision, frequency, and scale of satellite-based economic measurement. Increased satellite data availability and rapid developments in machine learning processes allow a deeper understanding of the fundamental forces that form economic growth. Satellite knowledge has changed our way of understanding, tracking, and achieving sustainable development of land by using remote sensing tools. As satellite data become more available and popular, not only can changes to the Earth be better understood, but it is now also

Land Resource Mapping and Monitoring 123

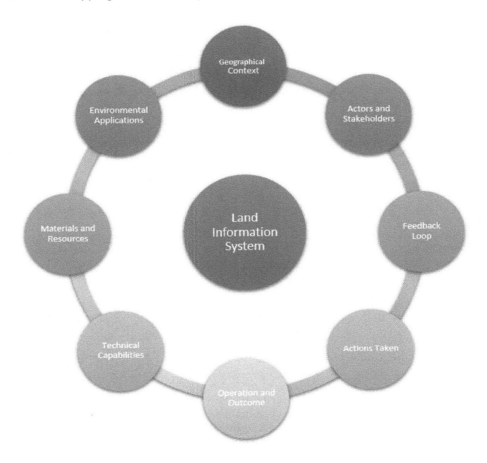

FIGURE 9.1 Life cycle of land information system.

possible to use this understanding to better determine and direct policies, provide services, and encourage better governance. For developing countries, which often do not regularly have accurate socio-economic or demographic data, satellites capture a significant number of physical, economic, and social characteristics of the Earth. Collaborating with powerful cloud storage and open source analyses, the increasing availability of "free" satellite data today has brought about the democratization of data innovation, which enables local government and local agencies to use satellite data to improve sector diagnostics, development indicators, program monitoring, and services delivery.

9.3 BIG (GEO) DATA

Data are often known as the "blood" of GIS. This is the same as Big Data, based on "information" and guided by "information." Big Data not only represents a huge volume of data but has now increased its characteristics from "3Vs (length, velocity, variety)" to "4Vs (+ Veracity)" and "5Vs (+ Value)" (Li and Li, 2014). For a number of years, huge data were the hot topic in GIS. Global engineering satellites are becoming stronger every day, generating new engineering impact opportunities. In different spatial, spectral, and temporal resolution, they capture millions of images from the Earth, producing data in increasing quantity, variation, and speed. The five categories of data (remote sensing data, large data survey and mapping, location-based data, social network data, and Internet-based data) are listed in Figure 9.2.

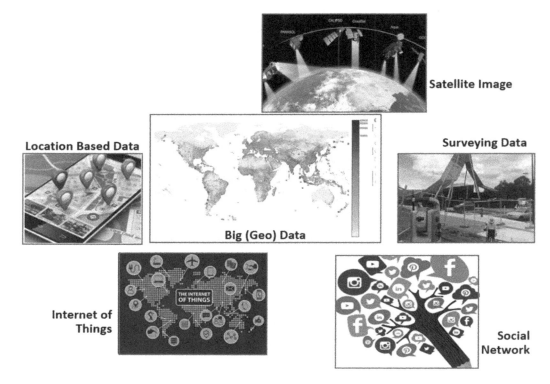

FIGURE 9.2 Classification diagram of big spatial data. (Modified from Yao, X. and Li, G., *Big Earth Data*, 2 (1), 108–129, 2018.) More and more satellite platforms are available, including aeronautical, aviation, and space. The space, time, and spectral resolution can be improved continuously and stored with PetaByte, which can be accomplished with the TerraByte stage. Four-dimensional surveys typically include regional conditions, geographical industries, and thematic mapping information, including digital products (DLG, DRG, DOM, DEM), use of land and other national fundamental surveys, and mapping data. The age of Big Data, such as point cloud and mobile mapping, has intensified in recent years with the development of new surveying equipment and mapping technology that enables the space distribution data for the measurement area to be collected quickly and efficiently. The location-based databases are known as geographic and human social data. These usually contain a spatial position and time identification. Location-based data are generated primarily by GPS, BDS, or other smart telephone positioning systems, field collection data, traffic trajectory data, etc., spatial internet information such as profile accounts, and social media information including WeChat, Facebook, Twitter, and other social apps. Social media data are currently playing a significant role in online public opinion, land monitoring, and sustainable land use planning. Data come from all kinds of sensors, such as environmental, meteorological, water, pipeline, watch, smart household, and so on. The Internet of Things produces higher frequency and greater variety compared with the traditional Internet open source remote sensing data.

Furthermore, land use maps based on remote sensor imagery play a crucial role in tracking human–environmental interactions such as changes in landscape, ecological services, and urban planning and administration (Sumarga and Hein, 2014; Hegazy and Kaloop, 2015). While land cover maps reflect the biophysical coverage observed on the surface of the Earth, land use maps define the structures, operations, and inputs people perform to create, change, or maintain them in a particular land cover form. Land cover reveals the natural and anthropogenic characteristics on the surface of the planet, e.g., deciduous forest, wetlands, developed/constructed areas, grasslands, water, etc. In comparison, land use defines activities that occur on the territory and reflect current land use. Change in land cover also plays a significant role in environmental change at the local and national levels. Land use/land cover (LULC) is local and site specific, and these changes ultimately lead to global environmental change. The annotated Open Sourced Datasets, the World Bank Open Data datasets, and other publicly available resources allow the processing and documentation of the data (e.g. Cumulus, label maker). Since its inception in the 1970s, the scale of the satellite imagery available has gradually increased. In the last couple of years, exposure to satellite imagery has been increasingly democratized. After the launch of the Earth Resources Technology Satellite (renamed Landsat 1) in 1972, remote sensed imagery has gradually been used to track Earth's habitats by quantifying land cover change (Hansen and Loveland, 2012), forest degradation (Asner et al., 2002), fossil reserves and pollution (Asner et al., 2010), habitat destruction and disease (Dennison et al., 2010), plant abundance (Hall et al., 2011), alien plants, habitat suitability Lahoz-Monfort et al. (2010), and animal populations. However, satellite imaging items differ in their spatial and spectral resolution, geographic and temporal coverage, cloud exposure, protection legislation, and quality—factors that can hamper their reliable conservation use.

9.3.1 COARSE RESOLUTION SATELLITE DATA (>100 m PIXEL SIZE)

Advanced High-Resolution Radiometer (AVHRR) data provide useful data sources for an almost constant wide coverage to enable a wide variety of work on environmental monitoring, involving weather forecasting, climate change, ocean currents, atmospheric sounding, ground cover tracking, search and rescue, forest fire identification, and several other tasks (Wu et al., 2020). Satellite systems from NASA complement operational satellites and have quantitative metrics for analyzing long-term patterns. For example, on the EOS Terra (AM) and Aqua (PM) platforms, the Moderate Resolution Imaging Spectroradiometer (MODIS) instruments have dramatically enhanced the functionality of the operational National Oceanic and Atmospheric Administration (NOAA)/MetOp Advanced Very High-Resolution Radiometer (AVHRR). These coarse resolution data are being used to identify and describe land cover globally and to identify regional-scale land cover transition. They often perform regular observation of terrestrial surface characteristics such as surface temperature, Leaf Area Index (LAI), and fire intensity, which are also indications of land cover transition. NASA allowed the creation of the Visible Infrared Imaging Radiometer Suite (VIIRS) instrument launched in 2011 as part of the Integrated System Office, which enabled a route from MODIS group discoveries into the operational domain. The NASA S-NPP Research Team is ensuring consistency of data between MODIS and VIIRS. The VIIRS instrument, together with MODIS, continued the long-term records of vegetation indices, land cover, and fire data. The Earth Observer 1 (EO1) platform offered a test bed for new sensor technologies and hyperspectral, spaceborne remote sensing. Similarly, for better surface identification and evaluation of MODIS thermal goods, the Advanced Spaceborne Thermal Emission and Reflection Radiometer (ASTER) sensor produced coarse resolution thermal data collocated with MODIS data.

9.3.2 MEDIUM RESOLUTION SATELLITE DATA (10–99 m PIXEL SIZE)

The United States Geological Survey (USGS) Earth Resources Observation and Science Center (EROS) maintains and disseminates data from Landsat satellites in collaboration with NASA, offering worldwide access to a continuously collected database of space-based land remote sensing data. Through the USGS website, data from the Landsat series of Earth observation satellites, which have been mapping the planet since the 1970s, became openly accessible to the public in 2008 (Wulder et al., 2019). They have also been geometrically modified and spatially adjusted and are usable in a ready-to-analyze format through the LEDAPS (Landsat Ecosystem Disturbance Adaptive Processing System) (Wulder et al., 2016). LEDAPS is a system that transforms raw satellite Landsat (Thematic Mapper [TM] and Enhanced Thematic Mapper Plus [ETM+]) data into surface reflectance (SR) properties, reducing much of the spectral variance induced by atmospheric effects (e.g., water vapor and aerosols) (Hansen and Loveland, 2012). Subsequently, SR products are generated using the Landsat 8 (Operational Land Imager [OLI]) Surface Reflectance Code

(LaSRC) algorithm (Rathnayake et al., 2020). In cooperation with the LCLUC Program, the NASA Harmonized Landsat Sentinel (HLS) Project has developed a co-located, unified Landsat 8 and Sentinel 2 package, which offers improved temporal frequency of moderate resolution coverage.

9.3.3 High Resolution Satellite Data (<10 m Pixel Size)

Google Earth and Microsoft Bing Maps provide access to satellite imagery in very high resolution (VHR), described in this chapter as pictures with spatial resolution of less than 5 m (Figure 9.3). As VHR images contain numerous detailed features and objects such as structures, roads, and individual woodland trees, the map-validation baseline sets are increasingly being improved by visual analysis of Google Earth images, while Google Earth producers and consumers use Google Map producers and consumers to collect data for validation (Biradar et al., 2009; Tsendbazar et al., 2016). Simultaneously, applications like Geo-Wiki use crowdsourcing for the collection of a reference knowledge set for the creation and validation of hybrid land maps based on the Google Earth and Microsoft Bing Map visual representations (See et al., 2015; Fritz et al., 2017) and Google Earth imagery for the collection of forest inventory data from the Earth collection tool. Google Earth offers access to historical images by archiving photos when connected to its network, unlike Microsoft Bing Maps. These historical images are a valuable knowledge source for analyzing landscape changes over time. Nevertheless, as the satellite images are displayed together in Google Earth and Microsoft Bing Maps, the impression may be that the data from satellites are constant and homogeneous in both time and space. It is furthermore necessary to note that not all VHR images are available from both providers, Google Earth and Microsoft Bing Maps, but only a subset

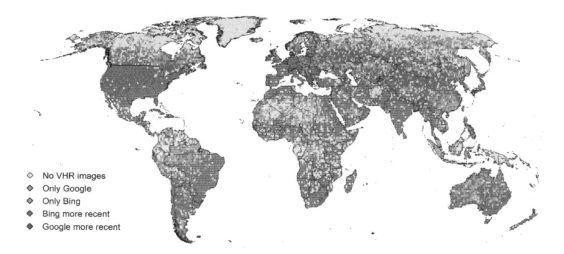

FIGURE 9.3 Comparison of the most recent VHR satellite imagery (<5 m resolution) available in Microsoft Bing Maps and Google Earth as of January 2017. (Lesiv, M., et al., *Land*, 7 (4), 118, 2018.) This imagery is often used to track statistically surveyed data collected via visual analysis of modern applications, such as Collect Earth. The worst coverage is in North America, with VHR pictures covering just half of the region, in Google Earth and Microsoft Bing Maps, and in Eastern Europe with only 39 percent VHR and 58 percent of Microsoft Bing Maps. This is possibly because these regions cover high latitudes in the north, where VHR images are less available. Google Earth has much higher coverage than Microsoft Bing Maps, by comparison, in northern Europe, Southeast Asia, and central Africa. In general, Google image is more recent than Microsoft Bing Maps, but in South America, Australia, New Zealand, and northern Europe, Microsoft Bing Maps complement Google. The richest collections of images exist in North America, southern Europe, South Africa, and Southeast Asia, while the majority of collections in eastern and northern Europe, central Asia, and north and central Africa consist of only one or two images.

of images negotiated through contracts. With Sentinel 2, a 10 m spatial resolution for the visual representation of many landscape features is readily available and may gradually replace the basic Landsat imagery on Google Earth.

9.4 WEB SEARCH ENGINE FOR FREE ACCESS REMOTE SENSING DATA

Any study of remote sensing revolves around the term "DATA" with a defined resolution, location, and sensor. There are plenty of websites providing free quality satellite images. Here are some top-notch useful examples of free satellite imagery from several important space agencies around the world: The Google Earth Engine (https://earthengine.google.org/), USGS Global Visualization Viewer (GloVis), NASA Earth Observation (https://neo.sci.gsfc.nasa.gov/), The USGS Earth Explorer (https://earthexplorer.usgs.gov/), The European Space Agency's (ESA) Sentinel data, NASA Earth Data (https://search.earthdata.nasa.gov/), NOAA (National Oceanic and Atmospheric Administration) Class, The Land, Atmosphere Near real-time Capability for EOS (LANCE) (https://earthdata.nasa.gov/), Earth Observation Link (EOLi), National Institute for Space Research (Instituto Nacional de Pesquisa Espaciais [INPE]), JAXA's Global ALOS 3D World, VITO Vision (www.vito-eodata.be/PDF/portal/Application.html#Home), NOAA Digital Coast (https://coast.noaa.gov/digitalcoast/), Global Land Cover Facility, BlackBridge/Geo-Airbus (www.intelligence-airbusds.com/satellite-image-gallery/), UNAVCO Research Data (https://web-services.unavco.org/brokered/ssara/gui), Natural Earth's Data (www.naturalearthdata.com/downloads/), and Bhuvan Indian Geo-Platform of Indian Space Research Organisation (ISRO) (https://bhuvan-app3.nrsc.gov.in/) are among the fast and simple Web search portals that offer complete, unrestricted, and open access satellite and aerial data. We can use land monitoring and mapping to various sources of satellite data. If you wonder which providers have the new satellite imagery, the highest resolution, or the biggest open data collection, we can find the answers. The important open source satellite data are described in Table 9.1.

9.4.1 OPEN SOURCE VECTOR DATA

High-precision and wide-range space vector knowledge has exploded internationally, including land range, social media, and other data sets, offering a strong chance of improving regional macroscopic decision-making, social networking, public safety, and emergency capabilities (Yao and Li, 2018). Over the last 10 years, Big Data has become a state-of-the-art technological concept in academia, industry, business, and politics. Big Data refers to rapid data growth that goes beyond conventional databases and electronic tools for collecting, storing, analyzing, and utilizing data (Chen et al., 2013). Possibly, one of the most productive fields of data growth and technological applications is broad space data (Lee and Kang, 2015). The theories and methods that have been implemented have slowly been introduced into the GIS area through the growth and improvement of new-generation high-performance computing technology, such as cloud computing, NoSQL databases, and others (Yang et al., 2017). Nonetheless, this is much less the case for large-scale big spatial vector data (BSVD) studies, because the vector data set also contains a variety of special factors, such as national economic protection or the building of other infrastructure, leading to difficulties in the sharing and acquisition of large-scale information.

Big BSVD Data (BSVD) is the set of spatial data types (Shekhar et al., 2014), represented by points, lines, or polygons (areas). BSVD usually includes both survey and mapping data, location data, social media data, and Internet of Things data. Within the data processing field, data storage, index, query, analysis, and analytical data are also considered as a large and more dynamic domain (Siddiqa et al., 2016). The current work on BSVD management has mainly emphasized certain features (volume, variety, or speed) of large space data and solved certain technical problems. The vector data model can be categorized as a model of geographic relationship and an object-oriented model based on the storage relationship between attributes and spatial data. Due to its advantages

TABLE 9.1
Free Satellite Imagery Data Sources

Satellite Data	Description	Pros	Cons	Website
Google Earth	While the highest-resolution images are actually taken by aircraft, Google Earth provides free access to some of the highest-resolution satellite images. In the last 3–4 years, the bulk of Google Earth data has been obtained. Google Timelapse is also a nice free resource if you are interested in seeing pictures of our changing planet.	Free to use, huge library (worldwide), no required account, easy to search, many resolutions available.	No recent images, specific licenses and prices for commercial applications may not be available, not downloadable.	https://earth.google.com/web/
Sentinel Hub	Sentinel Hub is one of the most widely used satellite data access portals. Users can access all Sentinel items by means of Sentinel Hub as soon as they become available. A collection of historic (archive) satellite images is also included. Includes access to Landsat and MODIS items by the Sentinel Center.Sentinel 2 provides the highest possible resolution for free satellite imagery, 10 m in red, green, blue and near-infrared resolution. Its 12 spectral groups extend from the shore to SWIR. Sentinel-2 covers the entire planet by two satellites that orbit the globe and provides new photos of an area every 5 days.	One of the common data sets currently available and the nearest people can come to accessing free, real-time satellite images at 5-day revisit rates. Great satellite communication forums and educational opportunities. ESA contracts for the management of the SNAP and SNAP toolbox forums are currently awarded to SkyWatch. They are amazing and absolutely free.	Access to just available data products. In many applications, the resolution does not even exceed 10 m, such as car count, construction control, tracking, etc.	www.sentinel-hub.com/
Earth Explorer	Earth Explorer is a free software application for satellite imagery access to Landsat. Landsat is the longest-running space-based archive of Earth's land in nature, a collaborative NASA/USGS program. Landsat-8, the newest mission satellite in 2013, has the second highest free-of-charge resolution for optical data.Earth Explorer also offers links to NASA Terra and Aqua land-data products MODIS, and U.S. and territorial level-1B ASTER data items. SPOT, IKONOS, and OrbView 3 also have some images. There are also Sentinel-2B products (since 2018), but Sentinel Hub or Copernicus Hub is recommended for access to Sentinel satellite data.	It provides direct access to Landsat's most recent satellite data. There are also available digital elevation models (ASTER and SRTM) with hyperspectral models (Hyperion).	Sentinel products have restricted access. Users will have to filter the images manually to select the best images to automatically access satellite data.	https://earthexplorer.usgs.gov/

(Continued)

TABLE 9.1 (CONTINUED)
Free Satellite Imagery Data Sources

Satellite Data	Description	Pros	Cons	Website
NOAA	Free GEOS-R and NOAA-20 data access is supported by NOAA. GOES-R data is updated every 15 minutes and is the nearest to satellite data in real time. However, GOES-R satellites, which are based in very high orbits, are geo-stationary satellites. This makes the data produced very low (250 m and higher) resolution.	Real-time satellite info, accessible free of charge from your browser, can be accessed directly without registering.	Restricted access, extremely low resolution, for commercial applications.	www.nesdis.noaa.gov/content/imagery-and-data
Copernicus Open Access Hub	The Sentinel Hub is close to available data sets, but it is designed for developers who need Earth Satellite Data with low resolution for their applications.	Open data sets, directly available for download through API, SAR data sets included.	Restricted use of low resolution for commercial applications.	https://scihub.copernicus.eu/
Earth on AWS	Amazon Web Services hosts several open data sets. Earth on AWS was inspired by AWS to include a direct approach to them. Data sets available for the Chino-Brazilian Earth Satellite (CBERS) processed by the INPE, including Sentinel-2, Landsat-8, GEOS, NOAA, Sentinel-1, and others.	The largest array of open data sets, which can be accessed directly by AWS.	The largest array of open data sets, which can be accessed directly by AWS.	https://aws.amazon.com/earth/
Zoom.Earth	Zoom Earth presents the latest satellite image and aerial views in a quick, zooming map, much like Google Earth. The platform retrieves data from the satellites NOAA GOES and JMA Himawari-8 every 10 minutes and from the EUMETSAT satellites every 15 minutes.	It does not need an account. The data are easy to browse, and different resolutions are quick to zoom in.	Probably not recent images. Data can't be downloaded.	https://zoom.earth/
Global Imagery Browse Services (GIBS)	Global Imagery Browse Services offers easy access, covering every part of the world, to over 900 satellite imagery items. Most images are updated every day, available a few hours after satellite monitoring. Some items span nearly 30 years. The data available via GIBS are the same as in Worldview (see above), but also, an API provides almost real-time access to satellite images.	One of the world's largest free satellite data libraries, which includes both recent and historic satellite imagery.	Open data only at low resolution.	https://earthdata.nasa.gov/eosdis/science-system-description/eosdis-components/gibs
Bhuvan	India Geo Platform of ISRO. It includes IMS-1, Cartosat, OceanSat, and ResourceSat (Hyperspectral).	The material comprises thematic maps of disasters, forestry, water resources, land cover, and ISRO satellite data that have been processed. The content includes thematic maps on disasters, agriculture, climate, land use, and satellite data collected by the ISRO.	However, most data are only for India.	https://bhuvan-app3.nrsc.gov.in/data/download/index.php

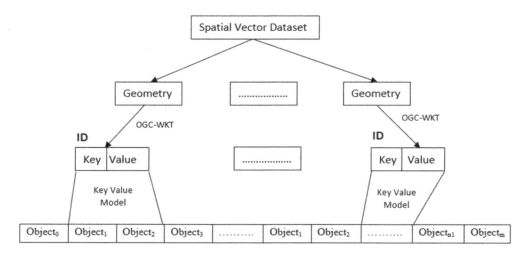

FIGURE 9.4 Structure of GeoCSV vector data. (Wang, L., et al., Paper presented at the 21st International Conference on Geoinformatics, Kaifeng, China, June 20–22, 2013.)

(Wojda and Brouyère, 2013) of being easy to understand, extend, represent, and enforce, the object-oriented model became very popular. While some common vector data structures (including KML, GML, WKT/WKB, and GeoJSON) are provided by Open Geospatial Consortium (OGC) standards (Castronova et al., 2013), not all of them are suitable for large data storage structures.

For a range of NoSQL applications, the central value model is now the standard storage model. Each key value model has two parts, also known as "key pairs," enabling simple data processing. Based on the simple key value model, spatial vector data can be imported and processed with new spatial vector data models on the Big Data platform, including HDFS and HBase (Wang et al., 2013). The OGC-WKT format is used to define spatial geometry using the basic key value storage model. CSV (comma-separated values) files are used to organize the space vector data such that each record is a one-space geometry object only. This is consistent with the cloud computing framework, which facilitates the segmentation, retrieval, and analysis of spatial data. This has various advantages on the basis of GeoCSV with simultaneous estimation, network transmission, and extension (Figure 9.4). Dispersed spatial vector data storage was focused on relational databases such as Oracle; PostgreSQL, a hot trend in the distributed spatial domain; and a long-term emphasis on the design and implementation of a distributed spacious database engine (SDE). Niharika (Yao and Li, 2018) is a distributed PostSQL and PostGIS database scheme that performs cloud-based spatial data partitioning, reading, and writing. The distributed spatial vector data management system is equipped with Oracle RAC (Real Application Clusters) and Oracle Spatial for parallel query (Hameurlain and Morvan, 2016). Sphinx (Eldawy et al., 2015) expands the Impala to provide spatial query support in the SQL language. LandQv1 (Yao et al., 2017) is designed to handle arable land-quality Big Data based on the SQLServer database and the ArcGIS server.

9.5 OPEN SOURCE SOFTWARE FOR LAND RESOURCE MAPPING AND MONITORING

In the past decade, open source geospatial software has made considerable progress, now vying with proprietary rivals in both ability and user-friendly graphical interfaces. There are a range of free open source raster analysis packages available, including ENVI, GRASS, White Box and ILWIS, and SAGA GIS for use in advanced landscape and hydrological analysis with free satellite imagery for natural resource mapping and tracking and digital elevation data.

GIS applications compile, interpret, and view two- or three-dimensional (2-D or 3-D) maps of geospatial data. As GIS offers useful data on land management, distribution transportation, infrastructure planning, population research, facilities management, and more, it will significantly support companies in all sectors. Data processing by GIS tools helps businesses to identify patterns and trends, thus allowing them to address future challenges and develop business processes. Some examples of open source GIS software are: QuantumGIS (https://qgis.org/en/site/), Geographic Resources Analysis Support System (https://grass.osgeo.org/), OpenJump (www.openjump.org/), uDig (http://udig.refractions.net/), SAGA GIS (www.saga-gis.org/en/index.html), Whitebox GAT (https://jblindsay.github.io/ghrg/Whitebox/), GeoDa (https://geodacenter.github.io/), gvSIG (www.osgeo.org/projects/gvsig/), and Diva GIS (www.diva-gis.org/). Many free GIS applications give the business the authority to do the work just like commercial software. We have assembled a list of the best GIS free/open source tools for geospatial research in Table 9.2.

Building local capacity for evidence-based policy in this sense will promote the integration of local knowledge and ultimately, contribute to more educated planning choices. The growing availability of free geospatial data and tools provides new incentives for greater openness and citizen participation in environmental policy. A significant barrier to the wider adoption of FOS GIS tools, however, is a lack of simple, detailed training content, especially with respect to the use of satellite and terrain analysis instruments. The future benefits of engaging in training content to enable local government departments to carry out their own planning and monitoring include:

- Creating credibility by involvement in the spatial analysis
- Supporting equity, integrity, and transparency through skills development
- Contributing to a single national database of teaching materials
- Legitimizing the use of FOS GIS software

To help decentralized applications for better LRM governance, training content needs to be generated that (1) uses free data, (2) is contextualized to local LRM problems, (3) results in data that meet local needs and monitoring standards of central government, and (4) is created in relevant languages.

9.6 CROWDSOURCE PLATFORM

Crowdsourcing enables efforts to allow more efficient use of space technologies for sustainable land creation through digital, Internet, and social networking platforms. Multiple NGOs linked to aid use crowdsourced image labeling to manually classify trends of affected areas and to automatically learn devices. Although remote sensing technology, in particular satellites, played an important role in supplying land coverage data sets, significant discrepancies were noted among the accessible products. In general, global land use is harder to map and cannot be sensed remotely in certain situations. In-house or ground-based data and high-resolution imagery are also a vital necessity for the development of reliable coverage and land use data sets (Fritz et al., 2017). The evaluation of global maps also restricted the effort to evaluate global goods in terms of both cost and practice. Crowdsourcing has demonstrated reliable, timely, and cost-efficient knowledge complementing conventional data collection approaches in order to tackle the shortage of reference data (Olofsson et al., 2012). Firtz et al. (2017) defined data from the Geo-Wiki (http://geo-wiki.org/) crowdsourcing platform across four campaigns as the global Earth cover and land use index. This global data set provides human impact information, land discrepancies, wildlife and land coverage, and land use (Figure 9.5). The Geo-Wiki app allows authorized users to interpret current spatial details, including land cover charts, overlaid by satellite images of high to medium resolution. It can also be used to train people, who then delegate different picture interpretation tasks for the collection of land coverage and land use information at certain locations all over the world.

TABLE 9.2
Basic Examples of Open Source GIS Software

Software	Description	Major Features	Pros	Cons	Source
QGIS	QGIS combines and extends its capabilities with other open source geographical information systems. It is primarily used to assess the area, raising the risk of disasters and map the environment.	• Data capturing • Overlaying • Spatial analysis • Create, edit, manage, and export data	Multiple plugins and tools that can be customized User-friendly interface	It can be made more image analysis friendly	
GRASS	GRASS refers to Geographic Resources Analysis Support System and is a land and environmental management tool. It provides tools and programs for many activities. It's an open source GIS app. It includes the processing of images, data management, spatial modeling, graphics, and visualization of data.	• Image processing • Raster analysis • Vector analysis • Geocoding	Image processing Raster analysis Vector analysis Geocoding	Not useful in cartographic design. Data management is complicated.	
OpenJump	Written in JAVA, this free GIS program can handle large sets of data. This function allows users to read and write from files, read different spatial databases, and interpret different vector formats. The ability to modify geometry is another striking feature of its selection. Users can analyze buffer, overlay, or vector data in OpenJump.	• Chart data • Layer editing • Geometry data	Editing geometry and attribute data Customizing the appearance of your information	Changes you make are often lost; thus, you need to keep backups	
uDig	uDig is most apt to map basics. For its Mapnik, base maps can be imported in the same tuning as provided by ArcGIS. The Rich Client (RCP) technology, developed with Eclipse, aims at building complex analysis data within a user-friendly frame.	• Drag and drop interface • Editing tools • Vector operations • Import base maps	It may work with RCP plugins as a separate application or extension This makes it powerful and strong through its catalog, its symbology, and its functionality	Limited tools to personalize the data Processing speed is quite slow	

(*Continued*)

TABLE 9.2 (CONTINUED)
Basic Examples of Open Source GIS Software

Software	Description	Major Features	Pros	Cons	Source
SAGA GIS	SAGA GIS specializes in digital geoscientific data. Because it started with field research, it is now filled with growing geoscience. It has the best morphometry tools, including a topographic wetness and topographical positioning index of SAGA, and is the ideal free GIS app for the geoscientific community.	• Intuitive graphical user interface (GUI) for data management, visualization, and analysis • Framework-independent function development • Object-oriented system design • Geo-referencing projections	It can handle vectors and knowledge in large quantitiesUse raster and spatial data in any format to benefit from the advantages	It isn't used exclusively in mappingObjects you add come separately, which is often annoying	
Whitebox GAT	LIDAR (Light-Detection and Ranging) data are used in Whitebox Geospatial Analysis Software, allowing users to explore environmental and manmade areas accurately. In addition, it has hydrodynamic modeling and storm surge modeling functions and coastal modeling features.	• LIDAR data • Image processing tools • Hydrology tools • GIS tools	Over 410 tools to clip, convert, analyze, buffer, manage, and extract geospatial informationOpen access and remote sensing GIS package	Cannot integrate it into other software and open source GIS projects	
gvSIG	No other software can beat gvSIG framework software when it comes to 3-D visualization. The computer-aided design (CAD) tools allow you to edit vertices, geometries, split lines, and polygons. In addition, the mobile application of gvSIG allows businesses to track the fieldwork with their GPS devices.	• 3-D and animation • Vector representation • Raster and remote sensing • Topology	It works with different vector formatsIt poses advanced tools for spatial analysis	Labeling vector information doesn't fit the presentationLimited options to change the layout of fields in the table	

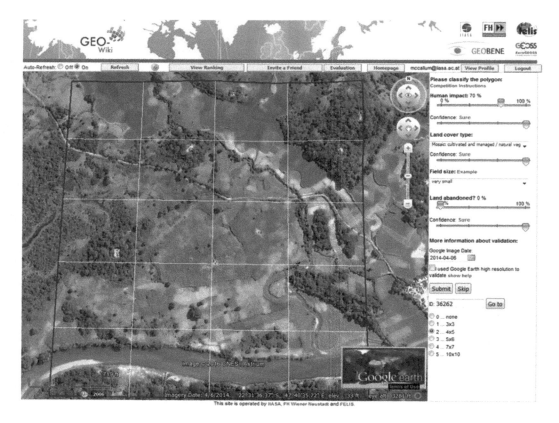

FIGURE 9.5 Screenshot of the human impact interface in Geo-Wiki. (Firtz et al., 2017). Such data are available both to scientists interested in analytical data concerning global satellite items and to those involved in the general monitoring of global terrestrial ecosystems. It tackles in particular a significant gap in global coverage and land use data and is the first global crowdsourcing-driven coverage and land use comparison data collection.

Global land use is much more difficult to map than land use, and in many cases, satellite imagery cannot be clearly viewed. This is an important catalyst for environmental change. The way the biophysical cover observed is used by human beings can be loosely defined. Timely reference data are therefore an important input for the creation of accurate land use maps (Cihlar et al., 2001). The incorporation of remote sensing technology with volunteered geographic information (VGI) data systems such as Open Street Map (OSM) and cloud computing systems such as Google Earth Engine (GEE) is a valuable tool for LULC and mapping and related research. In addition, a powerful tool to track, define, and measure the environment is the combination of remote sensing information and OSM. For researchers and policy makers analyzing land parcels, the product will be a significant source of information. A new way of collecting in situ data is needed in order to supplement Copernicus Sentinel's high spatial (10 m) and temporal (5 d) resolution. Sensitivity for Sentinel observations requires high-quality and timely training and validation data. The potential contribution of the opportunistic use of crowdsourced street-level imagery to the collection of large in situ high-quality data within crop monitoring was assessed by D'Andrimont et al. (2018).

VGI represents a new age of mapping and visualizing our environment, and in the last decade, its data and applications have been growing dramatically (Zook et al., 2010). OSM has been studied and applied in many disciplines and is one of the well-supported VGIs, but vast amounts of knowledge still remain to be investigated. In contrast to official geographical databases, OSM has unique advantages that allow its use to suit a wider range of applications. OSM will create superior maps by having "up-to-date" data to take into account temporary transition trajectories (Estima and Painho, 2013). From a geographical point of view, this is important, because the mapping and monitoring of the Earth was always a major challenge. OSM is an appealing choice for the research undertaken in this study to achieve the research goals because of the increasing coverage of broad space data and cloud computing (Zook et al., 2010). OSM offers a VGI forum through which many volunteers can contribute and collaborate, boosting their interest in and ability for selection and mapping. In addition, the accomplishments of OSM data sets allow analysis of multi-temporal trajectories based on past land changes (Neis et al., 2012). In addition, Web 2.0 offers a breakthrough in capturing, mapping, and analyzing geo-located data (Haklay and Weber, 2008). GEE is a newly built mapping and analysis tool for Earth Observation data sets, allowing large-scale spatial analysis through its special infrastructure and automated parallel calculations (Johansen et al., 2015; Patel et al., 2015). Because of its geo-location functionality and labels, OSM is an important new tool, which can be used in land cover mapping and evaluation. The combination of earth observations and OSM features will be a tool for large-scale mapping. Due to its specific workflow, it will provide a great opportunity to generate innovations to satisfy different map needs.

Several recent studies have demonstrated the potential of citizen science observatories as well as active or opportunistic crowdsourcing for collecting large, quality in situ data. People with no technical experience in remote sensing can be actively involved in the development and processing of large data sets. In conjunction with remote sensing, Kosmala et al. (2016) used citizen scientists to link ground-based phenology observations for scaling and validation purposes with the satellite sensor data. Data collection is carried out in mobile appliances in which people can take photos, observations, or ad hoc crop management applications in specific areas such as FotoQuest (2017). Search data from Google in the United States (Van der Velde et al., 2012) helps to track planting and harvesting dates for maize. Google Street View (GSV) is a well-known server. City images are taken by means of picture acquisitions from vehicles, motorcycles, or other forms of transport that are mounted on board. GSV coverage is broad, and every place has been collected, many times over the years in many cases (Juhász and Hochmair, 2016). GSV and its Street View Application (USGS, 2018) have been used by the Global Croplands project to gather knowledge about crops. A 30 m Africa cropland map was also built and validated with the approach (Xiong, 2017).

9.7 HIGH-QUALITY GROUND TRUTH AND LAND USE MANAGEMENT

The kind of auxiliary data that provide supporting and explanatory information are generally called the "ground truth" in a remote sensing picture. In the end, the goal of acquiring basic truth is to assist in calibrating and understanding remotely sensed information by analyzing the truths. The information sources for global positioning system (GPS) objects, direct observations, photo documentation, and a variety of charts, personal acquaintance, and other information will be given during the field review. The data are not densely accessible in many parts of the world, depending on the source. For example, Geograph has a set of images that reflect the grid square, which is very scarce for the urban object–level land use mapping mission. The accuracy of the geo-tags is subjective because many pictures either lack information on the camera's orientation and location or are inaccurate because the user enters a map with an incorrect zoom level. The pictures are not always accurate. It is much harder to obtain land use–related information from the picture strips when distinguishing different types of land cover groups based on their respective spectral signatures, as spectrum information from overhead images is not sufficient to distinguish the same materials (landscape) into various land use groups. Therefore, a land use class is sometimes composed of a collection of objects, perhaps of different materials. A traditional in situ set of truths is susceptible to sampling errors (Dickinson et al., 2010), and it is not possible to automate the integration using Big Data analysis. Classical field investigations often involve a significant organizational effort, which makes frequent re-sampling to determine improvements in agricultural dynamics difficult. Robust algorithms that analyze the entire world need specific training data and conventional architecture. Microdata are used, for example, to map urbanizing processes in computer training, verification, and calibration. Such in situ data are given in the EU by the LUCAS (Land Use and Land Cover Area Frame Survey). LUCAS is a 3-year harmonized in situ coverage and land use data collection activity covering the EU as a whole (Gallego and Delincé, 2010). For the development of crop type information in China, Wu and Li have created an individual system that combines GPS, video camera, and GIS. The same way of verifying paddy rice mapping items was used by (Singha et al., 2016). Srivastava et al. (2018) seek to recognize the use of a single object by means of visual indications found in GSV side-view images and to distinguish between schools, hospitals, and religious locations. Such photographs provide geo-referenced information not only about the composition of materials of objects but also about their actual use, which otherwise would be hard to obtain using other traditional data sources, including aerial images.

9.8 CLOUD COMPUTING

The "cloud" idea was created because businesses like Amazon had excess computing power, and they understood that when not in use (e.g., at night), they should "share their server space." A positive feature of cloud applications is that hardware resources are distributed from the cloud as software demands increase. Cloud computing workflows can simplify data collections and management for field personnel and make land management and conservation decisions timelier and more knowledgeable. The monitoring efficiency of these workflows is also substantially improved (Figure 9.6). Big Data technology in areas such as geographic data and technical data requirements of geological research organizations, as well as of related businesses and government agencies, must respond to various needs at different levels from individuals at different levels. These solutions can not only save time and money; they can also reduce possible mistakes caused by capturing on-paper field data and then inserting them manually into a computer database. Many choices are available for the storage of cloud data.

Cloud computing is a service to a network of end-users for processing and storage. A key problem in land management in the sense of large-scale distributed computing is the mapping of a collection of applications into a network where those applications can be applied and the local resources allocated for these applications are reserved for each of the applications running on the network. The resource allocation cycle must repeat itself over and over in response to such changes, in other words, be dynamic, to make sure it is possible to optimize the system utility at all times. Optimum distribution of resources is also computationally costly as a result of efficiency maximization. For example, in the case of grid computing, the problem of scheduling work on computers, which minimizes the total running time, can be formulated as the lowest-range scheduling problem, known as NP-hard. Second, the issue that applications have to be installed on machinery is sometimes referred to in the sense of cloud computing as a variant of the NP-hard knap-pack issue. In this work, we model the system as a dynamic set of nodes that represent the cloud machines. Cloud computing is a synthesis used to describe a variety of various types of computing concepts, including a large number of computers connected via an Internet-based real-time communication network. Cloud computing depends on resource sharing in order to ensure stability and cost savings, similarly to a power supply over a network.

Geospatial Web services are standards-based frameworks of the OGC for electronic access by geospatial information users, suppliers, and partners (Figure 9.7). The three key goals—data exploration, data analysis, and data access—are highlighted by geospatial Web services. It demonstrates an implementation scenario that supports the standard OGC Web services interface in a cloud environment with the selected open source components, such as GIS, GeoServer, ArcGIS, GRASS GIS, etc. (Vinothina et al., 2012). Figure 9.8 shows GeoServer application of geospatial data acquisition, monitoring, and management. Every user can collect field information electronically with his or her login credentials on your tablet through this ArcGIS Online account. With your Field iPad running

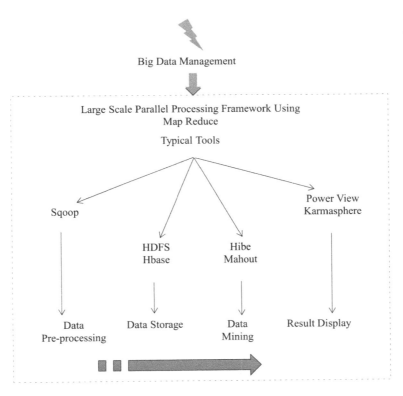

FIGURE 9.6 Schematic diagram of Big Data analysis for Land Information Science. A whole data connection is made by using advanced cloud computing technology, Internet of Things, and large-scale processing flows to link data, information and knowledge, and service.

ESRI Collector for ArcGIS, the field surveyor will reach your monitoring site and connect to an RTK GNSS receiver via Bluetooth. A tablet that interfaces exactly like your smartphone allows field users to collect all their data with one appliance (viewing information, geospatial data, and photos). Oversight managers can effectively monitor and communicate with other organizations, for example, the hypothetical animal control workers who handle predators. Managers get their basic data and breeding results in a timely manner, allowing quicker management decisions and overall budget savings. The combination of multiple data sets allows a timelier evaluation, and new techniques and frameworks for optimizing data handling and analytical analysis are needed. The provision of cloud computing services and tools like GEE, NASA Earth Exchange (NEX), Amazon Web Service (AWS), and the Descartes Labs Platform allows dimensions and depths to be measured and tracked that are difficult or impossible to achieve (Nemani, 2011). In addition to the Landsat data collection, the AWS also provides access to an overview of this data set in the cloud. GEE is an enhanced, high-performance computer platform that gives access and processing power to analyze a vast number of Earth observation data on a planetary scale.

The cost-effective development of LULC products could be possible on a low-cost and efficient basis through multi-temporal satellite observations and cloud-based analytical platforms such as GEE. The key benefits of using GEE include decreased processing times and increased capacity to access high-resolution data globally. But the use of GEE is difficult. Some obstacles include JavaScript and Python languages, a limited number of GEE functions built in, and a lack

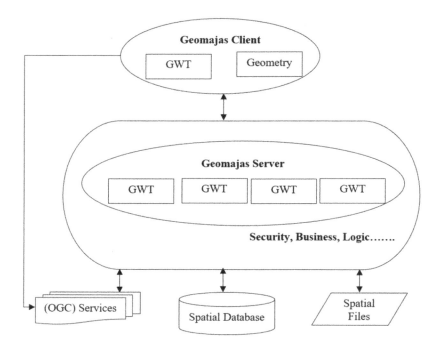

FIGURE 9.7 Components of Geomajas. The client is built upon the Google Web Toolit (GWT) and provides a complete mapping application programming interface (API) for the browser, similarly to Google Maps. On top of that, it also bundles a host of plugins, such as geometry editing or printing. By making use of GWT, applications using Geomajas are really light-weight, as the Javascript is generated and obfuscated at compile time. The Geomajas server can act as a "proxy" for many layer types. adding security or business logic or even influencing rendering mechanisms.

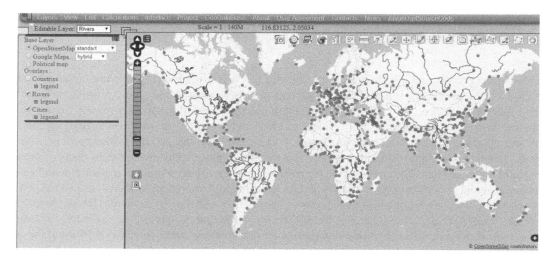

FIGURE 9.8 Open GeoServer application.

of integration with other geospatial analytic open source tools, such as R and QGIS for the existing GEE platform. Although a large Earth observation and geospatial information library is stored in GEE archives, data are not accessible in real time and may be of use only for certain operational applications. Table 9.3 shows open source Web mapping sources of geospatial data for land resource mapping and monitoring. Hansen et al. analyzed Landsat satellite data and machine learning for

TABLE 9.3
List of Open Source Web Mapping

Web Server	Computer Language	Major Features	Application	Source
GeoServer	Java		GeoServer is a software server based on Java, which allows users to access and edit geospatial information. GeoServer offers great flexibility in both mapping development and data sharing with open standards set by the OGC.	http://geoserver.org/
GeoMajas	Java	• Complete Web mapping API • Full HTML 5 support • Geometry editing framework • Printing services • Supported layer types: TMS, WMS, OSM	Geomajas is a library, software, and API set for a whole Web mapping solution, available FOS. The key components of the project are client and server applications built for close collaboration.	www.geomajas.org/
MapGuide	PHP, .NET, Java	• Interactive map viewing • Quality cartographic output • Built-in resource database for manageability • Uniform data access • Flexible application development • Extensive server-side APIs • Fast, scalable, secure server platform • Multiple platform support • OGC standards	A cloud-based platform that allows users to create and execute network mapping and geospatial Web services applications. MapGuide provides an interactive viewer that facilitates selection of elements, property inspection, map tips, and buffer, pick, and calculation operations. MapGuide offers the most common geospatial file formats, databases, and standard support in an XML contents management database.	http://mapguide.osgeo.org/
MapFish	Pylons, Javascript	• Compliant with the OGC standards • Complete RIA-oriented JavaScript toolbox • Built around an open HTTP-based protocol	MapFish is a versatile and detailed platform for developing rich applications for Web mapping. MapFish is based on the Web platform of Pylons Python. MapFish extends geospatially specific pylons. For instance, MapFish provides certain Web-based services that allow geographical objects to be queried and edited.	www.mapfish.org/

(Continued)

TABLE 9.3 (CONTINUED)
List of Open Source Web Mapping

Web Server	Computer Language	Major Features	Application	Source
MapServer	C	• Advanced cartographic output • Support for popular scripting and development environments (PHP, Python, Perl, Ruby, Java, and .NET) • Cross-platform support • Support of numerous OGC standards • A multitude of raster and vector data formats • Map projection support	MapServer is an open source website for publishing interactive Web mapping and spatial data. Originally developed in collaboration with NASA and the Minnesota Department of Natural Resources (MNDNR), the University of Minnesota (UMN) ForNet project developed MapServer. MapServer allows you to create "geographic image maps," i.e., maps that allow users to guide users to information, beyond browsing GIS data.	https://mapserver.org/
OpenLayers	Javascript	• Tiled layers • Cutting edge, fast and mobile ready • Vector layers • Easy to customize and extend	OpenLayers allows the dynamic map to be placed on every Web page. You can display map tiles, vector data, and loaded source markers. OpenLayers has been created to make use of geographical data of any type.	https://openlayers.org/
Open WebGIS	Javascript, HTML	• The Mozilla Firefox browser is tailored to our website. • The Coordinate Reference Scheme must be in WGS84 (EPSG 4326). • The pop-up windows on this site should be allowed in browser settings. • It takes some time for calculations to work with interpolation and raster. • The element separation symbol must be included in all decimal numbers.	OpenWebGIS is an open source GIS for both Web browsers and mobile phones, both online and offline. Users are able to extrude 2-D, 2.5-D, and 3-D images on a map with the aid of OpenWebGIS to achieve space and time distribution of earthquakes worldwide. The websites that use OWGIS may offer the following functions: the ability to download data (as a KML) for multiple languages, animation, vertical profiles, and transects.	http://opengis.dlinkddns.com/gis/opengis_eng.html#

mapping global forest cover for the 2000–2012 period as one of the first internationally relevant applications of the Landsat data collection. Midekisa et al. (2017) are using high-resolution Landsat satellite observations and GEE cloud computing platform to measure continental land coverage and impermanent surfactors over a long duration (2000–2015).

9.9 CONCLUSION

As satellite data become more and more available and popular, it is now possible not only to better understand how the environment shifts but also to make use of this knowledge to enhance decision-making, direct policy, and promote better governance. Artificial intelligence and computer education can allow users to quickly and efficiently analyze large quantities of Earth observation data. Revolutionary neural networks can be used for automating image recognition and classification tasks based on remotely sensed imagery with correct in situ observations. Earth observation data can therefore be analyzed in real time, reducing the required duration and effort of human analysts. The World Bank Group worked with partners in recent years to develop satellite measurement and tools through open source collaboration. The funding of the Big Data Innovation system and the World Bank's Development Economics Technical Support Group were the incubators of those developments. The aim was to analyze the value of street-level imagery in producing relevant information and recognize photos that could better view land plots. If this ability were realized, Copernicus Sentinels could complement and improve frequently observed data and with high resolution, taking this novel data source to advance the urgent need for knowledge in the fields of food security, product markets, monitoring of the environment, and support for the production and implementation of sustainable agricultural policies. The World Bank has formed a long-standing collaboration with the European Space Agency (ESA), currently part of the Earth Sustainable Development Observation (EO4SD) initiative, in the context of a widespread use of this technology and eventually fostering information transfers to country clients.

Wide geospatial data storage and management continue to be of high importance, including the optimization of numerous conventional cloud environment systems (e.g., MySQL and PostgreSQL) and modern database management systems (e.g., NoSQL, HDFS, SPARK, and HIVE). In order to leverage the spatiotemporal wide data mining system, information extraction and automation require real-time data processing and data extraction. More efficient methods of space-time mining should be built to take advantage of the cloud platforms' elastic storage and machine resources. Spatiotemporal methodologies for large geospatial data processing are crucial and must be further developed and formalized for the optimization of cloud computing. Further work is required to recognize and prevent attacks on the cloud platform in order to monitor and maintain trust.

REFERENCES

Asner, G.P., Keller, M., Pereira, R., and Zweede, J.C. (2002). Remote sensing of selective logging in Amazonia: assessing limitations based on detailed field observations, Landsat ETM+, and textural analysis. *Remote Sens Environ* 80: 483–496.

Asner, G.P., Powell, G.V.N., Mascaro, J., Knapp, D.E., Clark, J.K., Jacobson, J., Kennedy-Bowdoin, T., Balaji, A., Paez-Acosta, G., Victoria, E., Secada, L., Valqui, M., and Hughes, R.F. (2010). High-resolution forest carbon stocks and emissions in the Amazon. *Proc Nat Acad Sci USA* 107: 16738–16742.

Biradar, C.M., Thenkabail, P.S., Noojipady, P., Li, Y., Dheeravath, V., Turral, H., Velpuri, M. et al. (2009). A global map of rainfed cropland areas (GMRCA) at the end of last millennium using remote sensing. *Inter J Appl Earth Obser Geoinformation* 11 (2): 114–129. doi:10.1016/j.jag.2008.11.002.

Castronova, A.M., Goodall, J.L., and Elag, M.M. (2013). Models as web services using the Open Geospatial Consortium (OGC) Web Processing Service (WPS) standard. *Environ Model Softw* 41: 72–83.

Chen, J., Chen, Y., Du, X., Li, C., Lu, J., Zhao, S., and Zhou, X. (2013). Big data challenge: a data management perspective. *Front Comput Sci* 7 (2): 157–164.

Cihlar, J., and Jansen, L. (2001). From land cover to land use: a methodology for efficient land use mapping over large areas. *Prof Geogr* 53: 275–289.

D'Andrimont, R., Yordanov, M., Lemoine, G., Yoong, J., Nikel, K., and der Velde, M.V. (2018). Crowdsourced street-level imagery as a potential source of in-situ data for crop monitoring. *Land* 7 (4): 127. https://doi.org/10.3390/land7040127.

Dennison, P.E., Brunelle, A.R., and Carter, V.A. (2010). Assessing canopy mortality during a mountain pine beetle outbreak using GeoEye-1 high spatial resolution satellite data. *Remote Sens Environ* 114: 2431–2435.

Dickinson, J.L., Zuckerberg, B., and Bonter, D.N. (2010). Citizen science as an ecological research tool: challenges and benefits. *Annu Rev Ecol Evol Syst* 41: 149–172.

Eldawy, A., Elganainy, M., Bakeer, A., Abdelmotaleb, A., and Mokbel, M.F. (2015). *Sphinx: distributed execution of interactive SQL queries on big spatial data*. Paper presented at the 23rd ACM Sigspatial International Conference on Advances In Geographic Information Systems, Seattle, WA, November 3–6.

Estima, J., and Painho, M. (2013). Exploratory analysis of OpenStreetMap for land use classification. 2nd ACM SIGSPATIAL International Workshop on Crowdsourced and Volunteered Geographic Information, GEOCROWD 2013 – Orlando, FL, 5 Nov 2013, Association for Computing Machinery, ISBN 9781450325288, pp. 39–46.

Fisher, R. (2012). Tropical forest monitoring, combining satellite and social data, to inform management and livelihood implications: case studies from Indonesian West Timor. *Int J Appl Earth Obs Geoinf* 16: 77–84.

Fisher, R.P., and Myers, B.A. (2011). Free and simple GIS as appropriate for health mapping in a low resource setting: a case study in eastern Indonesia. *Int J Health Geogr* 10 (15). https://doi.org/10.1186/1476-072X-10-15.

FotoQuest Go. Available online: http://fotoquest-go.org/ (accessed on 24 November 2017).

Fritz, S., McCallum, I., Schill, C., Perger, C., Grillmayer, R., Achard, F., Kraxner, F., and Obersteiner, M. (2009). Geo-Wiki.Org: The Use of Crowdsourcing to Improve Global Land Cover. *Remote Sensing* 1 (3): 345–354. https://doi.org/10.3390/rs1030345.

Fritz, S., et al. (2017). A global dataset of crowdsourced land cover and land use reference data. *Sci. Data* 4:170075. https://doi.org/10.1038/sdata.2017.75.

Fritz, S., See, L., Perger, C., McCallum, I., Schill, C., Schepaschenko, D., Duerauer, M., Karner, M., Dresel, C., Laso-Bayas, J.-C., et al. (2017). A global dataset of crowdsourced land cover and land use reference data. *Sci. Data* 4: 170075.

Fuller, D.O. (2006). Tropical forest monitoring and remote sensing: a new era of transparency in forest governance? *Singap J Trop Geogr* 27 (1): 15–29.

Gallego, J., and Delincé, J. (2010). The European land use and cover area-frame statistical survey. In (eds.) R. Benedetti, M. Bee, G. Espa, F. Piersimoni, *Agric Surv Methods*, pp. 149–168. ISBN: 9780470665480; https://doi.org/10.1002/9780470665480.ch10

Hall, K., Reitalu, T., Sykes, M.T., and Prentice, H.C. (2011). Spectral heterogeneity of QuickBird satellite data is related to fine-scale plant species spatial turnover in semi-natural grasslands. *Appl Veg Sci* 15: 145–147.

Hansen, M.C., and Loveland, T.R. (2012). A review of large area monitoring of land cover change using Landsat data. *Remote Sens Environ* 122: 66–74.

Haklay, M., and Weber, P. (2008). Openstreetmap: user-generated street maps. *IEEE Pervas Comput* 7 (4): 12–18.

Hallett, S.H., and Caird, S.P. (2017). Soil-Net: development and impact of innovative, open, online soil science educational resources. *Soil Sci* 182 (5): 188–201.

Hameurlain, A., and Morvan, F. (2016). *Big data management in the cloud: evolution or crossroad?* Paper presented at the 12th International Scientific Conference on Beyond Databases, Architectures and Structures (BDAS), Ustron, Poland, May 31–June 03.

Hegazy, I.R., and Kaloop, M.R. (2015). Monitoring urban growth and land use change detection with GIS and remote sensing techniques in Daqahlia Governorate Egypt. *Inter J Sustainable Built Environ* 4: 117–124. https://doi.org/10.1016/j.ijsbe.2015.02.005.

Johansen, K., Phinn, S., and Taylor, M. (2015). Mapping woody vegetation clearing in Queensland, Australia from landsat imagery using the google earth engine. *Remote Sens Appl Soc Environ* 1: 36–49.

Juhász, L., and Hochmair, H.H. (2016). User contribution patterns and completeness evaluation of Mapillary, a crowdsourced street level photo service. *Trans GIS* 20: 925–947.

Kosmala, M., Crall, A., Cheng, R., Hufkens, K., Henderson, S., Richardson, A.D. (2016). Season spotter: using citizen science to validate and scale plant phenology from near-surface remote sensing. *Remote Sens* 8: 726.

Lahoz-Monfort, J.J., Guillera-Arroita, G., Milner-Gulland, E.J., Young, R.P., Nicolson, E. (2010). Satellite imagery as a single source of predictor variables for habitat suitability modelling: how Landsat can inform the conservation of a critically endangered lemur. *J Appl Ecol* 47: 1094–1102.

Lee, J.-G., and Kang, M. (2015). Geospatial big data: challenges and opportunities. *Big Data Res* 2 (2): 74–81.

Lesiv, M., See, L., Bayas, J.C.L., Sturn, T., Schepaschenko, D., Karner, M., Moorthy, I., McCallum, I., and Fritz, S. (2018). Characterizing the spatial and temporal availability of very high-resolution satellite imagery in Google Earth and Microsoft Bing maps as a source of reference data. *Land* 7 (4): 118. https://doi.org/10.3390/land7040118.

Li, Q., and Li, D. (2014). Big data GIS. Wuhan Daxue Xuebao (Xinxi Kexue Ban)/Geomatics and Information Science of Wuhan University. 39(6): 641–646. doi:10.13203/j.whugis20140150.

Midekisa, A., Holl, F., Savory, D.J., Andrade-Pacheco, R., Gething, P.W., Bennett, A., et al. (2017). Mapping land cover change over continental Africa using Landsat and Google Earth Engine cloud computing. *PLoS ONE* 12 (9): e0184926.

Moreno-Sanchez, R. (2012). Free and open source software for geospatial applications (FOSS4G): a mature alternative in the geospatial technologies' arena. *Trans GIS* 16 (2): 81–88.

Neis, P., Zielstra, D., and Zipf, A. (2012). The street network evolution of crowdsourced maps: openstreetmap in Germany 2007–2011. *Future Internet* 4: 1–21.

Nemani, R. (2011). Nasa earth exchange: next generation earth science collaborative. *Int Arch Photogramm* 38–8 (W20): 17.

Olofsson, P., et al. (2012). A global land-cover validation data set, part I: fundamental design principles. *Int J Remote Sens* 33: 5768–5788.

Patel, N.N., Angiuli, E., Gamba, P., Gaughan, A., Lisini, G., Stevens, F.R., and Trianni, G. (2015). Multitemporal settlement and population mapping from Landsat using Google earth engine. *Int J Appl Earth Observ Geoinf* 35: 199–208.

Rathnayake, S.W.M., Jones, S., and Soto-Berelov, M. (2020). Mapping land cover change over a 25-year period (1993–2018) in Sri Lanka using Landsat time-series. *Land* 9 (1): 27. https://doi.org/10.3390/land9010027.

See, L., Fritz, S., Perger, C., Schill, C., McCallum, I., Schepaschenko, D., Duerauer, M., Sturn, T., Karner, M., Kraxner, F., et al. (2015). Harnessing the power of volunteers, the internet and Google Earth to collect and validate global spatial information using Geo-Wiki. *Technol Forecast Soc Chang* 98: 324–335.

Shekhar, S., Evans, M.R., Gunturi, V., Yang, Y., and Cugler, D.C. (2014). *Benchmarking spatial big data*. Paper presented at the 2nd Workshop on Specifying Big Data Benchmarks, Pune, India, December 17–18.

Siddiqa, A., Hashem, I.A.T., Yaqoob, I., Marjani, M., Shamshirband, S., Gani, A., and Nasaruddin, F. (2016). A survey of big data management: taxonomy and state-of-the-art. *J Netw Comput Appl* 71: 151–166.

Singha, M., Wu, B., and Zhang, M. (2016). Object-based paddy rice mapping using HJ-1A/B data and temporal features extracted from time series MODIS NDVI data. *Sensors* 17: 10.

Spiegel, S.J., Ribeiro, C.A., Sousa, R., and Veiga, M.M. (2012). Mapping spaces of environmental dispute: GIS, mining, and surveillance in the Amazon. *Ann Assoc Am Geogr* 102 (2): 320–349.

Srivastava, S., Muñoz, J.E.V., Lobry, S., and Tuia, D. (2018). Fine-grained landuse characterization using ground-based pictures: a deep learning solution based on globally available data. *Int J Geogr Inf Sci*. https://doi.org/10.1080/13658816.2018.1542698.

Sumarga, E., and Hein, L. (2014). Mapping ecosystem services for land use planning, the case of Central Kalimantan. *Environ Manage* 54: 84–97.

Tsendbazar, N.E., de Bruin, S., Mora, B., Schouten, L., and Herold, M. (2016). Comparative assessment of thematic accuracy of GLC maps for specific applications using existing reference data. *Inter J Appl Earth Obser Geoinform* 44: 124–135.

USGS. (2018). *Global croplands street view application*. Available online: www.croplands.org/app/data/street (accessed on 27 March 2018).

Van der Velde, M., See, L., Fritz, S., Verheijen, F.G., Khabarov, N., and Obersteiner, M. (2012). Generating crop calendars with Web search data. *Environ Res Lett* 7: 024022.

Vinothina, V., Sridaran, R., and Ganapathi, P. (2012). A survey on resource allocation strategies in cloud computing. *Int J Adv Comput Sci Appl* 3 (6): 97–104.

Wang, L., Chen, B., and Liu, Y. (2013). *Distributed storage and index of vector spatial data based on HBase*. Paper presented at the 21st International Conference on Geoinformatics, Kaifeng, China, June 20–22.

Wojda, P., and Brouyère, S. (2013). An object-oriented hydrogeological data model for groundwater projects. *Environ Model Softw* 43: 109–123.

Wu, X., Naegeli, K., and Wunderle, S. (2020). Geometric accuracy assessment of coarse-resolution satellite datasets: a study based on AVHRR GAC data at the sub-pixel level. *Earth Syst Sci Data* 12: 539–553. https://doi.org/10.5194/essd-12-539-2020.

Wulder, M.A., Loveland, T.R., Roy, D.P., Crawford, C.J., Masek, J.G., Woodcock, C.E., Allen, R.G., Anderson, M.C., Belward, A.S., Cohen, W.B., et al. (2019). Current status of Landsat program, science, and applications. *Remote Sens Environ* 225: 127–147.

Wulder, M.A., White, J.C., Loveland, T.R., Woodcock, C.E., Belward, A.S., Cohen, W.B., Fosnight, E.A., Shaw, J., Masek, J.G., and Roy, D.P. (2016). The global Landsat archive: status, consolidation, and direction. *Remote Sens Environ* 185: 271–283.

Xiong, J., Thenkabail, P.S., Gumma, M.K., Teluguntla, P., Poehnelt, J., Congalton, R.G., Yadav, K., and Thau, D. (2017). Automated cropland mapping of continental Africa using Google Earth Engine cloud computing. *ISPRS J Photogramm Remote Sens* 126: 225–244.

Yang, C., Yu, M., Hu, F., Jiang, Y., and Li, Y. (2017). Utilizing cloud computing to address big geospatial data challenges. *Comput Environ Urban Syst* 61: 120–128.

Yao, X., and Li, G. (2018). Big spatial vector data management: a review. *Big Earth Data* 2 (1): 108–129.

Yao, X., Yang, J., Li, L., Yun, W., Zhao, Z., Ye, S., and Zhu, D. (2017). *LandQv1: a GIS cluster-based management information system for arable land quality big data*. Paper presented at the 6th International Conference on Agro-Geoinformatics, Fairfax, VA, USA, August 7–10.

Zook, M., Graham, M., Shelton, T., and Gorman, S. (2010). Volunteered geographic information and crowdsourcing disaster relief: a case study of the Haitian earthquake. *World Med Health Policy* 2 (2): 7–33.

10 Introduction to Part II
Mapping, Monitoring, and Modeling of Water Resources

P. Das and Dipanwita Dutta

CONTENTS

10.1 Introduction 145
10.2 Individual Chapters 146
References 147

10.1 INTRODUCTION

Due to its many uses, the water resource is considered essential for all living beings on the Earth. The availability of water has made our planet livable and distinguishable from other planets. Access to clean water and sanitation is one of the major foci of the sustainable development goals (6th Goal of SDG, UN, 2019) adopted by the United Nations. Although three-quarters of our planet is covered by water bodies, only 3% of the total water resource is freshwater and suitable for human consumption. However, the spatial distribution of water is not equal over the Earth, and this is driven by climatic factors to a great extent. The scarcity of water is a major problem in several parts of the world, especially in regions with an arid or semiarid climate. The freshwater stress in these regions is expected to be more severe in the near future due to climate change and related issues. Moreover, increasing per capita consumption of water in the domestic, industrial, and agricultural sectors, along with pollution and unsustainable use of water, is limiting the availability of freshwater worldwide.

The availability of water is important for preserving ecosystems and habitats, biodiversity, and the biosphere as a whole. It plays a significant role in the economic development of a country due to its diverse uses in agriculture, manufacturing and other industries, transport, and commercial activities. All types of water bodies, including wetlands, provide important ecosystem services, e.g., the source of drinking water, fodder, water purification, flood protection, a suitable niche for wildlife, etc. (Foote et al., 1996). They help to maintain the surface and sub-surface water and energy balance and regulate the regional climate conditions. As a consequence of climate change and intense human activities, a significant reduction in wetlands has been observed in India in the past few decades (ISFR, 2017). Moreover, the land use land cover (LULC) changes via deforestation, cropping intensity, urban area expansion, and other development activities are imposing immense pressure on the surface water area and water quality (Bassi et al., 2014). Considering this, the sustainable management of water has become a matter of great concern for the increasing global water crisis, population pressure, and overconsumption of the resource.

Recent advancement in the field of geoinformation technology has opened up multi-dimensional potential to study water resource and address the critical issues associated with the global water crisis. Remotely sensed datasets obtained from active and passive sensors onboard different satellites have become essential for the mapping, monitoring, and modeling of hydro-meteorological parameters, water quality, flood and droughts; watershed management; river dynamics; irrigation planning; and estimating run-off and soil erosion. Availability of a wide range of satellite rainfall data (TRMM, CRU, CHIRPS, and GPM) and soil moisture data (AMSRE, SMAP, SMOS, and ESA-CCI) has enabled the accurate measurement of the hydro-meteorological characteristics of an area. Studies pertaining to hydrological modeling require datasets with diverse spatial and temporal ranges. Processing and analyzing such a huge volume of datasets requires efficient geographic information system (GIS) software equipped with tools for hydrological analysis. However, commercial proprietary software has limited access for researchers, and most of this software does not allow further modification or customization of the codes. In contrast, open source free software like Q-GIS, GRASS GIS, SAGA GIS, and ILWIS and programming languages like R, Matlab, Python, and GrADS are gaining popularity in the GIS community for their ability to edit, modify, publish, and share the codes.

10.2 INDIVIDUAL CHAPTERS

The second part of the book comprises 11 chapters focusing on the mapping, monitoring, and modeling of water resources with special emphasis on the application of free and open source software. The studies presented in this part highlight the status and nature of deterioration in multiple water resource pools via wetland, ground water, and soil moisture, and its consequences for agricultural productivity and biodiversity. Most of the studies included in this part explore the potential of freely available satellite datasets for mapping and modeling water resources.

In Chapter 11, Talukdar et al. (2021) employed the image fusion technique and artificial neural network (ANN) for automatic water body extraction. Chapter 12 (Paul and Chowdary, 2021) and Chapter 13 (Shakya et al., 2021) assessed the use of open source geospatial techniques as standalone software and Web-GIS platform in preparing suitable water resource development plans. In Chapter 14, Brahma et al. (2021) compared several spectral indices for surface water area identification. Chapter 15, written by a group of researchers (Patwary et al., 2021), highlighted the wetland ecosystem service value (ESV), wherein the highest contribution to the ecosystem function is estimated for water regulation, followed by water supply and waste treatment. In Chapter 16, Bramha et al. (2021) employed the Multi-Criteria Decision Making (MCDM) technique for delineating the groundwater prospect zone (GPZ) in Bilaspur district, Chhattisgarh, India. Chapter 17 (Mitra et al., 2021) assessed the potentiality of Standardised Precipitation-Evapotranspiration Index (SPEI) and RCP4.5 projected data for forecasting drought scenario in the Kangsabati River basin till 2100. Their study has indicated the two most severe drought years, 1982–1983 and 1988–1990 in the twentieth century, and has projected a rise in drought frequency, duration, and severity till 2050. Chapter 18 (Rajkonwar et al., 2021) deals with the characterization of the Himalayan Frontal Thrust (HFT) in Pasighat, Arunachal Pradesh, India by using Google Earth imagery and the vertical electric sounding (VES) method. They have reported a shallow to intermediate water table depth (<5 to 15 m) over 67% of the study area, where the aquifers are formed due to movement along the HFT. In Chapter 19, Bera and Dutta (2021) assessed agricultural droughts in the western part of West Bengal using satellite-derived vegetation condition index (VCI) and standardized precipitation index (SPI) along with ground-based rainfall anomaly index (RAI) and yield anomaly index (YAI). Their study has highlighted the consequences of extreme drought and flood events as the major causes of crop loss in this region. Chapter 20, written by Soni et al. (2021), describes the spatio-temporal changes in snow cover area in the Beas River basin and reports a falling trend above 4000 m from 2001 to 2016. Monitoring the dynamics of snow cover area in the Himalayan region is crucial as it produces a number of perennial rivers and facilitates fresh water supply to the inhabitants in the downstream region. In Chapter 21, Das et al. (2021) highlighted the use of optical data–derived spectral indices to characterize the changes in surface water area and water quality parameters. Moreover, they assessed the potentiality of the latest freely available Sentinel-1 microwave data for demarcating the water inundation area under cloudy conditions, which is essential for nearly real-time flood monitoring and developing suitable mitigation plans.

Integrated analysis of satellite remote sensing and various geospatial data layers in the GIS environment has facilitated the long-term monitoring and modeling of various water resource pools for the past few decades. The GIS environment enables the integration of multiple thematic layers and the development of scenarios based on observations and user-defined criteria, whereas the Web-GIS platforms are interactive and have user-friendly applications for data representation. The spatio-temporal mapping of surface water and its ecosystem services are important for wetland management, biodiversity conservation, and other allied water resource management activities. Moreover, studies pertaining to the mapping, monitoring, and modeling of water resources are crucial for preparing a suitable framework for water resource management, policy intervention, overall socio-economic development, and conservation of various ecosystems under the threats of climate alteration.

REFERENCES

Bassi, N., Kumar, M. D., Sharma, A., & Pardha-Saradhi, P. (2014). Status of wetlands in India: A review of extent, ecosystem benefits, threats and management strategies. *Journal of Hydrology: Regional Studies*, 2, 1–19.

Foote, A. L., Pandey, S., & Krogman, N. T. (1996). Processes of wetland loss in India. *Environmental Conservation*, 23 (1): 45–54.

State of Forest Report (ISFR). (2017). Forest Survey of India (FSI). Dehradun.

11 Improving Wetland Mapping Techniques Using the Integration of Image Fusion Techniques and Artificial Neural Network (ANN)

Swapan Talukdar, Shahfahad, Roquia Salam, Abdus Samad, Mohd Rihan, and Atiqur Rahman

CONTENTS

11.1 Introduction 149
11.2 Study Area 151
11.3 Materials and Methodology 152
    11.3.1 Materials 152
    11.3.2 Methods for Image Fusion 152
    11.3.3 Methods for Evaluating the Performances of Image Fusion Techniques 152
    11.3.4 Artificial Neural Network for Wetlands Mapping 152
    11.3.5 Validation of Wetland Maps 153
    11.3.6 Comparisons of Wetland Mapping Models 153
        11.3.6.1 Kendall Correlation 153
        11.3.6.2 Spearman's Correlation 153
11.4 Results and Discussion 153
    11.4.1 Analysis of Image Fusion Techniques 153
    11.4.2 Evaluation of the Image Fusion Techniques 154
    11.4.3 Wetland Mapping Using Artificial Neural Network 156
    11.4.4 Validation of the Wetland Modeling Models 157
    11.4.5 Comparison of the Association among Wetland Modeling Models 158
11.5 Conclusion 158
Acknowledgments 160
Conflict of Interest 160
References 160

11.1 INTRODUCTION

Wetlands are essential components of the hydrological cycle, which has significant effects on the various aspects of ecosystems and society (Everard, 2016; Moor et al., 2015). Wetland occupies less than 5% of the total ice-free terrestrial surface but has many significant impacts on the global carbon cycle, ecology, hydrology, and biogeochemistry. For example, wetlands are the largest natural source of atmospheric methane and contribute 20–40% of the methane emissions of the world (Ciais et al., 2013). Although wetlands are of significant importance, they are shrinking and degrading due to the expansion of agricultural land, conversion of wetland into urban areas, climate change, and disconnection of wetland from river channels in flood plains (Talukdar et al., 2020a; Ganaie et al., 2020; Talukdar and Pal, 2019a; Shahfahad et al., 2019; Nguyen et al., 2016; Li et al., 2010). Therefore, the precise mapping of wetlands and changes in their area and volume has great significance for understanding their changes in relation to climate change and evaluating the impacts of these changes on the economy, society, and ecosystems.

Despite their significance and their continuous degradation, comparatively less mapping and modeling of wetlands has been done due to the difficulties associated with the complexity of their distribution across vast and remote regions (Jensen et al., 2018; Pal and Talukdar, 2018). During the past few decades, remote sensing (RS) technology has proved to be an effective source of data for the mapping and monitoring of wetland dynamics in different parts of the globe (Kumar and Singh, 2020; Stagg et al., 2020; Wen and Hughes, 2020; Hess et al., 2015). RS is more effective and efficient in comparison to the traditional in situ measurement techniques due to its ability to monitor the globe continuously at multiple spatial and temporal scales (Huang et al., 2018; Wentz et al., 2014). Consequently, numerous studies have been done to map and monitor the wetland dynamics in different parts of the world using multispectral optical and microwave RS data (Talukdar et al., 2020b; Talukdar and Pal, 2019b; Jensen et al., 2018; Mahdianpari et al., 2018; Hess et al., 2015; Huang et al., 2014). Although both microwave and optical RS data have been used for the mapping of wetland dynamics, the optical sensor data have been most widely used for wetland monitoring due to their easy and free access at suitable spatial and temporal scales (Huang et al., 2018).

Over the last few decades, the advancement of RS technology led to the abundant availability of satellite data for wetland mapping, such as Landsat (TM, ETM+, and OLI/TIRS), moderate resolution imaging spectroradiometer (MODIS), advanced spaceborne thermal emission and reflection radiometer (ASTER), Sentinel-1 and 2, and Indian remote sensing (IRS) satellite data: Linear Imaging Self-Scanning Sensor 3 (LISS-III) and Linear Imaging Self-Scanning Sensor-4 (LISS-IV). The diverse availability of RS data has enabled scholars in different parts of the world to develop different methods and techniques for wetland delineation and mapping (Talukdar and Pal, 2020; Guo et al., 2017; Fang et al., 2016). Therefore, numerous models and techniques have been developed and applied for mapping and monitoring wetland dynamics by coupling geospatial techniques and machine learning algorithms (Rezaee et al., 2018), statistical techniques (Humphreys et al., 2019), parametric and non-parametric classification techniques (Talukdar et al., 2020c; Pantaleoni et al., 2009), etc., using optical and microwave RS datasets.

DeVries et al. (2017) adopted a fully automated algorithm for wetland modeling by applying dynamic surface water extent (DSWE) and sub-pixel water fraction (SWF) models. Huang et al. (2014) mapped wetland inundation by creating a sub-pixel inundation percentage using a LiDAR dataset and establishing its statistical association with the wetland area mapped using Landsat data. Similarly, Talukdar et al. (2020c) compared parametric and non-parametric models for the identification of different types of land use classes, such as wetlands and non-wetland areas. Further, a few studies of wetland dynamics have been done in the Indian sub-continent using traditional statistical and geospatial methods (Sun et al., 2017; Talukdar and Pal, 2017; Bassi et al., 2014), but there is a lack of studies using advanced and innovative techniques.

Previous literature reported that researchers in developed countries utilize high-resolution images, such as bird's eye view and SPOT, while in developing countries like India and Bangladesh, it is very difficult to obtain expensive high-resolution images to prepare high-precision wetland maps. Thus, in order to generate high-resolution images, very rarely, research has been carried out to prepare high-resolution images by integrating panchromatic (spatial resolution: 15 m) and multispectral bands (spatial resolution: 30 m). This study was the foundation work, as it integrated both high-resolution images and artificial intelligence to generate reliable wetland maps. In addition, geographic information systems (GIS) and statistical software are required to conduct good scientific research, and these are particularly expensive. It is very difficult to obtain that expensive software in developing countries as well as in new institutions. In these circumstances, open source software, such as QGIS, R studio, and Python, has attracted attention. Open source software can provide similar output to subscription software. There are huge numbers of online tutorials, which help make the open source software popular for conducting scientific research. Keeping the advantages of open source software in mind, in the present study, we used open source software such as R studio (version 3.5.3) for performing statistical analysis.

Sunamganj district of Bangladesh is the home of several large and medium-sized wetlands. In the last few decades, wetlands have experienced shrinking area coverage and conversion to other land use classes (Sun et al., 2017). Very rarely, studies were conducted to map and monitor the wetlands by integration of high-resolution images and artificial intelligence models for conservation and restoration. Therefore, we set three objectives for this study. The objectives are (i) to prepare high-resolution satellite images using image fusion techniques, (ii) to validate the fused satellite images using statistical techniques, RS techniques, and visual interpretation, and finally (iii) to map the wetlands by integrating fused images and artificial neural network (ANN) models.

11.2 STUDY AREA

The Sunamganj district of Bangladesh has been selected as the study area. It is located in the northeastern part of Bangladesh. The district is sub-divided into 11 sub-districts or *upazila* (microadministrative divisions). Sunamganj district is characterized by a low-lying topography and three medium-sized rivers: the Ratna, Kushitara, and Surma. Sunamganj is the home of a number of large and small-sized wetlands, locally known as *haor*. The Tanguar *haor* (wetland) is the largest wetland, which covers an area of about 100 km^2 (Figure 11.1). The government of Bangladesh declared the Tanguar *haor* to be an "Ecologically Critical Area" during the 1990s due to continuous encroachment and loss of biodiversity in the past few decades (Alam et al., 2013). In the Ramsar Agreement (2000), Tanguar *haor* was identified as a wetland of global importance. Besides Tanguar *haor*, other major wetlands (*haors*) of Sunamganj district are Shanir, Nullah, Dekhar, and Pasna *haors*. The district has been facing continuous degradation of wetland areas and their ecological significance due to overexploitation of the wetland for freshwater fish production and other reasons (Haque and Kazal, 2008). The total area of district is 3669.58 km^2, and the total population and population density of the district are 2,467,968 and 670/km^2, respectively. The climate of the district is characterized by the tropical monsoon type, with mean annual rainfall above 350 cm and an average annual temperature of about 25°C.

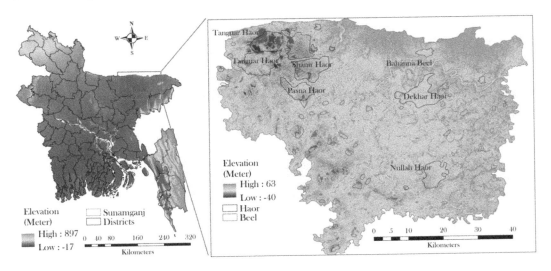

FIGURE 11.1 Location of the study area and major wetlands.

11.3 MATERIALS AND METHODOLOGY

11.3.1 MATERIALS

In the present study, we downloaded the Landsat 8 Operational Land Imager (OLI) image for the year 2019 from the website (https://earthexplorer.usgs.gov/) of the United States Geological Survey (USGS). For the present work, we used the blue, green, and near infrared (bands 2, 3, and 5) bands as multispectral bands and the panchromatic band (band 8) of Landsat 8 (OLI). The map of the Sunamganj district was extracted from the topographical map of Bangladesh, while the locations of the major and minor wetlands were identified using the Google Earth Pro domain. The image fusion techniques were performed in ERDAS Imagine (version 14) and ENVI software. For further image generation processes like wetland mapping, we used ArcGIS (version 10.5) software. The wetland mapping was done using the ANN in the TerrSet Geospatial Modelling Software. Finally, all the statistical analysis was conducted in R studio (version 3.5.3).

11.3.2 METHODS FOR IMAGE FUSION

In this study, four advanced image fusion techniques were employed on Landsat images: modified intensity hue saturation (IHS) and wavelet IHS, Gram–Schmidt, and Brovey image fusion techniques. The multispectral bands (2, 3, and 5) of the Landsat 8 OLI image were stacked. Subsequently, the stacked multispectral image was fused by the panchromatic band (band 8) of the Landsat 8 OLI image. First, the modified IHS technique, proposed by Siddiqui (2003), was applied to overcome the shortcomings of the traditional IHS technique. This technique can be applied to any combination of a multispectral and a panchromatic dataset.

The Gram–Schmidt image fusion technique was utilized to prepare a high-resolution image by simulating the multispectral and panchromatic bands in which the simulated panchromatic band is considered as the first input band. Furthermore, the high-resolution panchromatic band was replaced by the first Gram–Schmidt band before using an inverse Gram–Schmidt transformation to produce the pan-sharpened multispectral bands (Sarp, 2014).

The Brovey technique is a simple image fusion technique, which is based on the chromaticity transformation and used to fuse the data acquired by the different sensors. It preserves the spatial information of the PAN data, but the spectral distortion is very high in this technique. The Brovey was performed in this study following Wang et al. (2005).

Wavelet is based on multi-resolution analysis and is suitable for image fusion. In this study, the wavelet transformation technique was combined with the IHS to enhance the spectral character as well as spatial resolution. The combination of wavelet IHS fusions was applied to the multispectral and panchromatic bands by following Hong and Zhang (2003).

11.3.3 METHODS FOR EVALUATING THE PERFORMANCES OF IMAGE FUSION TECHNIQUES

To analyze the performance of the fused images, the Laplacian edge detection technique was used in this study. Numerous studies have utilized the Laplacian edge detection technique to assess the accuracy in geospatial studies using various earth observation satellite data (Bausys et al., 2020; Zheng and Zheng, 2019). In the present study, we applied Kendall's correlation technique to evaluate the differences of spectral values between the multispectral images and the fused images. For this, 5000 sample points were selected randomly from the different parts of the study area. The sample points were taken in equal proportion from both images. The plotting of Kendall correlation was done using the R software package version 4.0.2.

11.3.4 ARTIFICIAL NEURAL NETWORK FOR WETLANDS MAPPING

ANN is a widely used machine learning algorithm, which can work for big data analysis (Talukdar et al., 2020). It is a non-parametric classification algorithm, which is independent of the assumption of generally distributed data. The multilayer feed-forward feature is the basic form of neural networks. The structure of an ANN algorithm contains three layers: input layer, hidden layer, and output layer. All the layers are interconnected by neurons. In this study, the multilayer perceptron (MLP)–based ANN algorithm was applied for the mapping of wetland using TerrSet Modeling Software. The MLP-based neural network algorithm is a more general technique and is easy to perform (Amer and Maul, 2019). The detail of the MLP-ANN algorithm applied in this study, which is a back-propagation algorithm, is presented in Richards and Jia (2006).

11.3.5 Validation of Wetland Maps

The validation of a classified image describes the accuracy level of the technique used as well as the output generated using the technique (Shahfahad et al., 2020; Rashid and Aneaus, 2019). For the validation of wetland maps generated using ANN, the Kappa coefficient and root mean square error (RMSE) techniques were used in this study. First, the Kappa coefficient was applied using 5000 randomly selected points in such a way that the entire study area was fairly represented. The number of sample points was similar for both multispectral satellite imagery and the classified map. Subsequently, the cross validation of the points was done by field survey as the wetland area extracted from the Sentinel 2 satellite imagery. On the other hand, the RMSE was used to verify the results of experimental analysis. It is the standard deviation of the errors, which depicts how accurate the results are from the ground truth (Sun et al., 2015). A higher value of RMSE shows a low level of accuracy, and a lower value of RMSE depicts the higher accuracy of a classified output image.

11.3.6 Comparisons of Wetland Mapping Models

11.3.6.1 Kendall Correlation

The Kendall correlation technique, also known as Kendall's "T," is a non-parametric correlation technique that describes the strength of relationships between two variables (Valz and McLeod, 1990). This can be used as an alternative to Spearman's rank correlation technique. In this study, the Kendall rank correlation techniques were used to compare the image fusion models used for wetland mapping using the randomly selected sample points from the different parts of Sunamganj district as well as the areal variation of the land use land cover (LU/LC) types. The sample points were selected in a way that covers almost every part of the study area as well as covering all land use types. This method was performed in R studio software using the "performanalytics" package.

11.3.6.2 Spearman's Correlation

Spearman's correlation technique calculates the strength of the link between two sets of variables, either continuous or ordinal. In this kind of relationship, the variables are likely to change together, although not necessarily with a constant rate (Bishara and Hittner, 2017). Spearman's correlation technique is based on the rank values for the variables rather than the unprocessed data. The sample points selected from different parts of the Sunamganj district were used to calculate the Spearman's rank correlation in order to compare the association between the models used for wetland modeling in this study. This method was calculated using the "performanalytics" package in R studio software.

11.4 RESULTS AND DISCUSSION

11.4.1 Analysis of Image Fusion Techniques

Researchers have already recommended preparing high-precision wetland maps, which can be considered as a foundation step for conservation and restoration (Talukdar et al., 2020a; Talukdar and Pal, 2020). The study area belongs to a data-scarce region, and therefore, high-resolution images

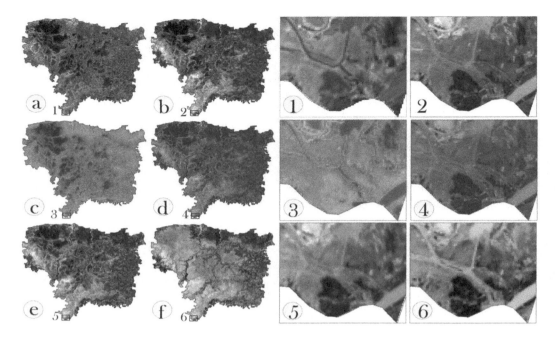

FIGURE 11.2 Image fusion of (a) multispectral bands and (b) panchromatic band (band 8) by using the (c) Modified IHS, (d) Brovey, (e) Gram–Schmidt, and (f) Wavelet IHS techniques.

are not always available. Consequently, high-precision and highly accurate wetland mapping is very difficult. Therefore, to prepare high-quality wetland maps in data-scarce regions, image fusion is an essential step to enhance the resolution of the available and free Landsat images. Hence, in this study, we fused the Landsat 5 Thematic Mapper (TM) multispectral bands (2, 3, and 4) having a spatial resolution of 30 m by the panchromatic band (band 8) having a spatial resolution of 15 m. In the present study, we used several image fusion techniques—modified IHS, Brovey, Gram–Schmidt, and wavelet IHS—for improving the spatial resolution of the images (Figure 11.2). The effect of image fusion can be found in Figure 11.2. It shows that the highlighted part (river and water body) of a1 had a lower resolution and a more unclear image than the fused images (c3–f6). Therefore, it can be concluded that spatial resolution was significantly enhanced by using these four techniques.

11.4.2 Evaluation of the Image Fusion Techniques

The performances of fused images should be evaluated before use for wetland mapping and other purposes (Colditz et al., 2006). Therefore, to analyze the performance of the image fusion techniques employed in this study, the edge detection technique and the correlation between spectral values of multispectral image and fused images were calculated. The edges of the fused images were assessed by 3×3 grid and Laplacian kernel–based edge detection technique. The previous studies suggested that if the edges of different features of the fused images were detected clearly, it can be inferred that the performance of the image fusion techniques would be highly satisfactory (Indhumathi et al., 2021; Asokan et al., 2020). The results showed that a river and associated features (Figure 11.3 a1) of multispectral images were identified to detect the edges of fused images (Figure 11.3 b2–e5). Figure 11.3 shows that the image fused by the Brovey technique detected the river more clearly than the other fused images. Therefore, it can be concluded that the edges of the whole area can be detected most clearly by the Brovey image fusion technique, followed by modified IHS, wavelet IHS, and Gram–Schmidt.

Improving Mapping Techniques Using ANN

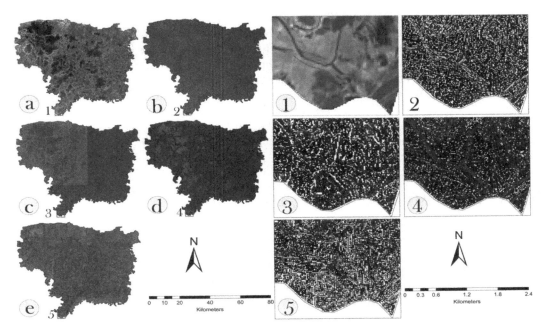

FIGURE 11.3 Laplacian image detection for (a) multispectral, (b) Modified IHS, (c) Gram–Schmidt, (d) Brovey, and (e) Wavelet–IHS.

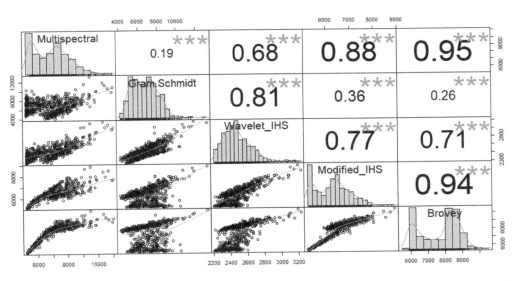

FIGURE 11.4 Spectral correlation among original image and fused images using the Kendall correlation technique.

Further, Kendall's correlation technique was applied for selecting the best image fusing techniques statistically. For this, spectral values of the original image and the fused image were extracted by the randomly selected 5000 points. Based on the extracted points from the original image and the fused images, the correlation technique was performed. Figure 11.4 shows the correlation between the original image and the fused images. It shows that Brovey had the highest agreement with the original image (r = 0.95), followed by modified IHS (r = 0.88), wavelet IHS (r = 0.68), and Gram–Schmidt (r = 0.19) at the 0.01 significance level in the case of spectral values. The correlation between the Brovey and modified IHS image fusion techniques was very high (r = 0.94), followed by modified IHS and wavelet IHS (0.77) at the significance level of 0.01.

Based on the results of the edge detection techniques and spectral correlation, it can be stated that the resolution was increased significantly for fused images, while the fused images had very similar spectral values except for the Gram–Schmidt technique. Therefore, high-resolution fused images by Brovey can be used for high-precision wetland mapping. However, in the present study, we applied an image classifier on all the fused images and multispectral images for preparing high-precision wetland mapping and comparing the accuracy level of wetlands maps among the fused images and the original image.

11.4.3 Wetland Mapping Using Artificial Neural Network

The wetland modeling of Sunamganj district was done using the ANN technique, and the whole area was classified into eight classes. Among these, five classes belonged to the wetland class (very deep water body, deep water body, shallow water body, saturated soil, and floating vegetation), while three classes belonged to non-wetland classes (sand bar, vegetation, and others). Figure 11.5 shows that the maximum area under wetland classes using the Gram–Schmidt–ANN classification technique was followed by Modified IHS–ANN and Brovey–ANN, while multispectral–ANN and Wavelet IHS–ANN had the minimum area under wetlands. On the other hand, the area under the non-wetland classes was maximum for the Wavelet IHS–ANN, followed by multispectral–ANN, Brovey–ANN, and Modified IHS–ANN, while the Gram–Schmidt–ANN had the lowest area under non-wetland.

Table 11.1 shows the area classified under different wetland and non-wetland classes using ANN and different image fusion imageries. Sun et al. (2017) worked on the Tanguar *haor* and found that more than half of the area of the district was wetland in the form of either waterlogged or swamp land. The LU/LC classification by the integration of all the fusion imageries and the ANN model showed that more than half of the district was under wetland (Table 11.1). The classified image generated by the Gram–Schmidt–ANN technique had the maximum area under very deep and

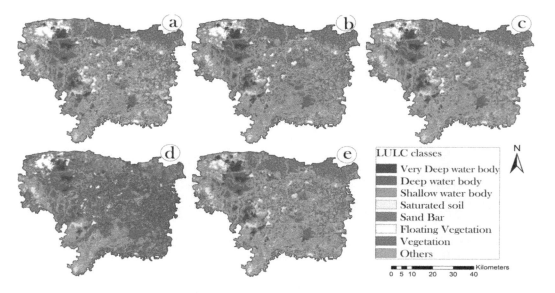

FIGURE 11.5 Wetland mapping using ANN for (a) multispectral, (b) Modified IHS, (c) Brovey, (d) Gram–Schmidt, and (e) Wavelet IHS.

TABLE 11.1
Area under Wetland and Non-Wetland Classes Using Different Image Fusion Images

	Area (km²)				
LU/LC Types	Multispectral–ANN	Brovey–ANN	Gram–Schmidt–ANN	Modified IHS–ANN	Wavelet IHS–ANN
Very deep water body	92.93	103.40	188.70	93.03	114.57
Deep water body	564.05	522.54	1289.41	592.84	508.41
Shallow water body	862.35	914.19	753.93	968.20	858.48
Saturated soil	69.41	72.64	82.45	72.99	90.92
Floating vegetation	356.50	304.61	231.93	247.68	302.55
Sand bar	20.76	17.98	7.22	14.32	22.08
Vegetation	778.52	649.68	673.79	667.44	659.76
Others	961.26	1118.38	475.98	1046.92	1146.64

deep water body (188.70 and 1289.41 km², respectively), while the multispectral–ANN image had the lowest area under very deep water body (92.93 km²). The Wavelet IHS–ANN had the lowest area under the deep water body (508.41 km²), followed by Brovey–ANN (522.54 km²) and multispectral–ANN (564.04 km²). The area under shallow water body was maximum for modified IHS (968.20 km²), followed by Brovey–ANN (914.19 km²), while it was minimum for the Wavelet IHS–ANN (858.48 km²), followed by multispectral–ANN (852.35 km²). Further, the area under saturated soil was maximum for the Wavelet IHS–ANN (90.92 km²), while the minimum area was predicted by multispectral–ANN (69.41 km²). Furthermore, the area under floating vegetation was maximum for the multispectral–ANN image (356.50 km²), while the lowest area was predicted by the Gram–Schmidt–ANN (231.93 km²).

Among non-wetland classes, the sand bar had the minimum area coverage in the Sunamganj district, while other LU/LC classes had the maximum area covered. The maximum area under sand bar was predicted by Wavelet IHS–ANN (22.08 km²), while the minimum area was classified by Gram–Schmidt–ANN image (7.22 km²). On the other hand, vegetation had the maximum area predicted by the multispectral–ANN image (778.52 km²), and the minimum area was classified by the Brovey fusion image (649.68 km²), while other LU/LC classes had the maximum area predicted by the Brovey–ANN image (1118.38 km²), and the minimum area was predicted by the Gram–Schmidt–ANN (475.98 km²).

11.4.4 Validation of the Wetland Modeling Models

In this study, two validation techniques, Kappa coefficient and RMSE, were used to evaluate the performances of the ANN model and fused images for classifying the wetlands. Studies reported that the fused imageries have better accuracy in the land use classification than the multispectral images (Vibhute et al., 2020; Luo, 2016; Colditz et al., 2006). This accuracy increases when the classification has been done using machine learning techniques (Vibhute et al., 2020). In this study, the highest overall accuracy was achieved by the Brovey–ANN image (Kappa coefficient: 89.23%, RMSE: 0.12), followed by Modified IHS–ANN (Kappa coefficient: 85.72%, RMSE: 0.22) and Wavelet IHS–ANN (Kappa coefficient: 85.5%, RMSE: 0.25), while the lowest accuracy was obtained by the Gram–Schmidt–ANN (Kappa coefficient: 80.13%, RMSE: 0.63) and multispectral–ANN image (Kappa coefficient: 82.6%, RMSE: 0.34) (Table 11.2). The previous study by Colditz et al. (2006) showed that all the image fusion techniques are effective in the LU/LC classification; however, in their study, the accuracy of the Brovey and IHS image fusion techniques was comparatively lower than the others. In this study, the accuracy of Brovey was found to be maximum.

TABLE 11.2
Validation of Wetland Mapping Models

Sl. No.	LU/LC model	Overall Kappa (%)	RMSE
1	Multispectral	82.6	0.34
2	Brovey	89.23	0.12
3	Gram–Schmidt	80.13	0.63
4	Modified IHS	85.72	0.22
5	Wavelet IHS	85.5	0.25

11.4.5 COMPARISON OF THE ASSOCIATION AMONG WETLAND MODELING MODELS

The association among the integrated image fusion and ANN models used for wetland modeling in this study was compared in order to analyze how much the models could be related to each other as well as their effectiveness. Kendall and Spearman's correlation techniques were used to compare the association between the models used. The results showed that the correlation among all the fusion models used for wetland modeling was positive and very high, while the correlation between fusion models and multispectral image was moderate and positive. The correlations between Brovey–ANN and Modified IHS–ANN (0.94 in Kendall and 0.95 in Spearman) as well as between Brovey–ANN and Wavelet IHS–ANN (0.94 in Kendall and 0.96 in Spearman) were maximum, while the correlations between all other models were also very high (Figure 11.6). The high association among the outputs of different image fusion techniques reflects that all models performed similarly and well (Manviya and Bharti, 2020). Therefore, it can be said that all the integrated image fusion and ANN models performed well in the wetland modeling, although there existed minor variation among their performances and associations.

The correlation among the wetland mapping models was also analyzed using the Kendall correlation technique to assess the classification similarity among the models. The result showed that the correlations among almost all the models are perfect and positive (Figure 11.7). The Gram–Schmidt–ANN model had a positive and very high correlation with all other models, including multispectral (0.79).

11.5 CONCLUSION

This study was carried out to analyze the performance of various image fusion techniques in increasing spatial resolution as well as for the precise mapping of the wetlands of Sunamganj district of Bangladesh. The wetland modeling was done using the ANN technique in eight classes (very deep water, deep water, shallow water, sand bar, vegetation, etc.). The study shows that about half of the district was under wetland in 2019 in all the integrated fused and ANN images as well as multispectral imagery. The areas under very deep and deep water bodies were maximum in the Gram–Schmidt–ANN image technique (188.70 and 1289.41 km^2, respectively), while the area under very deep water was minimum for multispectral–ANN imagery (92.93 km^2) and under deep water for Wavelet IHS–ANN image (508.41 km^2). For the non-wetland classes, sand bar had lower area coverage in the Sunamganj district than vegetation cover and others. The validation, using Kappa coefficient and RMSE, of the wetland modeled using ANN showed that the Brovey–ANN image model had better accuracy in wetland modeling than the others. Further, a literature survey showed that the accuracy of fusion images increases when integrated with machine learning and artificial intelligence. However, the comparison from the literature survey showed that the performance of Brovey was poor in comparison to the other image fusion techniques, although none of the studies compared the image fusion techniques for a specific land use modeling such as wetland. Therefore, it can be concluded that the performance of different image fusion techniques varies

Improving Mapping Techniques Using ANN

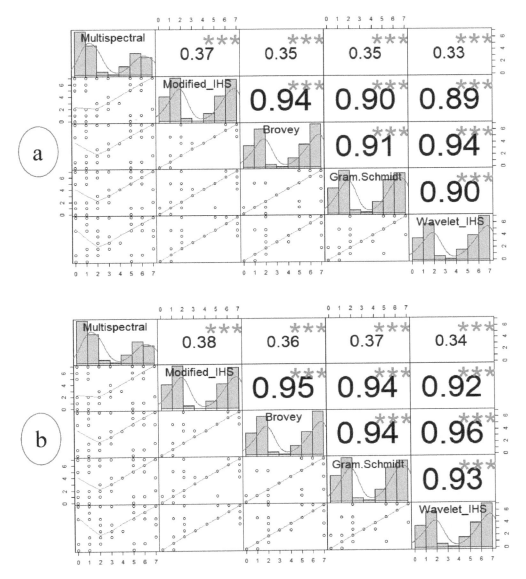

FIGURE 11.6 Correlation among wetland mapping models based on spectral values of classified maps by using (a) Kendall and (b) Spearman.

with the kind of study in which it is used as well as the approach of the study and the techniques and software used.

To conduct scientific research, not only in developing countries but also in developed countries, the utilization of open source software has become essential and an active topic. Therefore, the environmental researcher should use open source software more to overcome infrastructural issues, which will open the door to conducting scientific research. In the case of RS applications, the future will be the use of big data in the open source Python-based Google earth engine platform, which may be a game changer for future researchers. It provides a huge amount of data without downloading and can be used for not only wetland monitoring but also land management and urban studies. The management plan will be accurate and precise. Finally, it can be stated that research using open source software will be the future of the scientific community.

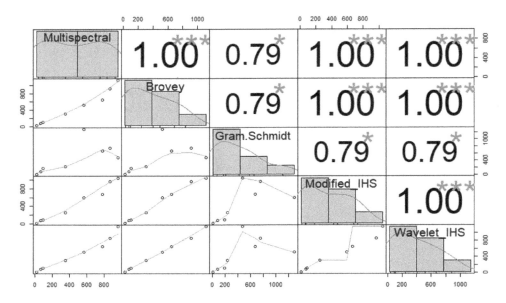

FIGURE 11.7 Correlation using Kendall among wetland mapping models based on the areal variation.

ACKNOWLEDGMENTS

The authors are thankful to the USGS for making satellite data freely accessible for use. The authors also acknowledge the Department of Geography, University of Gour Banga, Malda, West Bengal, India for providing laboratory facilities and other support to carry out this research work.

CONFLICT OF INTEREST

The authors declare that there is no conflict of interest on any financial or other issues.

REFERENCES

Alam, M.S., Hossain, M.S., Monwar, M.M. and Hoque, M.E. (2013). Assessment of fish distribution and biodiversity status in Upper Halda River, Chittagong, Bangladesh. *International Journal of Biodiversity and Conservation* 5(6): 349–357. DOI:10.5897/IJBC2013.0555.

Amer, M. and Maul, T. (2019). A review of modularization techniques in artificial neural networks. *Artificial Intelligence Review*, 52: 527–561.

Asokan, A., Anitha, J., Ciobanu, M., Gabor, A., Naaji, A. and Hemanth, D.J. (2020). Image processing techniques for analysis of satellite images for historical maps classification – an overview. *Applied Sciences*, 10(12): 4207.

Bassi, N., Kumar, M.D., Sharma, A. and Pardha-Saradhi, P., (2014). Status of wetlands in India: A review of extent, ecosystem benefits, threats and management strategies. *Journal of Hydrology: Regional Studies*, 2: 1–19.

Bausys, R., Kazakeviciute-Januskeviciene, G., Cavallaro, F. and Usovaite, A. (2020). Algorithm selection for edge detection in satellite images by neutrosophic WASPAS method. *Sustainability*, 12(2): 548.

Bishara, A.J. and Hittner, J.B. (2017). Confidence intervals for correlations when data are not normal. *Behavior Research Methods*, 49: 294–309.

Ciais, P., Sabine, C., Bala, G., Bopp, L., Brovkin, V. Canadell, V., Chhabra, A., DeFries, R., Galloway, J., Heimann, M., Jones, C., Le Quéré, C., Myneni, R.B., Piao, S. and Thornton, P. (2013). Carbon and other biogeochem cycles. In Stocker, T.F., Qin, D., Plattner, G.-K., Tignor, M., Allen, S.K., Boschung, J., Nauels, A., Xia, Y., Bex, V. and Midgley, P.M., (Eds.) *Climate Change 2013: The Physical Science Basis*. Cambridge University Press, Cambridge, UK; New York, NY, pp. 465–474. https://doi.org/10.1007/978-94-007-6172-8_254-4

Colditz, R.R., Wehrmann, T., Bachmann, M., Steinnocher, K., Schmidt, M., Strunz, G. and Dech, S. (2006). Influence of image fusion approaches on classification accuracy: a case study. *International Journal of Remote Sensing*, 27(15): 3311–3335.

DeVries, B., Huang, C., Lang, M.W., Jones, J.W., Huang, W., Creed, I.F. and Carroll, M.L. (2017). Automated quantification of surface water inundation in wetlands using optical satellite imagery. *Remote Sensing*, 9: 807.

Everard, M. (2016). Supporting services for wetlands: an overview. In Finlayson, C. et al. (Eds.) *The Wetland Book*. Springer, Dordrecht, pp. 1–10, https://doi.org/10.1007/978-94-007-6172-8_254-4.

Fang, C., Tao, Z., Gao, D. and Wu, H. (2016). Wetland mapping and wetland temporal dynamic analysis in the Nanjishan wetland using Gaofen One data. *Annals of GIS*, 22(4): 259–271.

Ganaie, T.A., Jamal, S. and Ahmad, W.S. (2020). Changing land use/land cover patterns and growing human population in Wular catchment of Kashmir Valley, India. *GeoJournal*. https://doi.org/10.1007/s10708-020-10146-y.

Guo, M., Li, J., Sheng, C., Xu, J. and Wu, L. (2017). A review of wetland remote sensing. *Sensors*, 17(4): 777.

Haque, A.K.E. and Kazal, M.H. (2008). Rich resources, poor people: The paradox of living in Tanguar Haor. *Int. Union Conserv. Nat. Gland Switz. Tech. Rep.*

Hess, L.L., Melack, J.M., Affonso, A.G., Barbosa, C., Gastil-Buhl, M. and Novo, E.M.L.M. (2015). Wetlands of the lowland Amazon basin: extent, vegetative cover, and dual-season inundated area as mapped with JERS-1 synthetic aperture radar. *Wetlands*, 35: 745–756.

Hong, G. and Zhang, Y. (2003). High resolution image fusion based on wavelet and IHS transformations, Proceedings of the IEEE/ISPRS Joint Workshop on Remote Sensing and Data Fusion over Urban Areas, Berlin, 99–104. Accessed November 20, 2016 from www.isprs.org/proceedings/XXXV/congress/comm4/papers/474.pdf.

Huang, C., Chen, Y., Zhang, S. and Wu, J. (2018). Detecting, extracting, and monitoring surface water from space using optical sensors: a review. *Reviews of Geophysics*, 56: 333–360.

Huang, C., Peng, Y., Lang, M., Yeo, I.Y. and McCarty, G. (2014). Wetland inundation mapping and change monitoring using Landsat and airborne LiDAR data. *Remote Sensing of Environment*, 141: 231–242.

Humphreys, J.M., Mahjoor, A., Reiss, K.C., Urbie, A.A. and Brown, M.T. (2019). A geostatistical model for estimating edge effects and cumulative human disturbance in wetlands and coastal waters. *Journal International Journal of Geographical Information Science*. https://doi.org/10.1080/13658816.2019.1577431

Indhumathi, R., Nagarajan, S. and Abimala, T. (2021). A comprehensive study of image fusion techniques and their applications. In Priya, E., Rajinikanth, V. (Eds.) *Signal and Image Processing Techniques for the Development of Intelligent Healthcare Systems*. Springer, Singapore. https://doi.org/10.1007/978-981-15-6141-2_8

Jensen, K., McDonald, K., Podest, E., Rodriguez-Alvarez, N., Horna, V. and Steiner, N. (2018). Assessing L-band GNSS-reflectometry and imaging radar for detecting sub-canopy inundation dynamics in a tropical wetlands complex. *Remote Sensing*, 10(9), 1431. https://doi.org/10.3390/rs10091431

Kumar, G. and Singh, K.K. (2020). Mapping and monitoring the selected wetlands of Punjab, India, using geospatial techniques. *Journal of the Indian Society of Remote Sensing*, 48: 615–625.

Li, Y., Zhu, X., Sun, X. and Wang, F. (2010). Landscape effects of environmental impact on bay-area wetlands under rapid urban expansion and development policy: a case study of Lianyungang, China. *Landscape and Urban Planning*, 94(3–4): 218–227.

Luo, H. (2016). *Classification precision analysis on different fusion algorithm for ETM + remote sensing image*. 2nd International Conference on Electronics, Network and Computer Engineering (ICENCE, 2016), pp. 983–990.

Mahdianpari, M., Salehi, B., Mohammadimanesh, F., Brisco, B., Mahdavi, S., Amani, M. and Granger, J.E. (2018). Fisher linear discriminant analysis of coherency matrix for wetland classification using PolSAR imagery. *Remote Sensing of Environment*, 206: 300–317.

Manviya, M. and Bharti, J. (2020). Image fusion survey: a comprehensive and detailed analysis of image fusion techniques. In Shukla, R., Agrawal, J., Sharma, S., Chaudhari, N., Shukla, K. (Eds.) *Social Networking and Computational Intelligence. Lecture Notes in Networks and Systems*, vol. 100. Springer, Singapore. https://doi.org/10.1007/978-981-15-2071-6_53

Moor, H., Hylander, K. and Norberg, J. (2015). Predicting climate change effects on wetland ecosystem services using species distribution modeling and plant functional traits. *AMBIO*, 44 Supplement 1: 113–126.

Nguyen, H.H., Dargusch, P., Moss, P. and Tran, D.B. (2016). A review of the drivers of 200 years of wetland degradation in the Mekong Delta of Vietnam. *Regional Environmental Change*, 16: 2303–2315.

Pal, S. and Talukdar, S. (2018). Drivers of vulnerability to wetlands in Punarbhaba river basin of India-Bangladesh. *Ecological Indicators*, 93: 612–626.

Pantaleoni, E., Wynne, R.H., Galbraith, J.M. and Campbell, J.B. (2009). Mapping wetlands using ASTER data: a comparison between classification trees and logistic regression. *International Journal of Remote Sensing*, 30(13): 3423–3440.

Rashid, I. and Aneaus, S. (2019). High-resolution earth observation data for assessing the impact of land system changes on wetland health in Kashmir Himalaya, India. *Arabian Journal of Geosciences*, 12: 453.

Rezaee, M., Mahdianpari, M., Zhang, Y. and Salehi, B. (2018). Deep convolutional neural network for complex wetland classification using optical remote sensing imagery. *IEEE Journal of Selected Topics in Applied Earth Observations and Remote Sensing*, 11(9): 3030–3039.

Richards, J.A. and Jia, X. (2006). *Remote Sensing Digital Image Analysis: An Introduction*. Springer-Verlag, Berlin, Heidelberg. 10.1007/3-540-29711-1

Sarp, G. (2014). Spectral and spatial quality analysis of pan-sharpening algorithms: a case study in Istanbul. *European Journal of Remote Sensing*, 47: 19–28.

Shahfahad., Kumari, B., Tayyab, M., Hang, H.T., Khan, M.F. and Rahman, A. (2019). Assessment of public open spaces (POS) and landscape quality based on per capita POS index in Delhi, India. SN. Applied Sciences, 1:368.

Shahfahad; Mourya, M., Kumari, B., Tayyab, M., Paarcha, A., Asif. and Rahman, A. (2020). Indices based assessment of built-up density and urban expansion of fast growing Surat city using multi-temporal Landsat data sets. *GeoJournal*. https://doi.org/10.1007/s10708-020-10148-w

Siddiqui, Y. (2003). *The modified IHS method for fusing satellite imagery.* In Proc. ASPRS Annual Conf., Anchorage, AK (CD).

Stagg, C.L., Osland, M.J., Moon, J.A., Hall, C.T., Feher, L.C., Jones, W.R., Couvillion, B.R., Hartley, S.B. and Vervaeke, W.C. (2020). Quantifying hydrologic controls on local- and landscape-scale indicators of coastal wetland loss. *Annals of Botany*, 125(2): 365–376.

Sun, C., Zhen, L. and Miah, M.G. (2017). Comparison of the ecosystem services provided by China's Poyang Lake wetland and Bangladesh's Tanguar Haor wetland. *Ecosystem Services*, 26: 411–421.

Sun, H., Qie, G., Wang, G., Tan, Y., Li, J., Peng, Y., Ma, Z. and Luo, C. (2015). Increasing the accuracy of mapping urban forest carbon density by combining spatial modeling and spectral unmixing analysis. *Remote Sensing*, 7(11): 15114–15139.

Talukdar, S., Mankotia, S., Shamimuzzaman, M., Shahfahad. and Mahato, S. (2020c). Wetland-inundated area modeling and monitoring using supervised and machine learning classifiers. In Pandey, P.C., Sharma, L.K. (Eds.) *Advances in Remote Sensing for Natural Resource Monitoring*. Wiley-Blackwell, New Jersey.

Talukdar, S. and Pal. S. (2017). Impact of dam on inundation regime of flood plain wetland of Punarbhaba river basin of Barind tract of Indo-Bangladesh. *International Soil and Water Conservation Research*, 5: 109–121.

Talukdar, S. and Pal., S. (2019a). Effects of damming on the hydrological regime of Punarbhaba river basin wetlands. *Ecological Engineering*, 135, 61–74.

Talukdar, S. and Pal., S. (2019b). Wetland habitat vulnerability of lower Punarbhaba river basin of the uplifted Barind region of Indo-Bangladesh. *Geocarto International*, 35(8): 857–886.

Talukdar, S. and Pal, S. (2020). Modeling floodplain wetland transformation in consequences of flow alteration in Punarbhaba River in India and Bangladesh. *Journal of Cleaner Production* 261: 120767.

Talukdar, S., Pal, S., Chakraborty, A. and Mahato, S. (2020a). Damming effects on trophic and habitat state of riparian wetlands and their spatial relationship. *Ecological Indicators*, 118: 106757.

Talukdar, S., Singha, P., Mahato, S., Shahfahad., Pal, S., Liou. Y.A. and Rahman, A. (2020b). Land-use land-cover classification by machine learning classifiers for satellite observations – a review. *Remote Sensing*, 12(7): 1135.

Valz, P.D. and McLeod, A.I. (1990). A simplified derivation of the variance of Kendall's rank correlation coefficient. *The American Statistician*, 44(1): 39–40.

Vibhute, A.D., Kale, K.V., Gaikwad, S.V., Dhumal, R.K., Nagne, A.D., Varpe, A.B., Nalawade, D.B. and Mehrotra, S.C. (2020). Classification of complex environments using pixel level fusion of satellite data. *Multimedia Tools and Applications.* https://doi.org/10.1007/s11042-020-08978-4

Wang, Z., Ziou. D., Armenakis. C., Li, D. and Q. Li, (2005). A comparative analysis of image fusion methods. *IEEE Transactions on Geoscience and Remote Sensing*, 43(6): 1391–1402.

Wen, L. and Hughes, M. (2020). Coastal wetland mapping using ensemble learning algorithms: a comparative study of bagging, boosting and stacking techniques. *Remote Sensing*, 12(10): 1683.

Wentz, E.A., Anderson, S., Fragkias, M., Netzband, M., Mesev, V., Myint, S.W., Quattrochi, D., Rahman, A. and Seto, K.C. (2014). Supporting global environmental change research: a review of trends and knowledge gaps in urban remote sensing. *Remote Sensing*, 6(5): 3879–3905.

Zheng, J. and Zheng, C. (2019). Recursive measures of edge accuracy on digital images. In Zheng, J. (Eds.) *Variant Construction from Theoretical Foundation to Applications*. Springer, Singapore. https://doi.org/10.1007/978-981-13-2282-2_12

12 Open Source Geospatial Technologies for Generation of Water Resource Development Plan

Arati Paul and V. M. Chowdary

CONTENTS

12.1 Introduction 165
12.2 Study Area 167
12.3 Data Used.. 168
12.4 Methodology... 169
    12.4.1 WRDP Generation Using Open Source Desktop GIS.. 170
    12.4.2 WRDP Generation Using Web-Enabled Open Source GIS 170
        12.4.2.1 Creation of Database and Web Services.. 171
        12.4.2.2 Development of Customized Web GIS Application for
                WRDP Generation 171
12.5 Results and Discussion 173
    12.5.1 Desktop Interface.. 173
    12.5.2 Web Interface.. 174
12.6 Conclusions... 176
Acknowledgments.. 176
References.. 176

12.1 INTRODUCTION

Open source software (OSS) refers to a type of computer software for which the source code is made available to users for modification as per their requirements (Andrew and Laurent, 2004). It promotes open exchange, collaborative participation, rapid prototyping, transparency, meritocracy, and community-oriented development (Open Source, 2020). The advantages of using OSS include: (i) low cost: OSS is generally free to use or costs less than its proprietary counterparts; (ii) enhanced security and stability: as the source code is open to users, OSS is continuously evolving, with more security and stability resulting from the large community; and (iii) control: OSS can be customized for specific requirements, which is difficult in the case of proprietary software (Mondal et al., 2018). Since most attention is paid to enhancing its functionality rather than its user interface, the general disadvantages of OSS include lack of user-friendliness. However, the strong user community and related documentation help to overcome this difficulty.

In recent years, advances in free and open source software (FOSS) for geographic information systems (FOSSGIS) have helped stakeholders to develop various open source tools for spatial data analysis in different technological domains (Steiniger and Hunter, 2012). It has also enabled administrators and policy makers to reach scientifically and technically sound decisions. Thus, open source GIS creates the platform for utilization of geospatial technologies for managing natural resources. It encourages non-GIS professionals to use open source–based GIS platforms for spatial data analysis when financial and technical constraints limit the implementation of professional and commercial GIS software. The open source platform also provides the ability to develop customized solutions to meet user requirements (Brovelli et al., 2017). Open Geospatial Consortium (OGC) (OGC, 2020) is an international, voluntary consensus standards organization and is taking an active role in the development of standards for geospatial and location-based services. It addresses interoperability challenges by defining software product standards that enable the seamless use of FOSS. As a result, the number of GIS applications using FOSS is increasing across different sectors (Brovelli et al., 2017): natural resource management (Paul et al., 2014), dissemination of green park information (Paul et al., 2016a, 2016b, 2019a), biodiversity conservation (Mose et al., 2018), estate property management (Paul et al., 2019b), plantation (Chakraborty et al., 2016), and facility mapping (Mondal et al., 2018). FOSSGIS is also used in land and water resource management (Paul et al., 2017; Kumar et al., 2018; Paul and Chowdary, 2021). FREEWAT is a free and open source water resource modeling platform integrated into the Quantum GIS (QGIS) software, which is being successfully used for water resource modeling (Cannata et al., 2018) and water quantity and quality management (Criollo et al., 2019).

Water is an important natural resource to be used judiciously, and sustainable planning and management is the need of the hour. Remote sensing (RS) and GIS-based techniques have been successfully utilized in managing this natural resource in many ways (Garg and Eslamian, 2017; Paul et al., 2018). Geospatial techniques are widely adopted for generating water resource development (WRD) plans in different regions (Chowdary et al., 2009; Jha et al., 2014; Singh et al., 2017). A GIS-based multi-criteria decision-making method is used to identify potential runoff storage zones within a watershed (Rana and Suryanarayana, 2020) for the planning of rainwater conservation (Singh et al., 2020). Socio-economic factors like distance from drainage networks, roads, urban areas, faults, farms, and wells are utilized along with GIS-based multi-criteria analysis for the identification of potential rainwater harvesting zones (Toosi et al., 2020). RS and GIS-based techniques are also used for the identification of prospective rainwater harvesting sites/zones (Sayl et al., 2016; Steinel et al., 2016). Analytic hierarchy process (AHP)–based geospatial techniques are used to integrate thematic layers in order to identify rainwater harvesting sites (Wu et al., 2018, Balkhair and Rahman, 2019, Karimi and Zeinivand, 2019).

The adoption of a sustainable WRD plan (WRDP) involves the identification of suitable locations for check dams, percolation tanks, and farm ponds. These water harvesting structures are widely implemented in India at the watershed scale as a measure of rainwater harvesting (Kaliraj et al., 2015). These structures not only serve as rainwater harvesting structures but are also useful for augmenting ground water recharge. Generation of a WRDP is a complex process, that involves aggregation of multiple spatial layers based on some criterion at watershed level. Hence, a customized solution is required to increase its application among users using FOSSGIS. Two case studies are discussed in this chapter for the development of customized WRDP generation software using FOSSGIS. The first is about developing a standalone desktop-based software tool, whereas the second deals with the FOSSGIS for developing a WRDP generation application in the web environment. MapWindow GIS, a standalone open source GIS component, has been customized in the first case study to develop a WRDP generation tool. This tool includes automatic processing of input data and produces output in the required format. It can be successfully adopted in cases where input data is available to the user. A web-enabled customized WRDP generation tool is also presented in this chapter using open source geospatial technology. The web GIS version includes data as well as processing knowledge that enables the user to generate a WRDP for a selected study area. It has no dependency or prerequisites from the user side other than an Internet browser.

In both these cases, WRD action plans are generated using a customized package that employed a set of logical conditions for integration of multiple spatial layers for a selected study area. Data pertaining to the West Sighbhum, Saraikela, and East Sighbhum districts of Jharkhand, India, has been

Water Resource Development Plan

used to demonstrate the open source platform. The WRDP generated by the customized software tools helps to identify suitable zones for taking up locally specific activities, such as the development of a check dam, percolation tank, underground barrier, etc., in the study area. Local area–specific activities refer to a certain type of water resource developmental activities that are recommended for implementation in a specific area. Thus, the customized tool combines spatial data, analyzes them, and produces meaningful information to support the planning and decision-making process.

The specific objectives of the present study include: (i) utilization of FOSSGIS for developing a customized geospatial tool for a specific purpose, i.e., generation of a WRDP for an area that combines complex processes involving the aggregation of multiple spatial layers based on several criteria, and (ii) demonstration of FOSSGIS capabilities in developing customized desktop-based as well as web-based applications to meet specific requirements. The significance of the study is that it incorporates user-friendliness in the developed applications to achieve complex geospatial operations using FOSSGIS, which otherwise may seem difficult for novice/non-GIS professionals. Hence, this chapter not only highlights the utility of FOSSGIS for achieving complex geospatial tasks; it also demonstrates the significance of customization for overcoming the general drawbacks of OSS.

12.2 STUDY AREA

West Sighbhum, East Singhbhum, and Saraikela districts, covering the southern part of the Jharkhand state, India, has been taken as a study region for demonstration of the customized software. The index map of the study area is shown in Figure 12.1. The West Sighbhum district extends between 21°58′ and 23°36′N latitude and between 85°0′ and 86°54′E longitude. The district is situated at a height of 244 m above sea level and has an area of 5351 km². East Singhbhum district, extending from 22°12′ to 23°01′N latitude and 86°04′ to 86°54′E longitude, is situated at the southeast corner of Jharkhand, covering an area of 3533 km². The Saraikela district is approximately situated between 22°29′ and 23°09′N latitudes and 85°30′ and 86°15′E longitudes and covers an area of 2724 km². The elevation ranges from 178 to 209 m. The study area is characterized by undulating topography, dense forests, and rivers. The area is rich in minerals. The major crop of this area in the *kharif* season is Paddy. Natural resources data pertaining to the study area are considered for demonstration of the open source geospatial software tools for the generation of a WRDP.

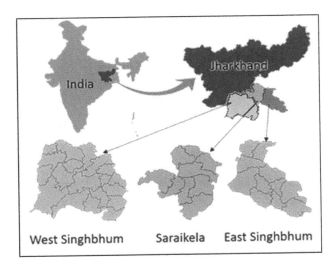

FIGURE 12.1 Index map of the study area.

12.3 DATA USED

Multispectral data, viz. LISS-3 and LISS-4 with a spatial resolution of 23 and 5.6 m, respectively, and panchromatic 2.5 m CARTOSAT-1 data were used to generate the spatial thematic layers. The spatial layers that are used for the generation of the WRDP include land use/land cover (LU/LC), soil, slope, drainage order (buffer layer), and geology (structural layer) pertaining to the study area. Multi-temporal data were considered to protect against loss of information and improve the quality of the spatial layers. The effects of image perspective and relief were removed from satellite images by ortho-rectification to generate a planimetrically correct image. Thematic layers are integrated with their respective administrative boundaries (state, district, and block) in GIS to perform location-specific query and analysis.

The ortho-rectified satellite images enabled accurate measurement of distances, angles, and areas. Hence, these images were used to generate the thematic layers. The base information for developmental planning of an area was obtained from the existing LU/LC and its spatial distribution in that area. To capture the seasonal variations, three-season LISS-III data were considered for mapping the LU/LC of the study area. A geomorphological map at 1:50,000 scale was prepared using morphological expressions in the satellite image. A digital elevation model (DEM) was generated using CARTOSAT-1 stereo pair images along with rational polynomial coefficients (RPC) at 10 m resolution. DEM is an important input for water resources planning and developmental activities and is an essential parameter for slope map generation. The guidelines provided in Integrated Mission for Sustainable Development guidelines (IMSD, 1995) were followed to categorize the slope map into eight classes. Strahler's method–based stream ordering was performed on the drainage network generated through visual interpretation of IRS LISS 4 and CARTOSAT-1 merged product (Strahler, 1964). This helps in planning artificial recharge structures. These spatial layers were used as inputs to produce decision maps either by direct translation from a single layer or by combining two or more thematic layers, as given in Tables 12.1 and 12.2, respectively.

TABLE 12.1
Spatial Layers and Sources for WRD Planning

Data/Map	Source
DEM	CARTOSAT-1 stereo pair image
Structures/Lineaments	Rajiv Gandhi National Drinking Water Mission project
Soil	National Bureau of Soil Survey and Land Use Planning (NBSS & LUP), India
Land use/ Land cover	Natural Resource Census
Drainage network	IRS LISS 4 + CARTOSAT-1 Merged product

TABLE 12.2
Derived Spatial Layers Required for WRD Planning

Derived Map	Theme Map	Remarks
Slope	DEM	Derived from DEM
Drainage order	Drainage network	Strahler method

Water Resource Development Plan

12.4 METHODOLOGY

GIS is a framework for capturing, storing, managing, and analyzing spatial data. It provides a platform to perform spatial queries and generate output in a spatial format, which enables understanding spatial patterns and relationships among data pertaining to a common geographical area. Information from diversified sources is combined in GIS (process/function) to produce output themes, as given in Equation 12.1. It also helps in describing the underlying process and developing models for prediction to support decision making.

$$\text{Output layer} = \text{Function}(\text{input layers}) \quad (12.1)$$

In this study, WRD plans were generated in the GIS environment by integrating different thematic layers based on multiple criteria. This highlights the suitability of an artificial recharge activity in the study area. Figure 12.2 depicts the overall concept of WRDP generation. Considering the availability of data sources, the present GIS application adopted decision criteria whereby the LU/LC, soil permeability, slope, and drainage order layers are considered as mandatory input, and the geology (structure) layer is treated as optional (Chowdary et al., 2009, 2010; Jha et al., 2014; Singh et al., 2017). The spatial layers are integrated in the GIS environment using the logical conditions presented in Table 12.3. However, the entire process that includes logical conditions is not readily available as a tool in any GIS package. Hence, in the present chapter, the process of developing a customized WRDP generation tool using open source geospatial technology is discussed.

Initially, the desktop-based WRDP generation tool is discussed, and next, the development of a web-based WRDP generation tool is summarized. Open source MapWindowGIS, where the Activex component of MapWindowGIS is customized in the dot.net environment, is used to develop the desktop-based tool. In the present study, VB.Net is used for developing the tool, while the geospatial functionalities are inherited from the MapWindowGIS. In the case of the web-based tool development, OSS is used as database, web server, GIS server, and libraries. The Postgres database

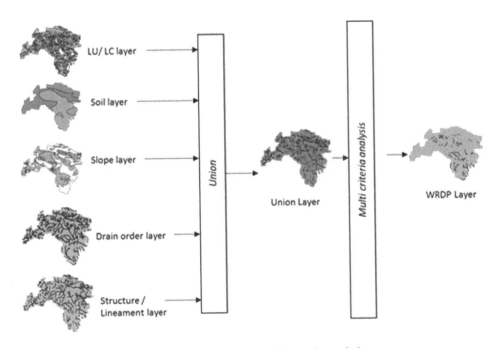

FIGURE 12.2 Process of WRD plan generation using different thematic layers.

TABLE 12.3
Criteria Adopted for WRD Plan Generation

Drain Order	Soil Permeability	Slope (%)	Land Use	Structure	Action Plan
Second/third	Moderate to high	<10	Agricultural areas		Check dam (in agricultural areas)
Second/third	Moderate to high	<10	Forest areas		Check dam (in forest areas)
Second/third	High	<3	Waste lands	Lineament and fractures preferred	Percolation tank (across streams)
	High	<3	Waste lands	Same as above	Percolation tank (in wasteland areas)
	Low	0–3	Agricultural areas	Lineament and fractures should be avoided	Farm pond (without seepage control)
	Moderate to high	0–3	Agricultural areas	Same as above	Farm pond (with seepage control)
Fourth–seventh	Sandy/Gravel river bed	0–3			Subsurface dykes (underground barrier)

Source: Chowdary, V.M., et al., *Water Resources Management*, 23, 1581–1602, 2009.

is used with the PostGIS extension for the spatial data repository, while GeoServer is used to serve the spatial layers in the web environment. PHP is used for querying the Postgres database, and JavaScript is used for dynamic web application development. Finally, the application is hosted using the open source Apache web server. The detailed methodology is presented in the following two sections using open source desktop GIS and the web GIS environment, respectively.

12.4.1 WRDP Generation Using Open Source Desktop GIS

MapWindow is an open source spatial data viewer and GIS that can be modified and extended using plugins (MapWindow, 2020). MapWindowGIS is an application program interface (API) that can be integrated into the dot.net programming environment to develop a customized GIS application or tool. In the present chapter, a customized WRDP generation tool developed using the Activex component is presented.

A list of thematic layers, as mentioned in Tables 12.1 and 12.2, is considered for WRDP generation. These need to be provided as inputs by the user. Subsequently, these layers are integrated using the "Union" operation of MapWindowGIS, whereby all the layer attributes are preserved. This union layer was used for suitability evaluation of an area for implementing a particular action plan. The set of logical conditions as given in Table 12.3 was applied on this layer to generate a suitable action plan for each polygon. Subsequently, the "Dissolve" operation was performed to reduce the number of contiguous polygons of the same category and generalize features. Finally, the output thus generated, can be saved as a shape file in the designated place for future use. A map area was also embedded in the WRDP generation tool to visualize input as well as output layers. The process flow for the development of the WRDP tool is shown in Figure 12.3.

12.4.2 WRDP Generation Using Web-Enabled Open Source GIS

WRDP generation can also be achieved using open source web GIS. However, unlike desktop-based GIS, data are integrated into the system and are stored in a spatial database in the web GIS environment.

Water Resource Development Plan 171

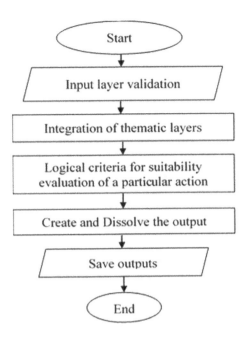

FIGURE 12.3 Process flow for development of desktop-based WRD planning tool.

Thus, the overall methodology for developing a WRDP generation application is broadly divided into: (i) creation of database and web services and (ii) development of the application.

12.4.2.1 Creation of Database and Web Services

The spatial thematic layers are augmented with state, district, and block names. This enables the selection of the target area for WRDP generation. All these layers are then integrated into a single layer using Quantum GIS (QGIS) software (Menke et al., 2015). However, a spatial database is required to store and use this union layer for WRDP generation in web environment. PostgreSQL/PostGIS is a powerful open source database (Douglas and Douglas, 2003; Kraft, 2018; PostgreSQL, 2020) and is used in the development of the application. It can be executed in most of the major operating systems, viz. Linux, UNIX, and Windows, for managing spatial data. In addition, it is reliable, stable, and extensible. PostgreSQL uses the PostGIS extension to provide support for geographic objects (PostGIS, 2020). This enables the PostgreSQL database to perform location queries in SQL. Hence, a spatial database is created in PostgreSQL/PostGIS, where the union layer is stored. Further, the data needs to be converted into Web Map Service (WMS) in order to share the spatial data as a map over the Internet. The open source Java-based GeoServer (Iacovella, 2014) is used to accomplish this task. GeoServer is an open source mapping server that provides the platform to share, edit, and display geospatial content in the web environment (GeoServer, 2020). It publishes data using open standards from any major spatial data source: PostGIS, ArcSDE, Oracle, and DB2. The use of OpenLayers, a free mapping library, in GeoServer enables very quick and easy map generation. The WMS of Bhuvan is used for visualization of base maps and satellite images of the study area (Arulraj et al., 2015; Bhuvan, 2020).

12.4.2.2 Development of Customized Web GIS Application for WRDP Generation

The development of a web-enabled open source application for the generation of a WRDP is envisaged using a three-tier architecture. The client and the web GIS application reside in the first and the second tier, respectively. The GIS and the spatial database server reside in the third tier. The system architecture of the application is presented in Figure 12.4. The GIS server is implemented using

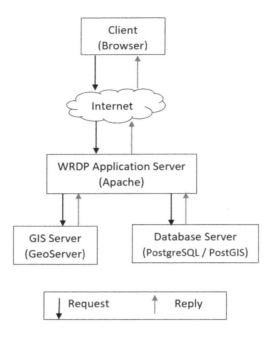

FIGURE 12.4 Architecture diagram of the web GIS application.

GeoServer, Postgres/PostGIS is was used as the spatial database server, and the open source Apache web server is used as the application server (Vukotic and Goodwill, 2011), where the WRDP application resides. It communicates with the client through hypertext transfer protocol (HTTP). The requests from clients are accepted and analyzed here. Subsequent inputs are collected from the GIS server and the database server to accomplish the task. Finally, the reply is communicated to clients by the application server using HTTP.

The interactive web GIS application for WRDP generation is developed using HTML and Java scripts, while OpenLayers4 (Santiago, 2014) is used to develop GIS functionalities. OpenLayers is a Java-based mapping library used to put a dynamic map in the web page (Hazzard, 2011). It enables the development of feature-rich web GIS applications. It can render tiled layers from OGC-compiled sources: OSM, Bing, MapBox, Stamen, etc. It also supports a range of vector data formats, including GeoJSON, TopoJSON, KML, GML, Mapbox vector tiles, etc. OpenLayers provides functions/methods for operations, map rendering, interactive drawing and editing, for which it does not require plugin at the client end. PHP (Hypertext Preprocessor) is a fast and flexible general-purpose scripting language that is very popular and useful in web development. It can be easily embedded in a HTML page. It is a server side script and is hence used to interface the database in the application development. In this study, PHP (Lurig, 2008) is used to query the database and fetch results as per the logical conditions presented in Table 12.3. The results are dynamically analyzed and managed using AJAX (Asynchronous JavaScript And XML) (Darie et al., 2006) and jQuery (Chaffer and Swedberg, 2013). AJAX is a web development technique that combines a group of existing web technologies, including HTML, CSS, JavaScript, DOM, XML, and the XMLHttpRequest object. It enables sending and retrieving data from a server in the background asynchronously without reloading the web page. jQuery is a feature-rich JavaScript library that simplifies event handling and AJAX and hence is used in the present development. A multi-tiered, reliable, and secure web GIS application is developed using OSS and libraries. The visual abstract of the process flow is given in Figure 12.5.

Water Resource Development Plan

12.5 RESULTS AND DISCUSSION

The desktop and the web application developed using FOSSGIS are demonstrated in this section. The interfaces of the developed packages were designed in a user-friendly manner. User actions are guided using proper messages during interactions.

12.5.1 Desktop Interface

The WRDP generation tool, developed using MapWindowGIS, contains buttons for input layer selection and a map area as shown in Figure 12.6. Each button enables selection of a particular input layer, while the map area enables visualization of input and output layers. In order to generate a WRDP for any selected area, the user needs to provide the required spatial layers as input using the designated buttons. Validation is performed for each input layer for correctness/availability of its attributes. In the absence of correctness, suitable messages are generated, and system halts for the appropriate input. The formats of input layers are described in the "HELP" document, which is embedded in the tool. On receipt of correct input layers, the system processes them to generate the WRDP layer. Subsequently, the output is "DISSOLVED" and stored in a predefined location. On completion of the entire process, the path of the output layers is notified on the tool's screen, and the dissolved WRDP layer is displayed in the map area, as depicted in Figure 12.6.

The WRDP tool integrates the input layers sequentially one after another using "UNION" operation. This is a computationally intensive process and is time and resource consuming. Hence, to optimize the performance, the process may be executed for a smaller region, viz. micro-watershed (<1000 ha), block, village, etc. The conventional method for generating a WRDP involves many controlling parameters, which are individually derived and integrated. This tedious process can be avoided by using this kind of open source GIS-based WRDP tool, which involves the demarcation of areas suitable for artificial recharge quite effectively and efficiently.

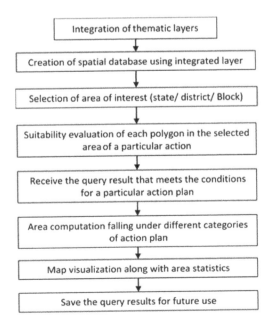

FIGURE 12.5 Process flow of the web-based application.

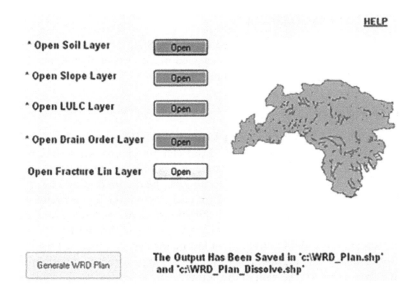

FIGURE 12.6 WRD plan generation tool developed using MapWindowGIS.

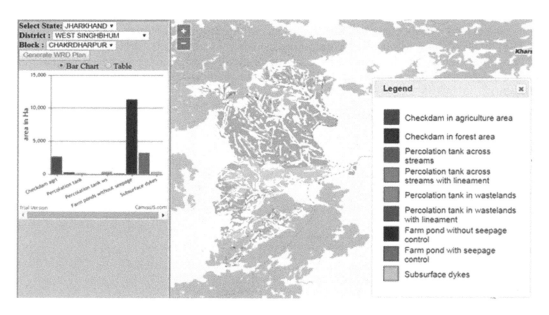

FIGURE 12.7 User interface of web-enabled WRD planning application with column chart and legend.

12.5.2 WEB INTERFACE

The open source–based web-enabled WRDP generation tool is an interactive and user-friendly GIS application. The user interface of the application contains two major components: one is the user interaction panel, and the other is a map viewer, as shown in Figure 12.7. The user interaction panel enables the user to choose the area of interest for which the WRDP is to be generated. The user can select the name of the intended state/district/block from the drop-down lists provided in the panel. Subsequently, the WRDP is automatically generated by querying the database and analyzing the suitability of the resultant polygons using the logical conditions. The area statistics, i.e.,

Water Resource Development Plan

the total area in hectares, were also calculated under different action plans. These are shown in tabular format as well as with a column chart in the left panel of the application. Once the WRDP layer is generated, it is displayed on the map and can be interpreted by using the legend, as depicted in Figure 12.7. A map on-click functionality is also available, which enables the user to know the action plan for a selected polygon on the map. The WMS of high-resolution satellite imagery can also be used as a base map to visualize the output, as shown in Figure 12.8. This helps the user to know the actual ground features present in that area.

The broader categories of action plans in this study include check dam, percolation tank, farm ponds, and subsurface dykes. The suggested area of a specific action plan provides a clue about the utilization of natural resources for the development of water resources. However, the specific location of implementing an action plan depends upon many other factors: owner's agreement, financial ability, government policy for granting permission, etc. Ground water recharge by artificial location-specific activities is required in regions where natural recharge is insufficient because of the lower permeability of the underlying surface, especially in hard rock areas. A percolation tank is an artificial reservoir, which is generally constructed across a stream to collect surface water runoff and facilitate percolation within the permeable land. It is an effective method of groundwater recharge in areas that are characterized by highly permeable soils rather than in areas of low permeability. Check dams are constructed upstream (second- and third-order streams with low to moderate slopes) to retain water. This enables water to percolate and recharge the ground water. Additionally, it helps to improve the irrigation potential of the area and reduce the denudation process. A farm pond is an important rainwater harvesting structure to accumulate surface runoff in a farm area. In addition to support irrigation, farm ponds also help in fish culture, duck farming, and ground water recharge. Apart from conservation of water, farm ponds are useful for conserving soil and nutrients (Reddy et al., 2012). Subsurface dykes (or underground barriers) are constructed across higher-order streams to retard the baseflow and improve ground water recharge through infiltration. This structure is generally planned across fourth–seventh-order wide streams where sufficient flow of water is available throughout the year. It is an effective way of providing water supply where other artificial recharge structures are difficult to implement because of adverse geomorphologic conditions or natural constraints.

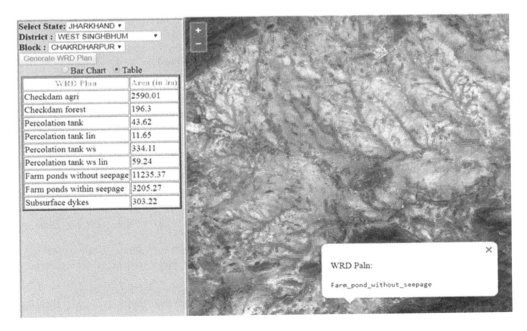

FIGURE 12.8 WRDP overlaid on satellite image with tabular statistics.

The desktop-based WRDP generation tool can be used independently when the requisite layers, as mentioned in Tables 12.1 and 12.2, are available with the user. The free availability of remotely sensed data and processing software from different sources enables these layers to be easily generated for the study area. Hence, this system is good for smaller areas where the number of users is limited. However, the web-enabled WRDP generation application removes this dependency. It includes the required data in essential format and produces geospatial output along with statistics. As the system is web based, the user only needs an Internet connection to generate the WRDP for the intended area without involving any additional software or hardware infrastructure. Thus, the web-enabled solution is good for larger areas with a huge number of stakeholders. Unlike the desktop-based solution, the web-based solution includes deployment infrastructure. As the system is developed using FOSSGIS, the deployment cost of the system is also minimal.

The case studies demonstrated in this chapter produce action plans using FOSSGIS by applying a set of criteria resulting from the geospatial analysis of scientific factors. However, for better understanding of the study area in terms of cause–effect analysis of a particular problem or limitation, the results may be integrated with social factors along with field inspection data.

12.6 CONCLUSIONS

In recent years, geospatial technologies are being effectively utilized in the planning, conservation, development, and management of natural resources. These technologies have proved to be faster and more efficient than conventional methods and are widely used across the world. The availability of FOSSGIS has made the technology more affordable and popular among stakeholders. In the present chapter, the potential of FOSSGIS has been successfully demonstrated by developing customized WRDP generation tool as a standalone desktop package and also as a web GIS application. In both cases, the complex processing knowledge is encoded in customized software that enables stakeholders with limited geospatial knowledge to generate a WRDP for their intended area for efficient water resource management. The present work may be further extended by increasing the study area and including local constraints to refine the results. The OSS used in the present study can be used for developing similar customized geospatial tools, such as generation of land resource development plans, in future. The open source web technologies discussed here can also be used for developing customized web GIS for the monitoring and management of natural/geospatial resources.

ACKNOWLEDGMENTS

Jharkhand Space Application Centre (JSAC), Jharkhand, India is acknowledged for the data that are utilized to demonstrate the results in the present chapter. The authors also acknowledge the support of GM and RRSC-East and the inspiration of CGM, RCs, and NRSC for accomplishing the present work.

REFERENCES

Andrew M. and Laurent St. 2004. *Understanding Open Source and Free Software Licensing*, O'Reilly Media Inc., USA. p. 4. ISBN 9780596005818.

Arulraj M., Aggarwal S., Naresh N., Kalyandeep K. and Mobina S. 2015. *Bhuvan User Hand Book*, NRSC-DPPAWA-GWGSG-AUG-2015-TR-728.

Balkhair K.S. and Rahman K.U. 2019. Development and assessment of rainwater harvesting suitability map using analytical hierarchy process, GIS and RS techniques, *Geocarto International*. doi: 10.1080/10106049.2019.1608591.

Bhuvan. 2020. *Documentation on Bhuvan*. https://bhuvan.nrsc.gov.in (accessed April 04, 2020).

Brovelli M.A., Minghini M., Moreno-Sanchez R. and Oliveira R. 2017. Free and open source software for geospatial applications (FOSS4G) to support future earth, *Int. J. Digital Earth*, 10(4):386–404. doi: 10.1080/17538947.2016.1196505.

Cannata M., Neumann J. and Rossetto R. 2018. Open source GIS platform for water resource modelling: FREEWAT approach in the Lugano Lake, *Spat. Inf. Res.*, 26(3):241–251. doi: 10.1007/s41324-017-0140-4.

Chaffer J. and Swedberg K. 2013. *Learning jQuery*, Packt Publishing, Birmingham, UK, ISBN 978-1-78216-314-5, 4th Edition.

Chakraborty D., Paul A., Dutta D., Jayaram A., Kumar T.P.G., Bajpai A., Mirza K., Soni V. and Thirani D. 2016. Indian tea garden information system – a WebGIS enabled solution, *Asian J. Geoinf.*, 16(1):8–16.

Chowdary V.M., Ramakrishnan D., Srivastava Y.K., Vinuchandran R. and Jeyaram A. 2009. Integrated water resource development plan for sustainable management of Mayurakshi Watershed, India using remote sensing and GIS, *Water Resour. Manag.*, 23:1581–1602.

Chowdhury A., Jha M.K. and Chowdary V.M. 2010. Delineation of groundwater recharge zones and identification of artificial recharge sites in West Medinipur district, West Bengal, using RS, GIS and MCDM techniques, *Environ. Earth Sci.*, 59(6):1209–1222.

Criollo R., Velasco V., Nardi A., Manuel de Vries L., Riera C., Scheiber L., Jurado A., Brouyère S., Pujades E., Rossetto R and Vázquez-Suñé E. 2019. AkvaGIS: an open source tool for water quantity and quality management, *Comput. Geosci.*, 127:123–132. doi: 10.1016/j.cageo.2018.10.012.

Darie C., Brinzarea B., Chereches-Tosa F. and Bucica, M. 2006. *AJAX and PHP*, PACKT Publishing, Birmingham, 1st Edition.

Douglas K. and Douglas S. 2003. *PostgreSQL – A Comprehensive Guide to Building, Programming and Administering PostgreSQL Databases*, Sams Publishing, Indianapolis.

Garg V. and Eslamian S. 2017. Monitoring, assessment, and forecasting of drought using remote sensing and the geographical information system. Ch. 14 in *Handbook of Drought and Water Scarcity*, Vol. 1: Principles of Drought and Water Scarcity, eds. Eslamian S. and Eslamian F., Francis and Taylor, CRC Press, USA, 217–252.

GeoServer. 2020. *Documentation on Geoserver*. http://geoserver.org/ (accessed April 04, 2020).

Hazzard E. 2011. *OpenLayers 2.10: Beginner's Guide*, Packt Publishing, Birmingham.

Iacovella S. 2014. *GeoServer Cookbook*, Packt Publishing, UK, ISBN 978-1-78328-961-5, 1st Edition.

IMSD. 1995. *Integrated Mission for Sustainable Development Technical Guidelines*. National Remote Sensing Agency, Department of Space, Govt. of India.

Jha M. K., Chowdary V.M., Kulkarni Y. and Mal B.C. 2014. Rainwater harvesting planning using geospatial techniques and multicriteria decision analysis. *Resour. Conserv. Recycl.*, 83:96–111.

Kaliraj S., Chandrasekar N. and Magesh N.S. 2015. Evaluation of multiple environmental factors for site-specific groundwater recharge structures in the Vaigai River upper basin, Tamil Nadu, India, using GIS-based weighted overlay analysis. *Environ. Earth Sci.*, 74:4355–4380.

Karimi H. and Zeinivand H. 2019. Integrating runoff map of a spatially distributed model and thematic layers for identifying potential rainwater harvesting suitability sites using GIS techniques. *Geocarto Int.*:1–20. doi: 10.1080/10106049.2019.1608590.

Kraft T.J., Mather S.V., Corti P., et al. 2018. *PostGIS Cookbook*, Packt Publishing, UK, ISBN: 9781788299329, 2nd Edition.

Kumar N., Singh S.K., Mishra V.N., Reddy G.P.O. and Bajpai R.K. 2018. Open-source satellite data and GIS for land resource mapping. In *Geospatial Technologies in Land Resources Mapping, Monitoring and Management. Geotechnologies and the Environment*, eds. Reddy G. and Singh S., vol. 21. Springer, Cham. doi: 10.1007/978-3-319-78711-4_10.

undefinedLurig, M. 2008. *PHP Reference: Beginner to Intermediate PHP5*. Lulu.com, ISBN: 978-1-4357-1590-5. 1st Edition. Available at: http://cdn.phpreferencebook.com/wp-content/uploads/2008/12/php_reference_-_beginner_to_intermediate_php5.pdf.

MapWindow. 2020. *Documentation on MapWindow*. www.mapwindow.org/ (accessed March 25, 2020).

Menke K., Smith Jr. R., Pirelli L. and Hoesen J.V. 2015. *Mastering QGIS*, Packt Publishing, UK, ISBN 978-1-78439-868-2.

Mondal R.S., Chakraborty D., Paul A. and Dafadar K.D. 2018. WebGIS enabled facility mapping and identification – a cost effective solution, *Int. J. Comput. Appl.*, 180(38):41–44. doi: 10.5120/ijca2018917019.

Mose V.N., Western D. and Tyrrell P. 2018. Application of open source tools for biodiversity conservation and natural resource management in East Africa, *Ecol. Inform.*, 47:35–44. doi: 10.1016/j.ecoinf.2017.09.006.

OGC. 2020. www.ogc.org/ (accessed April 04, 2020).

Open Source. 2020. *Documentation on Open Source*. https://opensource.com/resources/what-open-source (accessed April 04, 2020).

Paul A., Chakraborty D. and Dutta D. 2016a. Open source geospatial technology for developing web enabled Park GIS, *CSI Commun.*, 40(9):12–13, ISSN: 0970-647X.

Paul A., Chakraborty D., Dutta D., et al. 2016b. Park information system for Kolkata – a low cost web enabled solution, *J. Current Trends Info. Tech.*, 6(1):1–5.

Paul A. and Chowdary V.M. 2021. Application of web enabled open source geospatial technologies in generation of water resource development plan, *Int. J. Hydrol. Sci. Tech.* 11(1): 76–87. doi: 10.1504/IJHST.2020.10023542.

Paul A., Chowdary V.M., Chakraborty D., Dutta D. and Sharma J.R. 2014. Customization of freeware GIS software for management of natural resource data for developmental planning – a case study, *Int. J. Open Inf. Technol.*, 2(4):25–29. ISSN: 2307-8162.

Paul A., Chowdary V.M., Dutta D. and Sharma J.R. 2017. Standalone open-source GIS-based tools for land and water resource development plan generation. In *Environment and Earth Observation*, vol. 2, eds. Hazra S., Mukhopadhyay A., Ghosh A.R., Mitra D. and Dadhwal V.K., Springer, Switzerland, 23–34. doi: 10.1007/978-3-319-46010-9_2. eISBN: 978-3-319-46010-9.

Paul A., Mal P., Gulgulia P.K., Srivastava Y.K. and Chowdary V.M. 2019b. Spatial progression of estate property management system with customized freeware GIS, *Int. J. Info. Tech.*, 11(2):341–344. doi: 10.1007/s41870-018-0135-y.

Paul A., Mondal R.S. and Chakraborty D. 2019a. Open source geospatial solution for disseminating green park information, *J. Inf. Technol. Comput. Sci.*, 4(1):57–63. doi: 10.25126/jitecs.20194194.

Paul A., Tripathi D. and Dutta D. 2018. Application and comparison of advanced supervised classifiers in extraction of water bodies from remote sensing images, *Sustain. Water Resour. Manag.*, 4(4):905–919. doi: 10.1007/s40899-017-0184-6.

PostGIS. 2020. *PostGIS Documentation*. http://postgis.net/ (accessed April 04, 2020).

PostgreSQL. 2020. *PostgreSQL Documentation*. www.postgresql.org/ (accessed April 04, 2020).

Rana V.K. and Suryanarayana T.M.V. 2020. GIS-based multi criteria decision making method to identify potential runoff storage zones within watershed, *Ann. GIS*, 26(2):149–168. doi: 10.1080/19475683.2020.1733083.

Reddy K.S., Manoranjan K., Rao K.V., Maruthi V., Reddy B.M.K., Umesh B., Ganesh Babu R., Srinivasa Reddy K., Vijayalakshmi and Venkateswarlu B. 2012. Farm ponds: A climate resilient technology for rainfed agriculture; Planning, design and construction. Central Research Institute for Dryland Agriculture, Santoshnagar, Saidabad, Hyderabad, Andra Pradesh, India, p. 60.

Santiago A. 2014. *The Book of OpenLayers – Theory and Practice*, 3, Lean Publishing.

Sayl K.N., Muhammad N.S., Yaseen Z.M. and El-shafie A. 2016. Estimation the physical variables of rainwater harvesting system using integrated GIS-based remote sensing approach, *Water Resour. Manag.*, 30:3299–3313.

Singh L.K., Jha M.K. and Chowdary V.M. 2017. Multi-criteria analysis and GIS modelling for identifying prospective water harvesting and artificial recharge sites for sustainable water supply, *J Clean. Prod.*, 142:1436–1456.

Singh L.K., Jha M.K. and Chowdary V.M. 2020. Planning rainwater conservation measures using geospatial and multi-criteria decision making tools, *Environ. Sci. Pollut. Res.* doi: 10.1007/s11356-020-10227-y.

Steinel A., Schelkes K., Subah A., Himmelsbach T. 2016. Spatial multi-criteria analysis for selecting potential sites for aquifer recharge via harvesting and infiltration of surface runoff in north Jordan, *Hydrogeol. J.*, 24:1753–1774.

Steiniger S. and Hunter A.J.S. 2012. The 2012 free and open source GIS software map – a guide to facilitate research, development, and adoption, *Comput. Environ. Urban Syst.*, 39(2013):136–150. doi: 10.1016/j.compenvurbsys.2012.10.003.

Strahler A.N. 1964. Quantitative geomorphology of drainage basins and channel networks. In *Handbook of Applied Hydrology*, ed. Chow ByVenTe, McGraw Hill Book Company, New York, pp. 439–476.

Toosi A.S., Tousi E.G., Ghassemi S.A., Cheshomi A. and Alaghmand S. 2020. A multi-criteria decision analysis approach towards efficient rainwater harvesting, *J. Hydrol.*, 582(124501):1–14.

Vukotic A. and Goodwill J. 2011. *Apache Tomcat 7*, Springer, Apress, Berkeley CA, ISBN-13 (electronic): 978-1-4302-3724-2.

Wu R.-S., Molina G.L.L. and Hussain F. 2018. Optimal sites identification for rainwater harvesting in northeastern Guatemala by analytical hierarchy process, *Water Resour. Manag.*, 32:4139–4153.

13 Geo-Spatial Enabled Water Resource Development Plan for Decentralized Planning in India
Myths and Facts

Anand N. Shakya, Indal K. Ramteke, and Pritam S. Wanjari

CONTENTS

13.1 Introduction 179
    13.1.1 Water Resource Development Program in India 180
13.2 Materials and Methods 181
    13.2.1 Extent and Size of Water Resource Development Plans............................ 181
    13.2.2 Effect of Scale on Water Resource Development Plan 181
        13.2.2.1 Study Area... 182
        13.2.2.2 Remote Sensing/GIS Data Used... 182
        13.2.2.3 Remote Sensing / GIS Software Used.. 182
        13.2.2.4 Process Flow for Image Analysis.. 182
13.3 Results and Discussion 183
    13.3.1 Generation of Thematic Map... 183
    13.3.2 Slope... 183
    13.3.3 Land Capability 184
    13.3.4 Erosion 184
    13.3.5 Soil Depth 185
    13.3.6 Drainage... 185
    13.3.7 Land Use/Land Cover 185
        13.3.7.1 Land Evaluation and Land Use Plan 185
    13.3.8 Water Resource Action Plan Generation 185
    13.3.9 Weighted Overlay Techniques 186
    13.3.10 Resource Conservation Practices in India 187
        13.3.10.1 Water and Soil Conservation Activities in Maharashtra State............. 190
13.4 Conclusion 193
    13.4.1 Effect of Water Resources on the Community 194
References... 195

13.1 INTRODUCTION

The prosperity and growth of a region depend on the availability of infrastructure services. Water is the most crucial of these. Water and land are essential natural resources, which need to be monitored, assessed, and managed efficiently for any socio-economic development. Although adequate freshwater supplies are available globally to allow continued agricultural and industrial growth, there is growing concern about the long-term sustainable usage of water resources (Loucks and Beek, 2017).

Understanding the significance of remote sensing and geographic information systems (GIS) for providing cost- and time-effective resource databases, planning institutions such as the Ministry of Panchayati Raj and the Ministry of Rural Development proposed to develop atlases of district resources using geo-spatial technologies to strengthen the process of decentralized district-level planning through a coordinated approach with local self-government (Marble and Peuquet, 1983). The decentralization uses an information system comprising geo-spatial data to enable decentralized planning in the districts. Policymakers, scientists, and planners across the globe use the watershed approach for management (Tideman, 2000). The planning unit of water resource development is the watershed boundary because it includes all types of natural resources and the interrelationship between living beings. Watershed management tries to achieve equilibrium between the environment, resources, and living beings on the Earth. The big landscape is divided into small watershed units; this is more manageable for the purpose of decentralized planning, thereby benefiting the people living nearby. The watershed analysis includes conservation of land and water resources on a larger scale with resource planning aiming at sustainability, which results in rural reinforcement and development. Since 1980, the government has implemented this watershed approach to improve agricultural traditional practices and production. Scientists have conducted research on soil and water conservation measures using remote sensing and GIS. Advances in geo-spatial technologies encompass a wide range of alternatives for analysis for resources management.

GIS technologies have been productively used by experts for projects on water resources development, characterization, prioritization, and management (Suresh and Sudhakar, 2004). These techniques can be utilized for identifying suitable sites for soil and water conservation measures following the norms (Ravindran and Promod Kumar, 1992). Sankar and Hermon (2012) have devised a new approach to watershed prioritization and sustainable development through spatial techniques by integrating non-spatial data in order to formulate action plans for water resource development and conservation using land use planning. This research emphasizes the creation of water resource development plans (WRDP) using a weighted overlay method. The various GIS thematic layers have been chosen as input criteria for the identification of different water conservation zones, i.e., gully plugs, nala bunds, contour trenching, etc. In addition, the relevance of land use is studied using thematic layers at varied scales and their subsequent effects. The comparative analysis of the WRDP using different criteria on different scales provided a great deal of information for the researchers.

In a nutshell, the researchers tried to reveal myths and facts pertaining to the development of water resource plans and to use this information for decentralized grassroots planning at the Grampanchayats level. The effect of scale on WRDP using GIS databases at a higher scale was shown by the authors to be a myth. Facts about the significance of treatment plans for governmental development planning were also revealed.

13.1.1 Water Resource Development Program in India

Almost 142 million ha of total arable land (60 percent) in India is rainfed and therefore not highly productive, resulting in a low income. Due to fragile and marginal lands, a high incidence of poverty is observed (Joshi et al., 2008). The rainfall patterns in these areas are highly variable in terms of quantity and distribution. Such unbalanced conditions lead to moisture stress during the stages of crop production, making agricultural production susceptible to pre- and post-production risk. Since the early 1970s, the government of India has sponsored and implemented watershed development projects in the country. Watershed development programs like Drought Prone Area Program (DPAP), Desert Development Program (DDP), River Valley Project (RVP), National Watershed Development Project for Rain-fed Areas (NWDPRA), and Integrated Wasteland Development Program (IWDP) were launched subsequently in hydro-ecological regions affected by water stress and drought-like situations consistently. The entire watershed development program was mainly focused on a structurally driven compartmental approach to soil conservation and rainwater harvesting during the 1980s and before (Wani et al., 2005).

During the years from 1980 to 1990, a participatory approach was adopted for integrated water resource development programs. This strategy proved to be effective for raising the productivity of crops and improvement in the livelihoods of particular watersheds (Wani et al., 2006) by adopting soil and water conservation measures. Under the chairmanship of Prof. CH Hanumantha Rao, the central government appointed a committee, which strongly recommended migration from the conventional approach to bureaucratic planning without the involvement of local communities (Raju et al., 2008). Amendments were suggested in post-1995 guidelines about community participation, including involvement of stakeholders such as nongovernmental organizations (NGOs) and Panchayati Raj Institutions (PRIs) (Joshi et al., 2008). In 2001, further revised guidelines with the name "Hariyali" were issued in order to make community empowerment much simpler, involving PRIs in planning, implementation, and evaluation in an effective way (Raju et al., 2008) and later these guidelines were issued in 2003 (DOLR, 2003). Subsequently, in 2005, the Neeranchal Committee evaluated water resource development programs in India that were implemented by government, NGOs, and sponsors and suggested changing from a purely engineering and structural focus to a deeper concern with livelihood issues (Raju et al., 2008). The major emphasis of the watershed management program is on the conservation and utilization of natural resources, such as land, water, plant, animal, and human resources, in a harmoniously integrated manner to uplift the livelihood of rural India.

13.2 MATERIALS AND METHODS

Several scientists/scholars have demonstrated numerous methods using varied materials for water resource management across Indian territories. The contributors to this chapter focus precisely on the extent and effect of scale on the development of water resource plans in India. These researchers also include specially derived facts about water and soil resource planning, especially resource conservation strategies, that discriminate this view from traditional approaches.

13.2.1 Extent and Size of Water Resource Development Plans

Water resource development is an integral part of the watershed community, which exclusively accommodates people and animals inhabiting the watershed area. The size of the WRDP may vary from a few square kilometers to thousands of square kilometers. This is important if we consider the objective of the watershed management. For large irrigation projects, a watershed of thousands of square kilometers must be considered, whereas for a small storage structure in a farm (farm pond), consideration of only a few hectares of WRDP will be sufficient. The size is also affected by afforestation, grassland development, cultivation, etc. Many physiographic features such as valleys, undulating hillocks, and rugged hilly tracts influence the size of the water resource development jurisdiction. Larger watersheds can be selected in the plains or where afforestation and grassland development is the main objective, and smaller watersheds are chosen in the hilly areas where agricultural development is the main objective. In nutshell, a watershed is a physically existing socioeconomic system that intermingles with water resource development activities.

13.2.2 Effect of Scale on Water Resource Development Plan

With the availability of Survey of India (SOI) toposheets at 1:50,000 scale across India, a GIS thematic database was generated at the same uniform scale to serve the needs of geo-spatial communities for a variety of applications, water resource management being one of them. Later, a GIS database on higher scale was created in a piecemeal approach due to the availability of SOI toposheets at 1:25,000 scale and rarely, at 1:10,000 scale. With the advent of high-resolution remote sensing technologies, the Indian Space Research Organisation (ISRO) was able to generate an enormous GIS database on 1:10,000 scale at the national level. This unlocks the opportunities for researchers in India to investigate governmental planning at the grassroots level. In the context of water resources management development planning, Indian researchers were curious to know the effect of this changing scale on the development of a WRDP (Reddy et al., 2019). This fact was tested and verified with experimental results by virtue of research on one of the sub-watersheds in Maharashtra, India.

13.2.2.1 Study Area

WGKKC-2 watershed area lies on latitude 21°5′8.12″N/longitude 79°2′48.52″E, falling under three blocks of Nagpur District: Hingana, Kalmeshwar, and Katol. It covers an area of 360 km^2 and has a mean altitude of 310.5 meters above sea level. Climate conditions are tropical throughout the year. The area under study is situated on the Deccan Plateau of the Indian Peninsula and has a global identity as the "Orange City" of India due to the dominance of orchards in the region.

13.2.2.2 Remote Sensing/GIS Data Used

Satellite images of very high resolution of 2.5 m were used to generate thematic spatial layers and extract other required features necessary for research. The merged image of LISS IV MX and Cartosat-1D was created using the proper process of orthorectification defined by the National Remote Sensing Centre (NRSC), ISRO, India. The satellite sensor description is shown in Table 13.1. The multispectral LISS III satellite image is used for mapping GIS layers at 1:50,000 scale, i.e., land use/land cover. A thematic database at 1:50,000 scale was also derived from the SOI toposheet for better comparison among various datasets. The land use/land cover (LU/LC) for different scales of data has been generated using the merged satellite data product and LISS III imagery (Figure 13.1).

13.2.2.3 Remote Sensing / GIS Software Used

ILWIS 3.3 is used for the orthorectification process using the block bundled adjustment method. GIS software is used for further analysis in the GIS environment, and QGIS® (v.1.8.3) is introduced as an effective open source tool in place of proprietary ArcGIS functionalities. For orthorectifying the LISS IV MX data, the digital elevation model generated from Cartosat-1 stereo pairs and the orthorectified output are used. To accomplish this, projective transform is used. The projective transform requires elevation information and the orthorectified image references for rectifying the given LISS IV MX scene. The projective transform models are simulation models purely solved by ground control points (GCPs). The orthorectification procedure involves selecting the project properties, then the geometric model, and then projective transform. The process provides an accurate orthorectification of LISS IV MX imagery.

13.2.2.4 Process Flow for Image Analysis

The orthorectification of the LISS IV data is done using the Cartosat image, and then the merged image of both databases is prepared for the database generation and analysis. The SOI toposheet is referenced for generating thematic layers. The base layers, i.e., infrastructure, drainage, and LU/LC, are generated using high-resolution orthorectified data. Then, the different parameters are taken into consideration according to user-based priority for preparing a WRDP using the weighted

TABLE 13.1
Descriptions of Satellites Used in the Study

Satellite	IRS-P6	IRS-P5
Sensor used	LISS IV	Cartosat-1D
Pixel resolution (m)	5.8	2.5
Radiometry (bits)	7	10
Repeativity (days)	24	5
Swath (km)	23.9	30

Geo-Spatial Water Resource Development Plan

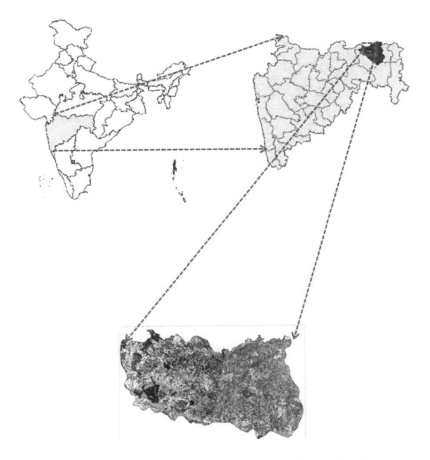

FIGURE 13.1 Study area covering watershed: WGKKC-2 belongs to Nagpur District.

overlay method. The methodology chart prepared is shown in Figure 13.2. The LU/LC is prepared using different images at different scales. The WRDP is prepared considering different scales of LU/LC. The output is then analyzed to see the effect of LU/LC and whether or not this should be considered. The comparative analysis and statistics are prepared to see the effect.

13.3 RESULTS AND DISCUSSION

13.3.1 GENERATION OF THEMATIC MAP

The GIS database was generated using merged satellite data (LISS IV and Cartosat-1D). The high-resolution dataset was used for feature extraction in different themes, such as LU/LC, drainage, and infrastructure. The orthorectified digital data were consequently imported into QGIS® (v.1.8.3) in order to digitize and generate various thematic layers: road, drainage, and LU/LC. Coding and standardization for all the thematic layers are done and stored for further analysis. All the themes are generated at 10,000 scale using orthorectified imagery, which was earlier produced in ILWIS 3.3.

13.3.2 SLOPE

The area exhibits a variety of slopes in different places. The slope classes can be identified as percentages, i.e., 3–5, 5–10, 10–15, and 15–35%. The slope criterion is a crucial input prior to proposing the harvesting structure.

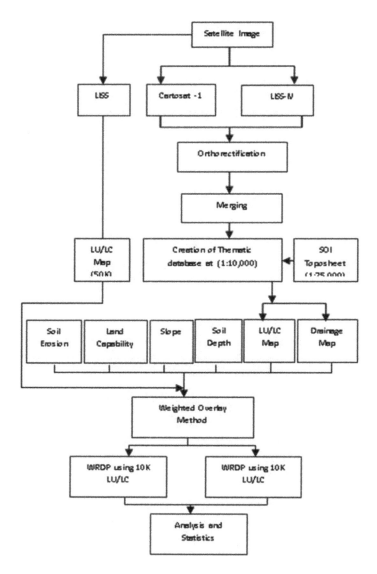

FIGURE 13.2 Methodology process flow.

13.3.3 LAND CAPABILITY

Soil is another important parameter for deciding the location of the conservation structure in a particular area. The soil classes detected in the study area comprise four land capability classes and eight subclasses, of which three are arable and one non-arable.

13.3.4 EROSION

This characteristic of soil was also included. The soil erosion has been classified into six classes, which are moderate in most of the area and severe in some places. Severe erosion is detected in hilly areas.

Geo-Spatial Water Resource Development Plan

13.3.5 Soil Depth

The soil depth is another important factor, which is detected in four classes for the area. The very deep condition of soil covers a large area. The soil depth and erosion are interdependent, so due to the very deep condition, the soil erosion is light in the plain areas.

13.3.6 Drainage

The drainage pattern is dendrite to sub-parallel in area. The area shows drainage to the greatest extent in most places and is enriched with crop production and scrub.

13.3.7 Land Use/Land Cover

Different features can be easily identified at 10,000 scale on the high-resolution orthorectified product. The LU/LC mapping done at this larger scale provides the basis for grassroots-level planning and monitoring the schemes. The area consists mainly of crop land due to the heavy drainage density.

13.3.7.1 Land Evaluation and Land Use Plan

Land evaluation is the assessment of land performance when used for specified purposes. As such, it provides a rational basis for making land use decisions based on analysis of relations between land use and land, giving estimates of required inputs and projected outputs. Land evaluation is an integral part of land use planning. After the scientific evaluation of the watershed land area, a suitable integrated land use plan may be generated and put into practice in the watershed. In this plan, each and every land, water, and human resource may be utilized in a sustainable manner towards farmers' income. Based on the land evaluation database, different land use options, such as agro-forestry, agri-fishery, sericulture, goat rearing, small-scale industries, farm industries, poultry, worm compost, agro-horticulture, floriculture, and vegetable cultivation along with main field crops, have been suggested in the watershed. This integrated approach saves the farmers if there is a loss due to any one component, and their income is not affected.

13.3.8 Water Resource Action Plan Generation

The spatial database layers are integrated in the GIS environment for better advanced analysis. A variety of tools and techniques can be used for the mapping and zonation of water conservation structures. The feasibility of various methods for the generation of a watershed action plan is validated subject to locally occurring constraints.

A decision tree is a popular tool specially devised for classification and prediction scenarios. It uses a rule-based inductive approach that recognizes discovered knowledge, and therefore, sufficient ground truth information is a must for deducing inferences. Collecting ground water samples and deriving information by means of testing would not be affordable anyway. More classes have been used for better accuracy. Hence, the decision tree approach is not suitable for addressing the problem of zonation of watersheds.

The pairwise comparison method compares two attributes at a time and gives a preference ratio between them. For example, the ratio w_i/w_j indicates how much attribute i is preferred to attribute j. However, it may be cumbersome to make such combinations if the attributes under considerations are greater in number.

Decision-making methods have three variants: the weighted sum model (Fishburn, 1967), the weighted product model (Miller and Starr, 1969), and multi-attribute utility theory (MAUT) (Stanney et al., 1994). The weighted sum model is chosen for analysis due to its simplicity, whereas the other two methods are complex in nature. The weighted sum model derives the optimal option as the "best" value of the weighted sum. The model is formulated for problems in which all variables have the same physical dimensions.

The analytic hierarchy process (AHP) is an incremental improvement over pairwise comparison methods because it uses the multiattribute decision analysis approach. In this study, the weighted overlay method is used, in which ranks are given to different classes of criteria. Then, after combining all criteria, the sum of these ranks is given. Accordingly, by analyzing the criteria, the range of summation of ranks is classified and assigned to different harvesting structures.

13.3.9 WEIGHTED OVERLAY TECHNIQUES

In QGIS, Weighted Sum Raster Overlay Analysis is a method for applying a common constraint of values to assorted and divergent inputs for integrated analysis by virtue of cumulative results. The overlay techniques allow the evaluation criterion to be ranked on input layers in order to determine the composite map layer (output maps). By doing this, priority areas have been suggested after considering soil and LU/LC parameters. Waterbody is considered as a constrained category for critical analysis purposes. Weighted Overlay accepts raster as well as vector data. Usually, digital number (DN) values of continuous raster images are grouped into clusters, like slopes or Euclidean distance outputs. Before it can be used in the weighted overlay analysis, each range must be assigned a single value, called ranking the criteria. With the accurate evaluation criterion chosen, simply add the input raster images to the Weighted Sum Raster Overlay Analysis dialog box. The cells in the raster will already be set according to suitability or preference. The output rasters can be weighted by rank to produce an output map.

Steps involved in Weighted Sum Raster Overlay Analysis and zonation map generation:

1. Define the set of evaluation criteria on input maps and standardize each criterion.
2. Define the criterion weights or rank; that is, a weight of relative preferences is assigned to each criterion. Assign drainage density using "V to Point" tool in QGIS (GRASS).
3. Generate the weighted standardized map layers by multiplying with corresponding weights.
4. Second, apply union of all criteria. After taking the sum of these ranks of all criteria and analyzing the criteria, the range of summation of ranks is classified and assigned to different soil/water conservation measures. Convert vectors into raster for such operations.
5. Rank the alternative outcomes according to the overall weighted scores; the alternative with the highest rank is the best option. Use the ordered weighted averaging tool box.
6. Generate the aggregate score for each alternative using the add overlay operation on the weighted standardized output map layers.

In the weighted overlay method, the layers are assigned weights according to the contribution of the different themes to the central theme. The slope is assigned higher weightage based on the importance of water harvesting structure. It reveals many facts about the physiographics of the land. It has been given the weightage of 0.4 at the scale of 0 to 1. Then, the soil erosion and land capability are given the same weightage of 0.2. A parameter like land capability shows the enrichment of soil to produce crops; indirectly, it indicates the percolation level and water table at a particular place. The soil erosion shows the amount of eroded soil in the land so that water flow can be minimized and erosion of soil can be stopped. The soil depth and LU/LC are given the weightage of 0.1. The soil depth indicates the magnitude of vegetation in that place, which in turn, reveals the soil's capability to hold the water flow. The relevance to the chosen classes of themes decides the rank to be assigned. The criteria for proposing water conservation structures are slope, soil depth, land capability, soil erosion, and LU/LC as presented in Table 13.2.

TABLE 13.2
Decision Rules for WRDP

Criteria Chart

Conservation Type	Land Capability	Erosion	Slope(%)	Depth	Land Use/ Land Cover
CCT	V, VI	Severe	15–35	Very Shallow	Not Suitable for Agriculture
GP	III, IV	Moderate to Severe	1–3/3–5	Shallow/Moderately Deep	Open Scrub
LBS	II	Slight to Moderate	5–10	Deep	Agriculture
ENB	II	Slight to Moderate	5–10	Deep	Agriculture
CNB	I	Slight	0–5	Very Deep	Agriculture

TABLE 13.3
Comparison between Different LU/LCs' Output

Water Conservation Structures	Area Using 10K LU/LC (km^2)	Area Using 50K LU/LC (km^2)
CCT	10.46	9.9197
CNB	92.9	97.661
ENB	189.417	186.46
GP	39.4	37.356
LBS	12.37	13.198

The WRDP is calculated using both LU/LC, i.e., at 10K and 50K, by the same overlay method, and the change in the output is detected. A comparative analysis between the outputs is shown in Table 13.3.

The area difference for the water conservation structures is small in percentage terms, as shown by the pie charts in Figure 13.3.

This can be better understood in terms of a graph. The output shows a diminutive difference in the area statistics between the outputs using different scales of LU/LC. We can conclude that the scaling of LU/LC doesn't affect the WRDP (Figure 13.4).

13.3.10 Resource Conservation Practices in India

Resource conservation primarily emphasizes soil and water conservation practices, which are treated as the primary approach towards water resource management. Conservation may be achieved by two means: in situ and ex situ management. The in situ type of management covers land/water conservation practices implemented within the agricultural domain, like terrace building, contour trenching, graded bunding, field bunding, broad bed and furrow practice, etc. Such conservation measures prevent land degradation, improve the water table results in groundwater availability, and promote healthy soil. Farm ponds, check dams, gully plug-ins, and excavation of pits in streams are a few examples of ex situ type management. This reduces peak discharge and harvests an extensive amount of runoff, which in return, raises the groundwater table and hence, the irrigation potential in the surrounding water resource area, resulting in an enriched watershed (Chalam and Krishnaveni, 1996).

For want of precise resource conservation strategies, it is sensible to identify geographically correct sites for the construction of soil and water conservation features. The capabilities of

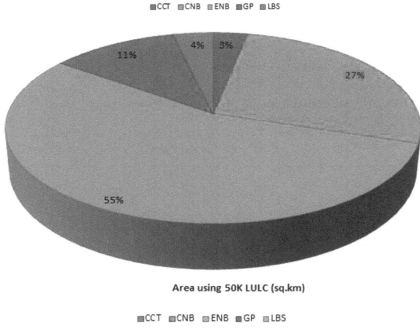

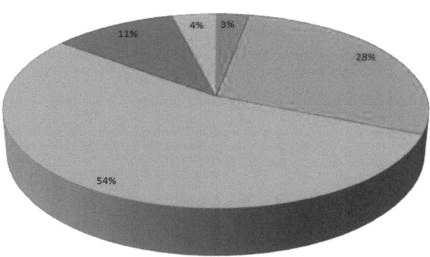

FIGURE 13.3 Pie chart showing area statistics: (a) using 10K, (b) using 50K.

geo-spatial technologies for soil and water conservation programs are commendable. The government of Maharashtra conducted such mission mode programs across the entire Maharashtra state considering the crucial parameters: drainage lines, elevation/contours, geomorphology, slope, soil erosion, soil texture, soil depth, and lineaments (Ramteke et al., 2017). The layers were amalgamated in the remote sensing/GIS environment using open source geo-spatial tools for the generation of village-wise "potential treatment maps" covering around 45,000 villages in Maharashtra.

Geo-Spatial Water Resource Development Plan

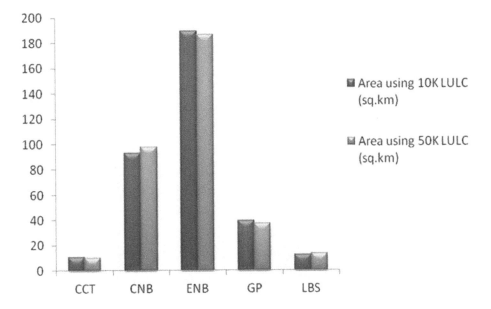

FIGURE 13.4 Variations in area of outputs at different scales.

The output treatment map shows the potential sites along "drainage line" and "area treatment." The drainage line treatments include gully plug-ins/loose bolder structure (LBS) and earthen nala bund (ENB)/cement nala bund (CNB). "Nala deepening" identifies existing waterbodies that need de-siltation. The area treatment activities comprise farm ponds, contour trenching, field/graded bunding, terracing, afforestation, etc. The map also helps in finding sites for development of new water resources.

The methodology is oriented to the concept of a "ridge to Valley" model of a watershed. A typical watershed consists of three major zones: the "runoff zone" (steep slope with a greater degree of erosion, high speed of water flow, early stage of a river), the "recharge zone" (moderate slope, gentle speed of flow, and middle stage of river course), and "storage zone" (no slope to gentle slope, silt deposition with thick soil layer, and more water percolation) (Chalam and Krishnaveni, 1996). The suitable locations for water conservation were determined using drainage line treatment and area treatment methodologies.

Drainage line treatment refers to the construction of 'LBS', 'gully plugs' are planned with early stream orders. The "earthen and gabion bunds" are planned in the recharge zone with secondary streams, and all the major water conservation structures like check dams and "river rejuvenation" are in the storage zone with stream order of three and above.

Area treatment covers activities like the construction of "contour trenching" and tree plantation that are planned in the "runoff zone." Pasture development, afforestation, and water absorption trenches (WAT) activities are planned in the recharge zone. The storage zone comprises activities like bund repairs, farm ponds, artificial well recharge, etc.

In Figure 13.5, existing WRDP activities are marked with gray shades, while proposed activities are indicated with dark gray shades. Village cadastral maps were overlaid on GIS database layers so as to conduct micro-level planning of water resources. It is clearly seen from the figure that drainage line treatments were done on LBS, ENB, or CNB. However, area treatments were carried out on dry land agricultural areas by means of contour bunding/farm ponds, horticultural/plantations, contour trenching, forest bunding, etc. The source information is derived from Vasundhara State

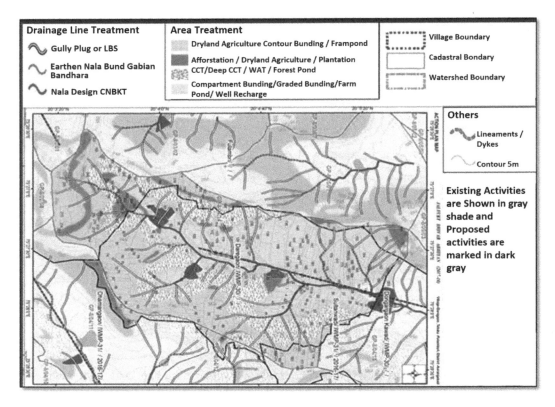

FIGURE 13.5 Drainage line treatment and area treatment map.

Level Nodal Agency (VSLNA), Maharashtra Remote Sensing Application Centre (MRSAC), and government of Maharashtra departments.

Figure 13.6 shows the output map of the potential sites along "drainage line" and "area treatment" belonging to Georai Kubri village of Aurangabad District, Maharashtra. The map was generated under the flagship program of the Hon. Chief Minister of Maharashtra during the year 2017–18. It was a combined effort by the government of Maharashtra, VSLNA, and MRSAC towards the utilization of geo-spatial technologies for devising a WRDP, which was intensively used by the State Agriculture Department and the State Water Conservation Department. It is obvious from the figure that the majority of the activities are proposed because of limitations of existing activities in the region so as to cater to the needs of the population in the targeted region.

13.3.10.1 Water and Soil Conservation Activities in Maharashtra State

The drainage line treatments included gully plugs/loose bolder structures, earthen nala bunds, cement nala bunds, underground nala bund, diversion bund, gabion structures, and "nala deepening" and recognized existing waterbodies that needed de-siltation. The area treatment activities were carried out in two zones: arable and non-arable. Arable area treatment activities include compartment bunding, graded bunding, farm pond, and bodi type. However, continuous contour trench, deep continuous contour trenching (CCT), stone bund/LBS/gabion, farm ponds, terracing, afforestation, WAT, etc., are all covered under non-arable area treatment activities. The map also helps in finding sites for development of new water resources (Figure 13.7).

A ground truth survey was carried out in watershed WGKKC-2, which comprises boundaries of Kalmeshwar, Katol, and Hingna Taluka of Nagpur District. It is evident that the majority of the area (approx. 310 km^2) is treated with compartment bunding and graded bunding. Due to less undulating terrain, CCT activities are limited to 3 km^2 approximately (Table 13.4).

Geo-Spatial Water Resource Development Plan

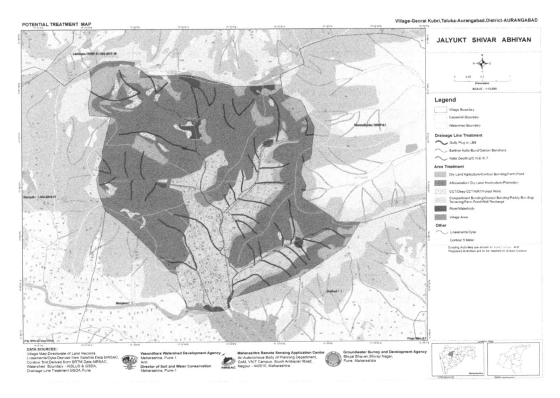

FIGURE 13.6 Potential sites for area/drainage line treatment map.

FIGURE 13.7 Water and soil conservation activities.

TABLE 13.4
Soil and Water Conservation Area Treatment Existing Activities in WGKKC-2 Watershed

Sl. No.	Soil and Water Conservation Existing Activities in WGKKC-2 Watershed	Total Area Treated (ha)
	Area Treatment	
1	Graded Bunding/ Compartment Bunding	30,968
2	Continuous Contour Trenching	286.58
	Total	*31,254.58*

Data Source: Soil and Watershed Management Dept., Commissionerate of Agriculture, Pune.

TABLE 13.5
Soil and Water Conservation Drainage Line Treatment Activities in WGKKC-2 Watershed

Sl. No.	Soil and Water Conservation Existing Activities in WGKKC-2 Watershed	Total Numbers
	Drainage Line Treatment	
1	Earthen Structure	10
2	Loose Bolder Structure	87
3	Mati Nala Bandh	35
4	Cement Nala Bandh	261
5	Farm Pond	184
	Total	*577*

Nala deepening activities of 58.11 km were also carried out in WGKKC-2 watershed.

Referring to Table 13.5, the authors concluded that CNB structures are abundant in number as compared with other drainage line treatment activities. Another reason is the quality of soil, which holds this type of structure well, and hence, this watershed falls mainly into the horticultural zone. Since WGKKC-2 watershed is covered under the rainfed zone, the majority of farm ponds are found in this region.

When the figures are scaled up to the extent of Maharashtra State, the inferences remain similar to those in WGKKC-2 watershed. Compartment bunding activities are carried out at the maximum scale because of the rainfed dominance across the State. Unlike Punjab or Haryana, 33 out of 36 districts in Maharashtra are covered under the rainfed region; it is known to be the second most rainfed state in India. As per Table 13.6, the area treated under compartment bunding is 3,966,000 ha, which comprises almost 50% of the total area treatment activities of 8,115,000 ha. The undulating geography of the state is another reason why bunding activities of this type are taken up on a large scale. A few districts of the eastern Vidharbha region and the coastal Konkan region contribute to paddy bunding activities in Maharashtra.

In total, the enormous number of 3,542,829 drainage line treatment activities were carried out in Maharashtra State by the Soil & Water Conservation Department, as described in Table 13.7. Considering the geomorphology of Maharashtra, LBS structures contributed more than 50% towards water management. As a cost-effective mode, the government has a preference for Mati/earthen nala bunds, which altogether make up fewer than 50% of the total activities carried out in

TABLE 13.6
Soil and Water Conservation Area Treatment Activities in Maharashtra State

Sl. No.	Soil and Water Conservation Activities	Total Area Treated (ha)
	Area Treatment	
1	Compartment Bunding	3,966,000
2	Graded Bunding	3,122,000
3	Continuous Contour Trenching	466,000
4	Deep Continuous Contour Trenching/ WAT	42,000
5	Paddy Bunding	518,000
Total		**7,596,000**

TABLE 13.7
Soil and Water Conservation Drainage Line Treatment Activities in Maharashtra State

Sl. No.	Drainage Line Treatment Activities	Total Numbers
1	Earthen Structure (ENB)	943,771
2	Loose Bolder Structure (LBS)	2,341,844
3	Mati Nala Bandh (MNB)	165,760
4	Cement Nala Bandh (CNB)	71,810
5	Gabion Structure	9,840
6	Underground Bandh	4,629
7	Diversion Bandh	5,175
Total		**3,542,829**

Data Source: Soil and Watershed Management Dept., Commissionerate of Agriculture, Pune.

Maharashtra. The most of the study (almost 50%) covered by forest, wastelands, and steep slopes. CNB is the first choice in such areas due to its runoff coefficients (Chalam and Krishnaveni, 1996). Except for the Konkan region, almost the whole of Maharashtra possesses black cotton soil, which is good for ENB/LBS with the constraint of soil depth.

In addition, the government promotes a few practices to elevate the economic standards of farmers. The agronomy practices usually followed in Maharashtra are contour cultivation, inter cropping, vegetative bandh, dead furrow, and broad base furrow technology.

13.4 CONCLUSION

Geo-spatial approaches provide an effective way for water resource management. In this study, an LU/LC map was generated using satellite data from different sensors, and a WRDP was prepared by the weighted overlay method using different constraints and overlaid on the merged product of LISS IV and Cartosat-1D. The WRDP was prepared by the Weighted Sum Raster Overlay Analysis method in QGIS. The current version of QGIS3 possesses a fantastic method named "Multi Criteria Overlay Analysis" for effective analysis in a similar way. The LU/LC scale was taken at 10K and 50K, and the WRDP was generated. Then, a comparative analysis was done using the variability in scaling of the LU/LC parameter, and statistics were prepared.

The need of the hour is to shift towards sustainable development. It is difficult to regenerate resources without conserving them. The process of conservation begins with better and effective management of primary resources. Land (soil) and water conservation are essential, and their interaction influences the quality of life or living standards. The only option to manage this interaction of primary resources is to work within a hydrologically defined unit, i.e., a watershed. All the processes within the limits of a watershed influence one another. Water resource management can be achieved through land evaluation and land use planning.

A WRDP can be used for rural empowerment and assists in employment generation. It reveals the facts needed for decentralized planning. The different criteria, slope, soil, and LU/LC, can be considered for preparing a WRDP using the weighted overlay method. The effect of variability in the scaling of LU/LC is generated. The output shows negligible difference in the zonation areas provided for water conservation structures.

Water conservation can be done through CNB, which are usually constructed in forest lands, wastelands, and areas with slopes. Water harvesting structures like LBS, ENB, and MLB are mainly put into operation on less deep soil and almost plain terrain, whereas water recharge activities can be carried out in low-lying areas. As inferred from statistics of Maharashtra, compartment/graded bunding area treatment activities were carried out at a large scale with the criterion of 0–3% slope area due to its undulating terrain. Drainage line water conservation activities were also carried out frequently, depending upon the characteristics of soil present in Maharashtra.

One myth about water resources management is that all watershed programs are successful, but unfortunately, this is untrue. Inadequate analysis of the physical, socio-economic, and environmental parameters of the watershed under study leads to different socio-economic difficulties for the farmers. A lack of involvement by farmers and no flexibility in technological options to suit farmers' needs against a background of meager resources may lead to the failure of such programs. Poor acceptance is observed for contour-based water conservation measures because of conflicts over ownership boundaries. Due to lack of continuous financial support, proper implementation of the WRDP programme has not been achieved. Conflicts of interest arise due to investment by one farmer towards the maintenance of water conservation structures like diversion drains, which benefits others but not the actual investor. Indigenously known soil and water conservation practices have often been disregarded due to personal interest. Lack of clear arrangements and understanding on sharing of the harvested water may lead to failure of water resource management programs. For example, the Kaveri river dispute is still a burning issue for two states in India sharing common boundaries.

13.4.1 Effect of Water Resources on the Community

Water resource management has a range of effects on livelihood on a large scale because soil, water, and vegetation are the most vital natural resources utilized by people. The production of food, fuel, fodder, fruit, and woods on a sustainable basis can be ensured by judicious and effective management of the water resources that impact human life.

With the help of such research methodologies, it is possible to reduce the frequency of water supply by tankers (for those districts facing water scarcity) by devising a WRDP. It is also possible to maintain double crop cultivation and potential irrigation of farms even during rain gaps. The successful implementation of potential treatment maps will lead to practicing transparency in scheme execution, generating awareness about water utilization among farmers, and an absolute increase in ground water level that leads to a sustainable water supply for future generations.

As the first of its kind, the "ridge to valley" concept was put into practice and broke the traditional myths of water resource management in Maharashtra. The departments were not able to cope with utilization of state of the art technologies like GIS/remote sensing. Capacity building of the villagers was done for identification suitable sites for water conservations. This leads to more transparency and less political interference in the system.

Water and soil resource conservation gives rise to numerous possibilities for developing strategies for the policy makers in India with minimal human effort, increasing areas under irrigation, crop productivity and livelihood, and sustainable development.

REFERENCES

Chalam B.N.S., Krishnaveni M., Karmegam M., 1996. Correlation of runoff with geomorphic parameters. *J Appl Hydrol.* IX(3–4): 24–31.

Fishburn, P.C., 1967. Additive utilities with incomplete product set: Applications to priorities and assignments. *Oper. Res.* 15: 537–542, www.jstor.org/stable/168461.

Joshi P.K., Jha A.K., Wani S.P., 2008. Impact of watershed program and conditions for success: a meta-analysis approach. Global Theme on Agroecosystems, Report 46. International Crops Research Institute for the Semi-Arid Tropics and National Centre for Agricultural Economics and Policy Research, New Delhi.

Loucks D.P., van Beek E., 2017. Water resources planning and management: an overview. In *Water Resource Systems Planning and Management.* Springer, Cham. https://doi.org/10.1007/978-3-319-44234-1_1

Marble D.F., Peuquet D.J., Boyle A.R., 1983. *Joint Geographic Information System and Remote Sensing, Manual of Remote Sensing.* Ed. Colwell R. N. American Society of Photogrammetry, Falls Church.

Miller D.W., Starr M.K., 1969. *Executive Decisions and Operations Research.* Prentice-Hall, Inc., Englewood Cliffs, NJ.

Raju K.V., Aziz A., Sundaram M.S.S., 2008. Guidelines for planning and implementation of watershed development program in India: a review. Global Theme on Agroecosystems Report 48. Andhra Pradesh, India.

Ramteke I.K., 2017. Development of geo-spatial database of land resources for alternate land use in Bali island of Sundarban Delta, West Bengal. PhD diss., Dr Panjabrao Deshmukh Krishi Vidyapeeth, Akola, India.

Ratna Reddy V., Geoffrey J. Syme, Chiranjeevi Tallapragada, 2019. Integrated approaches to sustainable watershed management in xeric environments. https://doi.org/10.1016/B978-0-12-815275-1.00002-4.

Ravindran K.R., Promod Kumar, Tiwari A.K., 1992. Integrated approach for resources planning using remote sensing and GIS: a case study of song watershed. *Proceedings of National Symposium on Remote Sensing for Sustainable Development*: 11–15.

Sankar P., Hermon R.R., Alaguraja P., 2012. Comprehensive water resources development planning Panoli village blocks in Ahmednagar district, Maharashtra using Remote Sensing and GIS techniques. *Int J Adv Remote Sens GIS.* 1(1): 40–58.

Stanney K.M., Pet-Edwards J., Swart W., Safford R., Barth T., 1994. The design of a systematic methods improvement planning methodology: Part II – A multiattribute utility theory (MAUT) approach. *International Journal of Industrial Engineering* 1(4): 275–284.

Suresh M., Sudhakar S., Tiwari K.N., 2004. Prioritization of watersheds using morphometric parameters and assessment of surface water potential using remote sensing. *J Indian Soc Remote Sens.* 32(3): 249–259.

Tideman E.M., 2000. *Watershed Management: Guidelines for Indian Conditions.* Omega Scientific, New Delhi.

Wani S.P., Ramakrishna Y.S., 2005. *Sustainable Management of Rainwater through Integrated Watershed Approach for Improved Rural Livelihoods, Watershed Management Challenges: Improved Productivity, Resources and Livelihoods.* International Water Management Institute, Colombo, Sri Lanka: 39–60.

Wani S.P., Ramakrishna Y.S., Sreedevi T.K., 2006. Issues, concepts, approaches and practices in the integrated watershed management: experience and lessons from Asia in integrated management of watershed for agricultural diversification and sustainable livelihoods in Eastern and Central Africa: lessons and experiences from semi-arid South Asia. *Proceedings of the International Workshop held at Nairobi*, Kenya: 17–36.

14 Automatic Extraction of Surface Waterbodies of Bilaspur District, Chhattisgarh (India)

Soumen Brahma, Gouri Sankar Bhunia,
S. R. Kamlesh, and Pravat Kumar Shit

CONTENTS

14.1 Introduction 197
14.2 Study Area 198
14.3 Methods 199
    14.3.1 Spectral Water Indexes 199
    14.3.2 Normalized Difference Water Index (NDWI)..199
    14.3.3 Modified Difference Water Index (MNDWI) 200
    14.3.4 Water Ratio Index (WRI) 200
    14.3.5 Automated Water Extraction Index (AWEI) 200
    14.3.6 Accuracy Assessment 201
    14.3.7 Correlation Analysis And Evaluation 201
14.4 Results and Discussion 201
14.5 Conclusion 207
References..208

14.1 INTRODUCTION

Water on the field, as a river, a lake, a wetland, and the ocean, means water on the surface of the Earth. Surface water observation is an ecological and hydrological research prerequisite. Surface freshwater bodies are important for both humans and environmentally sustainable structures and tools. They are essential to preserving all modes of life (Karpatne et al., 2016). Plenty of flora and fauna can be found in water, preserving the biodiversity of the ecosystems of rivers or wetlands (Vörösmarty et al., 2010). Surface freshwater is not only essential to ecosystems as a key component of the hydrological cycle; it also affects all aspects of our lives, such as water, agriculture, power generation, transport, and industries (Vörösmarty et al., 2000). Surface water systems, because of various natural and human causes, are changeable in nature as they grow, shrink or alter their appearance or flow over time (Karpatne et al., 2016). Variations in water systems have an impact on the environment and other natural resources and human assets. Changing the level of surface water typically leads to severe consequences (Huang et al., 2018). A rapid increase in surface water can lead to flooding in extreme cases. It is therefore important that surface water is efficiently detected, its volume extracted and quantified, and its dynamics controlled.

    Bilaspur district is very rapidly transforming its need for water from an agricultural society to an industrial civilization. The issues with our water supply have not gone away, and our observation and use of water have become one of Chhattisgarh's most important questions for the future (https://cgwrd.in/water-resource/ground-water-status.html). The cumulative impact of several individual effects of abstractions and pollution sources is showing that groundwater concerns are gradually increasing. The groundwater has failed to reach its previous optimum level in the past few years. In every rainy season, the level of groundwater in the state is about 5 to 6 meters below ground level, and about 50–60 meters below ground level in the central parts of the state during the summer (CGWB, 2012). In Bilaspur, the groundwater level is rapidly decreasing, and chemical pollution with fluoride, arsenic, and iron, particularly in the southern sections, and other heavy metals has increased significantly in many places throughout the country (Shrivastava et al., 2014).

Remote sensing provides a powerful means of tracking the dynamics of surface water. In comparison with traditional in situ measurements, the ability to continuously track the Earth's surface at multiple scales makes remote sensing far more effective. Remote sensing datasets provide spatially explicit and temporally frequent observational information on a number of physical attributes about the surface of the Earth that can be appropriately exploited to map the extent of water systems at a global or regional level and to regularly monitor their dynamics. Timely monitoring of surface water and the provision of data on surface water dynamics are important for policy- and decision-making processes (Frey et al., 2010). Measuring surface water has recently become a new age in the world of satellite-based optical remote sensors. The combination of remote sensing data with geographic information systems (GIS) has been used in the automatic or semiautomatic extraction and mapping of water bodies in recent years (Giardino et al., 2010). Multi-temporal images of water bodies with subpixel precision techniques were automatically retrieved from Landsat Thematic Mapper (TM) and Enhanced Thematic Mapper Plus (ETM+) (Verpoorter et al., 2012). Landsat is one of the most successful satellite series in history. It has been supplying medium-resolution images for over 40 years since the first mission was launched in 1972. Multispectral scanners (MSS) were the on-board sensors for early Landsat missions, later upgraded on Landsat-4 and Landsat-5 to TM and on Landsat-7 to ETM+. Land Imager Operation (OLI) is the new Landsat-8 optical sensor. This level of resolution is ideal for the dynamic detection of nearly every type of surface water body. Landsat-5 has been in service for an unforeseen long time. This makes Landsat TM image applications very common in surface water detection (Chen et al., 2014; Pardo-Pascual et al., 2012). Visual (Ryu et al., 2002) and hybrid digital interpretation (Suresh Babu et al., 2002) and supervised classification techniques were used to extract the surface water derived from remotely sensed data. Automatic extraction of water bodies (Xu, 2006) based on spectral indices has been tried, but the inclusion of a number of urban, field, and cloud shading pixels as water pixels restricts these procedures. Subramaniam et al. (2011) proposed a knowledge-based hierarchical decision tree algorithm for automated extraction of water bodies, and this was used for water pixel extraction. Further refinements were made to include the mixed pixels at the edges of the body of water. For post-monsoon periods (with maximum water spread), contiguous pixels are applied to the water body with positive values for either Normalized Difference Water Index (NDWI) or modified NDWI (MNDWI) (Acharya et al., 2018b) and values with predefined threshold values for the other function parameters such as turbidity, Normalized Difference Pond Index (NDPI), and band ratios (Sarpa and Ozcelik, 2017).

The present study focused on the recent progress in measurement of surface water using open source geospatial information in Bilaspur district of Chhattisgarh, India to demonstrate the automated extraction of surface water bodies.

14.2 STUDY AREA

Bilaspur is located in the east of Chhattisgarh and extends between 21°47′ and 23°08′ N latitude and between 81°14′ and 83°15′ E longitude. It is surrounded by Koriya in the north, Madhya Pradesh district in the south, Raipur in the east, and Champo in the west by Janjgir (Anonymous, 2003). It is mountainous in the north and in the south, leading to very cold and rainy conditions, respectively.

Automatic Extraction of Surface Waterbodies

The Bilaspur district has a mean temperature of 45 °C with an annual rainfall of 1220 mm. May and June are the hottest months, and the lowest temperatures are observed in December and January. Agaar, Maniyaar, and Arpa are the main rivers around Bilaspur. Bilaspur can be divided physiographically into two parts. The first portion is created from the highlands covering the north and central portion of the district (covering the blocks of Lormi, Kota, Gaurela, Pendra, and Marwahi). The second part is a lowland area covering the southern parts of the district (the blocks of Takhatpur, Mungeli, Pathariya, Belha, and Masturi). Roughly 38.78% of the total area is forest covered. The south side of the district is a plain region with gentle slopes that occupy 48% of the district's total geographical area. The topography ranges from 250 m above sea level in the southern plains to 1120 m above sea level in the northern hills. The southeastern part of the district is mostly filled with vertisol. In the eastern and northern areas of the region, the ultisol soil types are found, and they are red to yellow. Fertile leached soils found in wetlands where falling leaves form a dense humus layer every year are alfisol soils. The land is highly fertile and is primarily used for agriculture with no surface irrigation. Most of the northern part of the district is hilly and highly undulating; agriculture is confined to just a few areas. Almost 40% of the area is irrigated from all sources by the net sown area. The irrigation systems are provided by the irrigation network of the canals – almost 72% of the net irrigated area – and parts of the blocks of Bilaspur, Masturi, Mungeli, and Lormi. Surface water irrigation supports mainly the cultivation of Kharif (98.6%). The district population is 2,662,077 according to the census of 2011. The population density is 332 people/km^2, and the rural literacy rate is 72% in the study area.

14.3 METHODS

As a pre-processing step, radiometric calibration was performed, and digital numbers were converted into satellite reflection in the Environment for Visualizing of Images (ENVI). For economic data size and processing time, the sub-images containing the lake region were extracted using the ENVI Region of Interest method. Fast line-of-sight atmospheric analysis of Hypercube (FLAASH) atmospheric corrections were submitted for further study in images registered to Universal Transverse Mercator zone 45 north projections using WGS-84 datum.

14.3.1 Spectral Water Indexes

Automatic delineation of water bodies is a challenging method because of the nature of the land–water saturated environment. Several spectral water indices were developed for remotely sensed images to extract water bodies by generally estimating a standardized difference between two image bands and then using a reasonable threshold for dividing results into two groups (water and non-water). The NDWI, MNDWI, Water Ratio Index (WRI), and Automated Water Extraction Index (AWEI) satellite imagery were used in this analysis to take Landsat images. All water indices were applied in order to construct a final raster used in the study area with different thresholds to extract the water bodies of the study area. Once the map was finalized in line with the spectral index, thresholds were carefully selected for other extraction techniques based on:

a) The output of the spectral index with reference image as close as possible
b) Non-inclusion of features in the map of surface water characteristics such as cloud shadows, vegetation, and roads

14.3.2 Normalized Difference Water Index (NDWI)

The NDWI is used to separate water from dry soil or better, to map the body surface. In the visible infra-red wavelength range, water bodies show low radiation and high interoperability. NDWI uses remote sensing imaging in the range of infra-red and green bands. It will improve water knowledge efficiently in most cases. It's subtle in built-up land and often ends up overestimating water bodies. McFeeters (1996) suggested the NDWI for wetland identification and calculation of surface water level in wetland environments.

$$\text{NDWI} = \frac{\text{Band}_2 - \text{Band}_4}{\text{Band}_2 + \text{Band}_4}$$

This results in optimistic values for water characteristics and development. Vegetation and soil characteristics are typically zero or negative and suppressed.

14.3.3 Modified Difference Water Index (MNDWI)

In general, the digital number of TM band 5 is much greater than the average TM band 2 digitally reflecting MIR radiation (Xu, 2006). This also reduces built-up area characteristics, which in many indices are usually related to open space. Therefore, the built-up land will have negative values if a mid-infra-red (MIR) band is used in place of the near infra-red (NIR) band in the NDWI. On this basis, the NDWI will be changed by replacing the NIR band with the MIR. Xu (2006) describes a common MNDWI method, which is a strong index capable of extracting watercourses.

$$\text{MNDWI} = \frac{\text{Band}_2 - \text{Band}_5}{\text{Band}_2 + \text{Band}_5}$$

The findings show positive values due to their higher representation in band 2 than in band 5, and non-water characteristics have negative NDWI values. A threshold value may be set in two groups (water and non-water characteristics) for MNDWI (e.g., simply a zero value) category.

14.3.4 Water Ratio Index (WRI)

Another water index (Rokni et al., 2014) is the WRI:

$$\frac{(\text{Green} + \text{Red})}{(\text{NIR} + \text{SWIR1})}$$

It uses green, red, NIR, and short-wave infra-red (SWIR) bands to create a combination for ambient illumination reduction. Generally, WRI shows values greater than 1 for water bodies.

14.3.5 Automated Water Extraction Index (AWEI)

The AWEI's main aim is to improve the separability between water and water pixels by differentiating bands, applying different coefficients, and using them. Two separate equations are therefore proposed for efficient non-water pixel suppression and surface water extraction with increased accuracy.

$$\text{AWEI}_{nsh} = 4 \times (\rho\text{band}_2 - \rho\text{band}_5) - (0.25 \times \rho\text{band}_4 + 2.75 \times \rho\text{band}_7)$$

where ρ variables represent Landsat-5 TM spectral band values: band_1 (blue), band_2 (green), band_4 (red), band_5 (NIR), and band_7 (SWIR). In areas with urban backgrounds, AWEI_{nsh} is designed to effectively eliminate non-water pixels like dark built-up surfaces.

TABLE 14.1
Principle of the Confusion Matrix

		Predicted	
		Negative	Positive
Truth	Negative	a	b
	Positive	c	d

14.3.6 Accuracy Assessment

Since the purpose here is to make detailed adjustments over time without additional evaluation costs and time, the proposed approach should be tested for accuracy. Since images are historic and the surface of the water sometimes shifts, at this time, there is no database of water features. The on-screen digitalization of Landsat TM and OLI images through visual interpretation of expert knowledge has therefore been produced very thoroughly with the use of a panchromatic image and the derivative, i.e., the real-colored pan-sharpened image. The efficiency characteristics were compared with the overall accuracy (OA) and kappa coefficient calculated from the confusion matrix based on the index-based change maps and reference maps. Table 14.1 shows the principle of the typical confusion matrix, where the correct number of forecasts is the negative instance, "b" means the number of incorrect predictions that an instance is positive, "c" is the number of false predictions, and "d" is a positive correction (Congalton and Green, 2008).

The OA and kappa coefficient calculations are shown:

$$\text{Overall accuracy} = \frac{\text{Number of correct predictions}}{\text{Total predictions}}$$

$$\text{Kappa coefficient} = \frac{\text{Observed accuracy} - \text{Chance accuracy}}{1 - \text{Chance accuracy}}$$

The OA is the actual estimate and ranges from 0 to 1 with a value close to 1. However, the relationship between datasets that is attributable to chance alone cannot be taken into account. The kappa coefficient is therefore frequently used in combination with the OA as a tool to control this random agreement factor. The kappa coefficient can typically range from −1 to +1, where 0 is the amount of agreement predicted from random chance and 1 is a perfect agreement between raters. The kappa coefficient may be negative in very rare situations, which means that there is no successful agreement between the two ratters (McHugh, 2012).

14.3.7 Correlation Analysis And Evaluation

To measure the strength and association between two images, Pearson's correlation coefficient (r) test was used. The "r" value should be as close as possible to one. The difference between the Pearson r values will demonstrate just how well the spatial consistency is established (Sarp and Ozcelik, 2017).

14.4 RESULTS AND DISCUSSION

All indices with the selected threshold appear to have different patterns based on visual inspection. In Bilaspur district, the NDWI (Figure 14.1) and AWEI (Figure 14.2) indices appear to work well. First of all, WRI performed worst for the standard threshold. In flat and hilly areas, NDWI and AWEI seem to remove pixels of distilled water, but even WRI (Figure 14.3) and MNDWI (Figure 14.4) seem

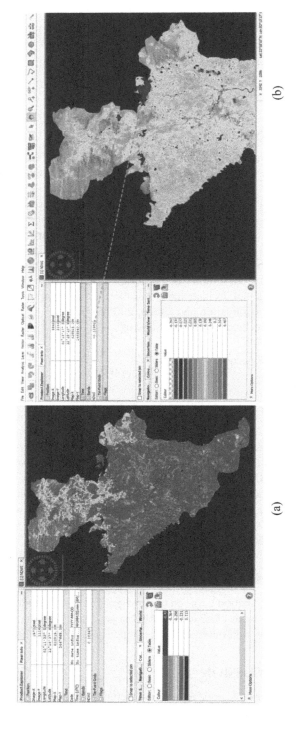

FIGURE 14.1 Extraction of surface water bodies through NDWI based on SNAP tool: (a) 1991, (b) 2019.

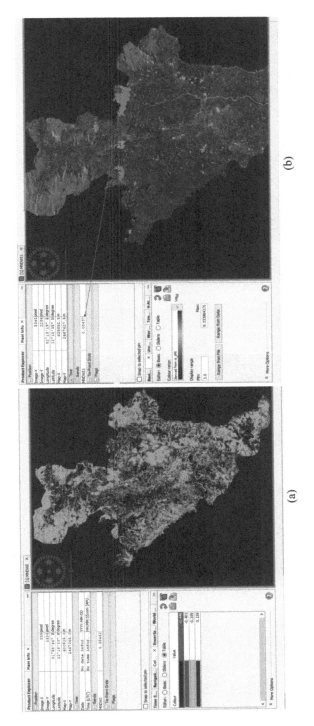

FIGURE 14.2 Extraction of surface water bodies through AWEI based on SNAP tool: (a) 1991, and (b) 2019.

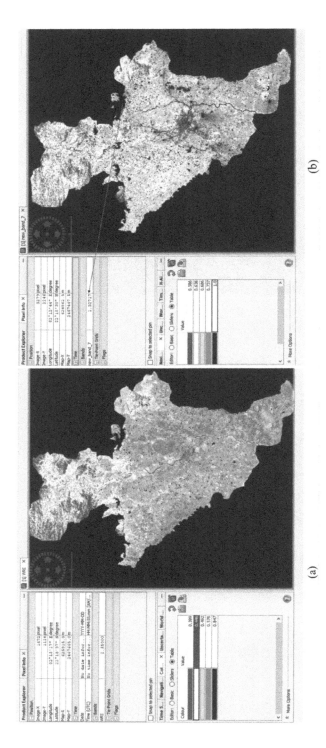

FIGURE 14.3 Extraction of surface water bodies through WRI based on SNAP tool: (a) 1990, (b) 2019.

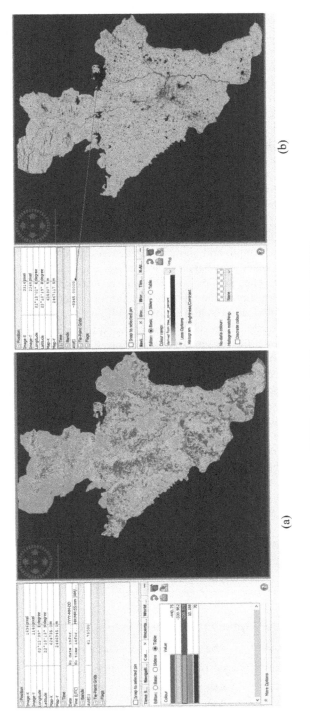

FIGURE 14.4 Extraction of surface water bodies through MNDWI based on SNAP tool: (a) 1991, (b) 2019.

to be enabled at the cost of mountain-shadowing mixed pixels. In the meantime, every derivative of the water index seems to have improved its optimum threshold. Visually very transparent, they reflect several water pixels, removing most shadow and non-water surface detection errors. NDWI and AWEI again appear to produce strong and consistent outcomes in flat and hilly areas for pure water pixels, but have left many mixed pixels for small and narrow water bodies. In the optimum range, all WRI and MNDWI variables appear unable to avoid the seamless water cover (Figure 14.4).

In Table 14.2, Producer Accuracy (PA) is higher for all indices with the standard threshold, which decreases with manual selection, but it is differing for User Accuracy (UA), OA, and kappa, which are increased. For both threshold instances, WRI and MNDWI were poor for performance. With over 80% of OA and a kappa of 0.60 and 0.54, respectively, NDWI and AWEI with optimum thresholds performed better. No spectral water indices showed OA above 88% in spite of the increase in OA and kappa with a balanced PA and UA at the optimal threshold. Most literature studies show that OA is more than 90%, even up to 99% (Acharya et al., 2016; Zhou et al., 2017). This demonstrates that these scenes require the combination of extraction and special conditions. The bulk of the experiments focused, as described in the original study, on scenes of large bodies of water or a subset of higher water content. The case scene was extremely complicated because it has only 0.85% of surface water. In fact, it was the wide disparity in height and surface water level and the prevalence of all kinds of ambient noise on the scene that were troubling problems. Water indexing methods are, in principle, simple and show high accuracy when using standard or optimum thresholds to extract surface water. Nevertheless, following assessment in this chapter, it is clear that this is not always the case. They are not feasible because various forms of water and noise exist.

Of the indices of regular thresholds, the higher PA is more evident. Around the same time, non-water surfaces were misclassified, and UA and kappa lower than 43% resulted. This means that non-water surfaces are only arbitrarily categorized when opposed to water surfaces. The classification of the shadow surface as water could also make this possible. Therefore, it is important to consider a shadow detection method to remedy this. A threshold that represents the actual surface water is needed to reduce such misclassification. In small areas, the vision of an expert and close interaction with high-resolution images provide better ideas about the optimal threshold (Acharya et al., 2018a). But in this optimal threshold selection, the other challenge in this study was that it covered so much, only a little of which was water with a lot of noise from the surroundings. For this reason, we have used test and error as a guide for our validation dataset to pick the highest accuracy threshold and kappa. NDWI and AWEI could only identify pure pixels for the maximal threshold values chosen. Mixed pixels along with ambient noise, resulting in lower UA, were observed by derived indexes from WRI and MNDWI. Each water detection algorithm has been evaluated by basically verified NDWI images with a threshold of 0.28, and the results are presented in Table 14.3. The results indicate that the performance of MNDWI and AWEI closely matches that of NDWI in terms of water detection efficiency.

The effect of the weather/atmosphere is another problem that has not been discussed but that may be important in the identification of surface water. The variations in geography and land coverage

TABLE 14.2
Accuracy Assessment for Optimum Threshold Values for Different Water Indices

Index	Producer Accuracy		User Accuracy		Overall Accuracy		Kappa^	
	1991	2019	1991	2019	1991	2019	1991	2019
NDWI	0.53	0.60	0.89	0.92	0.88	0.92	0.60	0.69
MNDWI	0.84	0.93	0.53	0.66	0.80	0.88	0.54	0.64
WRI	0.81	0.88	0.47	0.59	0.74	0.72	0.44	0.57
AWEI	0.77	0.85	0.69	0.78	0.84	0.86	0.56	0.65

TABLE 14.3
Percentage of Area Correctly Classified as Surface Water

Index	Threshold Demarcation	Correctly Classified (%)
NDWI	0.25	86.23
MNDWI	0.35	80.35
WRI	1.20	74.05
AWEI	−8320	85.43

TABLE 14.4
Pearson's r between the NDWI, MNDWI, WRI, and AWEI in 1991 and 2019

Water Body Indices	Pearson's r in 1991				Pearson's r in 2019			
	NDWI	MNDWI	WRI	AWEI	NDWI	MNDWI	WRI	AWEI
NDWI	1	0.82	0.65	0.86	1	0.92	0.85	0.96
MNDWI		1	0.58	0.57		1	0.68	0.78
WRI			1	0.60			1	0.80
AWEI				1				1

relate to the variations in weather. Depending on the elevation and wind conditions, the scene may be transparent, or there may be darkened or cloudy areas. Therefore, atmospheric effects should be carefully noted and pre-processed in order to reduce the effect to the fullest extent possible. By creating and using a shadow mask from the elevation data, we can solve the shadow problem. Based on the findings of this study, surface water extraction can be refined with smaller elevation classes using index methods, and these can be applied to any further characteristics that can divide a scene.

The spectral qualities of the water body indices were measured with Pearson's r, as depicted in Table 14.4. The maximum r value indicated strong correlation. The highest correlation was observed between the NDWI and AWEI in 1991 and 2019, with the values of 0.92 and 0.96, respectively.

14.5 CONCLUSION

In our research, in relation to the abovementioned methods, which used various algorithms and satellite data, we proposed an automated method to extract water bodies from remote sensing data. The proposed approach used Landsat data to distinguish water reliably and rapidly from non-water features. The characteristic properties of water were used as a basis in the analysis, such as low reflectance values in the SWIR band, a higher reflectance value in band 4 than in band 2, and a higher reflectance in band 2 than in band 5. The findings of this analysis were analyzed in a Landsat-8 scene in the district of Bilaspur: NDWI, MNDWI, WRI, and AWEI. With validation points, the optimal threshold was identified and compared with the standard threshold for each procedure. These properties, in addition to an NDWI, are used as the function vectors for improving classification. This model is competently used for Landsat data by the proposed perception model, and researchers in the global field of water science should apply this model for appropriate accuracy and precision in water supply systems. However, its sensitivity and efficiency should be checked by looking at resolution, satellite seasonal data in conjunction with land use, and patterns of land cover around water bodies not addressed in this report. In most situations, with higher OA but lower kappa, the optimal threshold increases stimulation. NDWI and AWEI were perfect for corresponding image identification and exclusion with maximal threshold, while MNDWI and WRI were able to distinguish mixed pixels from rivers and shallow ponds/water bodies. The proposed approach was shown to be effective in isolating large water bodies, and the model needs to be improved by highlighting a specific threshold, which can also isolate smaller water bodies, in light of the current global water crisis. In the light of this, more work in the field of perception models has a wide scope for designing and creating a potentially specific tool, which can use different satellite data and more rigorously extract water bodies, including smaller water systems. These results and suggestions are used to analyze the implementation of the new approach in many scenes and to compare it with state-of-the-art methods of machine learning.

The technique could also be used to identify patterns in any features if indices and thresholds are carefully selected in order to distinguish the subject matter from the background. The threshold for the identification of water bodies, sensor data, and locations using standard water indicators for accurate detection of changes must be adjusted accordingly for future research.

REFERENCES

Acharya, T. D., Lee, D. H., Yang, I. T., Lee, J. K. Identification of water bodies in a Landsat 8 OLI image using a J48 decision tree. *Sensors* 16 (2016): 1075. doi: 10.3390/s16071075.

Acharya, T. D., Subedi, A., Yang, I. T., Lee, D. H. Combining water indices for water and background threshold in Landsat image. *Proceedings* 2 (2018a): 143. doi: 10.3390/ecsa-4-04902.

Acharya, T. D., Subedi, A., Huang, H., Lee, D. H. Classification of surface water using machine learning methods from Landsat data in Nepal. In *Proceedings of the 5th International Electronic Conference on Sensors and Applications*, 15–30 November 2018b.

Anonymous. *Survey Report of Forest Department*. Chhattisgarh State, 2003.

CGWB (Central Ground Water Board). Ground water brochure of Bilaspur district, Chhattisgarh 2012–2013. Available at: http://cgwb.gov.in/District_Profile/Chhatisgarh-/Bilaspur.pdf

Chen, Y., Wang, B., Pollino, C. A., Cuddy, S. M., Merrin, L. E., Huang, C. Estimate of flood inundation and retention on wetlands using remote sensing and GIS. *Ecohydrology* 7 (2014): 1412–1420. https://doi.org/10.1002/eco.1467.

Congalton, R. G., Green, K. *Assessing the Accuracy of Remotely Sensed Data: Principles and Practices*. Boca Raton: CRC Press, 2008, Chap. 7.

Frey, H., Huggel, C., Paul, F., Haeberli, W. Automated detection of glacier lakes based on remote sensing in view of assessing associated hazard potentials. *Grazer Schriften Geogr. Raumforsch* 45 (2010): 261–272.

Giardino, C., Bresciani, M., Villa, P., Martinelli, A. Application of remote sensing in water resource management: The case study of Lake Trasimeno, Italy. *Water Resources Management* 24 (2010): 3885–3899.

Huang, C., Chen, Y., Zhang, S., Wu, J. Detecting, extracting, and monitoring surface water from space using optical sensors: A review. *Reviews of Geophysics* 56 (2018): 333–360. doi: 10.1029/2018RG000598

Karpatne, A., Khandelwal, A., Chen, X., Mithal, V., Faghmous, J., Kumar, V. Global monitoring of inland water dynamics: State-of-the art, challenges, and opportunities. In J. Lässig, K. Kersting, K. Morik (Eds.), *Computational Sustainability* (pp. 121–147). Cham: Springer International Publishing, 2016.

McFeeters, S. K. The use of the normalized difference water index (NDWI) in the delineation of open water features. *International Journal of Remote Sensing* 17 (1996): 1425–1432.

McHugh, M. L. Interrater reliability: the kappa statistic. *Biochemia Medica* 22 (2012): 276.

Pardo-Pascual, J. E., Almonacid-Caballer, J., Ruiz, L. A., Palomar-Vazquez, J. Automatic extraction of shorelines from Landsat TM and ETM+ multi-temporal images with subpixel precision. *Remote Sensing of Environment* 123 (2012): 1–11. https://doi.org/10.1016/j.rse.2012.02.024..

Rokni, K., Ahmad, A., Selamat, A., Hazini, S. Water feature extraction and change detection using multitemporal Landsat imagery. *Remote Sensing* 6(5) (2014): 4173–4189.

Ryu, J. H., Won, J. S., Min, K. D. Water-line extraction from Landsat data in a tidal flat. A case study in Gomso Bay, Korea. *Remote Sensing of Environment* 83 (2002): 442–456.

Sarpa, G., Ozcelik, M. Water body extraction and change detection using time series: A case study of Lake Burdur, Turkey. *Journal of Taibah University for Science* 11 (2017): 381–391.

Shrivastava, D. K., Yadav, S., Chandra, T. P. Evaluation of physico-chemical quality of drinking water in Bilaspur District of Chhattisgarh State. *International Journal of Scientific & Engineering Research (IJSR)* 3(6) (2014): 1267–1271.

Subramaniam, S., Suresh Babu, A. V., Roy, P. S. Automated water spread mapping using resources at-1 AWiFS data for waterbodies information system. *IEEE Journal of Selected Topics in Applied Earth Observation and Remote Sensing* 4(1) (2011): 205–215.

Suresh Babu, D. S., Hindi, E. C., Da Rosa Filho, E. F., Bittencourt, A. V. L. Characteristics of Valadares Island aquifer, Paranagua coastal Plain, Brazil. *Environmental Geology* 41 (2002): 954–959.

Verpoorter, C., Kutser, T., Tranvik, L. Automated mapping of water bodies using Landsat multispectral data. *Limnology and Oceanography: Methods* 10 (2012): 1037–1050.

Vörösmarty, C. J., Green, P., Salisbury, J., Lammers, R. B. Global water resources: Vulnerability from climate change and population growth. *Science* 289(5477) (2000): 284–288. doi: 10.1126/science.289.5477.284

Vörösmarty, C. J., McIntyre, P. B., Gessner, M. O., Dudgeon, D., Prusevich, A., Green, P., et al. Global threats to human water security and river biodiversity. *Nature* 467(7315) (2010): 555–561.

Xu, H. Modification of normalize difference water index (NDWI) to enhance open water features in remotely sensed imagery. *International Journal of Remote Sensing* 1(27) (2006): 3025–3033.

Zhou, Y., Dong, J., Xiao, X., Xiao, T., Yang, Z., Zhao, G., Zou, Z., Qin, Y. Open surface water mapping algorithms: A comparison of water-related spectral indices and sensors. *Water* 9 (2017): 256. doi: 10.3390/w9040256

15 Valuing Ecosystem Services for the Protection of Coastal Wetlands Using Benefit Transfer Approach
Evidence from Bangladesh

Muhammad Mainuddin Patwary, Md. Riad Hossain, Sadia Ashraf, Rabeya Sultana, and Faysal Kabir Shuvo

CONTENTS

15.1 Introduction 211
15.2 Materials and Methods 213
    15.2.1 Study Area 213
    15.2.2 Methodology 214
    15.2.3 Land Use Land Cover Classification 214
    15.2.4 Ecosystem Services Value (ESV) Estimation 215
    15.2.5 Spatial Analysis of ESV Flow 215
    15.2.6 Sensitivity Factor 217
15.3 Results 217
    15.3.1 Land Cover Classes in the Study Area 217
    15.3.2 Estimation of Total ESV 218
    15.3.3 Estimation of Ecosystem Function Value (ESV_f) 218
    15.3.4 Significance of Land Use Types for ESV 220
15.4 Discussion 221
15.5 Conclusion and Policy Implications for Sustainable Conservation 222
References 222

15.1 INTRODUCTION

Wetlands are one of the most productive habitats in the world, providing environmental, economic, and social benefits to human beings (Islam and Gnauck, 2007; Islam, 2010). It is estimated that wetlands cover 600 million to 1.2 billion hectares (ha) of the Earth's terrestrial surface globally (Mitra et al., 2003). Wetlands function as the Earth's "kidneys" and play a vital part in biodiversity conservation. They act as reservoirs for a diverse range of flora, fauna, and threatened species (Bai et al., 2013). In addition, wetland serves as a transition between the terrestrial and aquatic ecosystems, providing a secure livelihood for people living in the adjacent area (Rebelo et al., 2010). The global contribution of wetlands is estimated to be about US$70 billion per year (Schuyt and Brander, 2004). Therefore, conservation of wetlands is necessary because of their major role in sustainable development as well as poverty reduction (Finlayson et al., 2015). However, the multiple benefits of wetlands and their conservation priority are often overlooked by decision-makers, which could result in overexploitation of resources. Thus, the inclusion of the value of wetland conservation in environmental decision-making is critical.

Economic analysis is an essential quantitative tool for the assessment and policy implications of the environmental resources provided by the wetland ecosystem (Barbier et al., 1997). However, lack of sufficient knowledge of the economic importance of wetland services and lack of clear understanding of the possible revenue opportunities associated with them may contribute to other planning activities being taken up (Siew et al., 2015). These concerns have made it necessary for wetland valuations to measure the benefits of wetland resources in monetary units. There is a range of approaches available to measure global ecosystem services, but the most popular is the benefit transfer approach developed by Costanza et al., (1997) and later revised (Costanza et al., 2014). In this study, land use land cover (LULC) was used as a proxy estimation of various biomes that are proportional to land cover. After that, previously published value coefficients were adjusted locally to estimate the total ecosystem service value (ESV). Following the method, several studies have observed changes in the ESV of wetlands, including Msofe et al. (2020), Zhang et al. (2019), Camacho-Valdez et al. (2013), Zhao et al. (2005), Kreuter et al. (2001), and so on. There is, however, still limited research focusing on the comprehensive evaluation of the value of wetland ecosystem services worldwide (Zhang et al., 2019; Costanza et al., 1997).

In recent years, computational capacity and data availability enabling geographic information system (GIS) research have advanced significantly. Free and open source software has been increasingly used in ecological study in recent years (Steiniger and Hay, 2009; Boyd and Foody, 2011). Compared with commercial software, free and open source software provides several empirical advantages, such as impartial analysis and clarification of the study's computational and statistical background. Furthermore, open source applications facilitate ecological research and expertise transfer in developing countries, with financial constraints limiting access to confidential technologies (Steiniger and Hay, 2009). Leon and George (2008) reported that a variety of issues, such as shortage of resources, lack of skills, insufficient training capability, and reliance on foreign experts, impacted the modeling and decision support of water resource management in developing countries. Thus, open source software acts as a pathway to overcoming these challenges. A number of freeware GIS platforms, such as QGIS (Quantum GIS), GRASS GIS (Geographic Resources Analysis Support System GIS), SAGA (System for Automated Geoscientific Analyses), and gvSIG (Generalitat Valenciana Sistema de Información Geográfica), have been developed, which have features similar to the common ESRI commercial ArcGIS software (Luque and Zulian, 2017). Among them, the QGIS software has been found convenient by users due to its straightforward installation and wide range of functionalities (Chen et al., 2010).

The Ganges–Brahmaputra–Meghna (GBM) Delta constitutes the world's largest floodplain wetland area, which plays a major role in the socio-economic growth of coastal Bangladesh (Islam, 2016). However, different anthropogenic activities have caused a rapid decline of the wetlands of this region (Ahmed et al., 2008). According to Khan (1994), around 2.1 million hectares of freshwater wetlands have been depleted by development activities in this GBM delta. Beel Dakatia, a freshwater wetland in southwest coastal Bangladesh, experienced rapid degradation due to various natural and anthropogenic activities including waterlogging, siltation of river beds, poor maintenance of sluice gates, and conflicts among Beel users (Ali and Syfullah, 2017). Quantifying the ecosystem services in terms of monetary value has been recognized internationally for policy decisions on improved management of natural resources that support human well-being as well as promoting sustainable conservation (Pandeya et al., 2016). However, there is limited research focused on ecosystem services assessment in this region. Parvin et al. (2017) evaluated the implications of LULC changes on food and water supplies in the southwest delta of Bangladesh. Akber et al. (2018a) quantified the protective value of the Sundarbans mangrove forest in Bangladesh. A recent study conducted by Akber et al. (2018b) investigated the spatio-temporal changes of ESV in southwest coastal Bangladesh. Huq et al. (2019) used LULC as a proxy measure to quantify the values of Khulna-Gopalgonj Beel, a freshwater wetland in southern Bangladesh. However, no single study has been done so far in Beel Dakatia for assessing the environmental benefits in monetary units. Therefore, this study estimated the ESV of Beel Dakatia using a previously developed global ecosystem value coefficient. The main objective of the study was to estimate the current value of ecosystem services offered by Beel Dakatia by using a benefit transfer approach with the application of open source GIS software. The findings of the study will help policymakers to reduce the inefficiencies in the protection of coastal wetlands in Bangladesh.

15.2 MATERIALS AND METHODS

15.2.1 Study Area

Beel Dakatia is a freshwater floodplain located in the southwest coastal region of Bangladesh. It is the second largest Beel (smallest depression in floodplain) of Bangladesh. The total area of the Beel is about 17,400 ha or 174 km². The region is bounded by Dumuria and Phultala Upazila (sub-districts) at 89°20′E and 89°35′E longitude, 22°45′N and 23°00′N latitude of Khulna district (Rahman, 1995). The area is characterized by low elevation and almost flat topography. Solmari, Hamkura, and Salta are the three main rivers in this area that are interconnected with the Beel. The Beel Dakatia has been experiencing increased degradation due to waterlogging, inundation, river bed siltation, and other anthropogenic activities (Ali and Syfullah, 2017). Figure 15.1 shows the geographic location of Beel Dakatia.

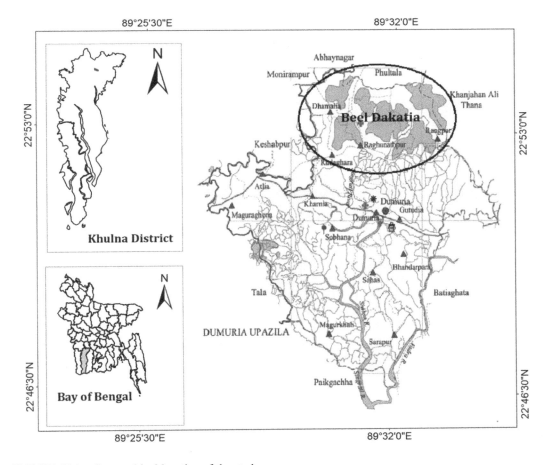

FIGURE 15.1 Geographical location of the study area.

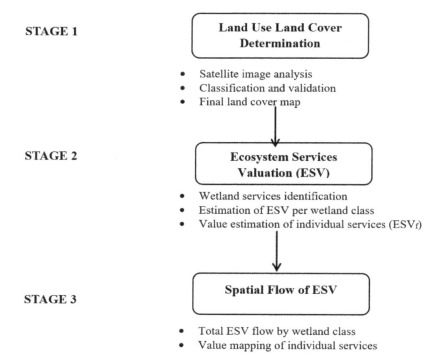

FIGURE 15.2 Flow chart of the overall methodological approach.

15.2.2 Methodology

In this study, the following methodological steps were used: (1) LULC classification of selected wetland through satellite image analysis; (2) estimation of land cover types of the wetland and the value of their services (ESV) based on the value transfer approach; and (3) analysis of total ESV flow and mapping of services values (Figure 15.2).

15.2.3 Land Use Land Cover Classification

LULC data were obtained by analyzing the remote sensing images from the United States Geological Survey (USGS) database (www.earthexplorer.usgs.gov). The medium resolution (30 m) of the Landsat 8 (Operational Land Imager [OLI]/Thermal Infrared Sensor [TIRS]) image for 2019 was used to determine the land use types. Images that stand for several stages taken in the same season are preferable for land cover change detection (Coppin et al., 2004). However, it is difficult to find cloud-free and well-matched satellite data during monsoon due to the cloudy condition of the area. It is therefore necessary to receive cloud-free data that represents the dry season images. Thus, the images acquired on February 2, 2020 were used for this study. The whole study area was covered by path 138 row 44 of the Landsat satellite. The images were projected into Universal Transverse Mercator (UTM), Zone 46 N, and WGS 84 Datum.

In this study, the LULC map was prepared using QGIS software (version 3.14 pi). The QGIS is an open source software platform with all the features of a modern GIS system, enabling user-based plugins to quickly extend their core features, which can be downloaded in a desktop setting (QGIS Development Team, 2013). The semi-automatic classification plugin (SCP) developed by Congedo (2017) is a significant extension of the QGIS that facilitates the supervised classification method by using multispectral satellite, aircraft, or drone images. The SCP involves a range of integrated raster processing techniques to create an automated workflow that enables faster and simpler land cover

TABLE 15.1
Detailed Descriptions of the Land Cover Types in the Study Area

Land Cover Type	Description
Agriculture	High land agriculture, cropland.
Aquaculture	Shrimp/Prawn, aquaculture ponds with saline water, agri-fisheries.
Settlements	Temporary and permanent settlements, road networks.
Vegetation	Trees, shrublands, grassland, and vegetable lands.
Water body	River, shallow water, canals and reservoirs, low-lying areas with marshy vegetation.

classification. The plugin offers several tasks, including pre-processing of input images by creating Band Set, classification accuracy assessment, raster manipulation using post-processing tools, spectral signature analysis, and raster calculation to accelerate the classification (Congedo, 2017).

Land use classification is a step-by-step process comprising image pre-processing, image mosaicking, image sub-setting, sample set preservation, supervised classification, and post-processing. The Semi-Automatic plugin offers several tasks including pre-processing of input images by creating Band Set, classification accuracy assessment, raster manipulation using post-processing tools, spectral signature analysis, and raster calculation to accelerate the land cover classification. Pre-processing images involves atmospheric correction, clipping the study region, creating spectral indices, and generating band collection. Reflectance correction and DOS1 (dark object subtraction) tools were used for atmospheric correction. Using published band combinations, various spectral responses were used for different land use practices to produce a region of interest (ROI). For each land use class, a total of 100 training samples was obtained using Google Earth aerial imagery. After training site development, the image was classified using a supervised classification approach with a maximum likelihood algorithm. In this study, five land use (agriculture, aquaculture, settlements, vegetation, and water body) practices were identified for the Beel Dakatia (Table 15.1). The accuracy of the land cover image was assessed using Google Earth images. A total of 200 random points was generated, and an accuracy assessment was performed for each land use class. Table 15.1 gives a detailed explanation of the currently practiced land use types in the Beel Dakatia.

15.2.4 Ecosystem Services Value (ESV) Estimation

The overall approach used in this study involved calculating ESVs and mapping their spatial distribution for the reference year 2020. The LULC dataset for the corresponding year was used as a proxy for calculating the ESVs. The present study used an ESV model developed earlier by Costanza et al. (1997) to assess the ESV for the Beel Dakatia. On the basis of this study, 5 land use practices were linked with 16 most related biomes, which are proportional to the various land cover forms used as proxies in the present study, namely, cropland biomes as "agriculture," tropical forests as "vegetation," lakes/river biomes as "water body," and urban as "settlement." Aquaculture represents the water body that produces fish and provides values of food production. However, this land use does not provide any freshwater or water regulation services and has no recreational or cultural value, unlike wetlands (Akber et al., 2018a). Therefore, this study used the food production service of the wetlands, mentioned earlier by Costanza et al. (1997), as a proxy for aquaculture.

15.2.5 Spatial Analysis of ESV Flow

The ESV of a specific category of LULC was calculated by multiplying the area of each LULC type by the respective value coefficients of the biome. The total ESV was calculated by summing the values estimated for each LULC for the reference year.

TABLE 15.2
LULC Classes and Their Equivalent Value Coefficients for Biomes (1994 US$ ha^{-1} y^{-1}).

LULC Type	Equivalent Biome	Value Coefficient (US$ ha^{-1} y^{-1})
Agriculture	Cropland	92
Aquaculture	Wetlands	256
Settlements	Urban	0
Vegetation	Tropical forest	2007
Water body	Lakes/River	8498

Source: Adapted from Costanza, R., et al., *Nature*, 1997, pp. 253–260.

TABLE 15.3
Value Coefficient (US$ ha^{-1} y^{-1}) of Ecosystem Service Functions for each LULC Type

	Ecosystem Function	Agriculture	Aquaculture	Settlement	Vegetation	Water Body
Provisioning	Food production	54	256	0	32	41
	Water supply	0	0	0	8	2117
	Raw material	0	0	0	315	0
	Genetic resources	0	0	0	41	0
Regulating	Water regulation	0	0	0	6	5445
	Waste treatment	0	0	0	87	665
	Erosion control	0	0	0	245	0
	Climate regulation	0	0	0	223	0
	Biological control	24	0	0	0	0
	Gas regulation	0	0	0	0	0
	Disturbance regulation	0	0	0	5	0
Supporting	Nutrient cycling	0	0	0	922	0
	Pollination	14	0	0	0	0
	Soil formation	0	0	0	10	0
	Habitat	0	0	0	0	0
Cultural	Recreation	0	0	0	112	0
	Cultural	0	0	0	2	230

Source: Adapted from Costanza, R., et al., *Nature*, 1997.

Table 15.2 presents the type of land use identified in the study area, its corresponding biomes, and coefficient values. The equivalent value coefficients for ecosystem function are presented in Table 15.3. The total ESV for each LULC was calculated using Equation 15.1:

$$ESV = \sum (A_k \times VC_k) \quad \quad (15.1)$$

where ESV is the ecosystem service value, A_K is the area of respective LULC (ha), and VC_K is the value coefficient (US$ ha^{-1} year^{-1}) for LULC type "k."

Equation 15.2 was used to quantify the individual ecosystem function values within the study landscape:

$$ESVf = \sum (A_k \times VCf_k) \quad \quad (15.2)$$

where ESV_f is the value of each ecosystem service function, A_k is the area of respective LULC in ha, and VC_{fk} is the value coefficient of the respective ecosystem function (US\$ ha^{-1} year^{-1}) for each LULC type.

15.2.6 Sensitivity Factor

For each LULC class, a sensitivity factor was determined to evaluate the value of land use classes with respect to their contribution to ESV (Aschonitis et al., 2016). The sensitivity factor was estimated by Equation 15.3:

$$CS_i = \frac{C_{bi} \times A_i}{ESV_b} \quad\quad\quad\quad\quad\quad\quad (15.3)$$

where CS_i = estimated sensitivity factor for land use type i; ESV_b = total ecosystem service value (s); C_{bi} = value coefficient for land use type i; A_i = area (ha) of the land use type i.

15.3 RESULTS

15.3.1 Land Cover Classes in the Study Area

Figure 15.3 demonstrates the spatial distribution of the five land use practices in Beel Dakatia, Bangladesh. Agriculture constituted the highest portion (32.42%) of land cover in the study area

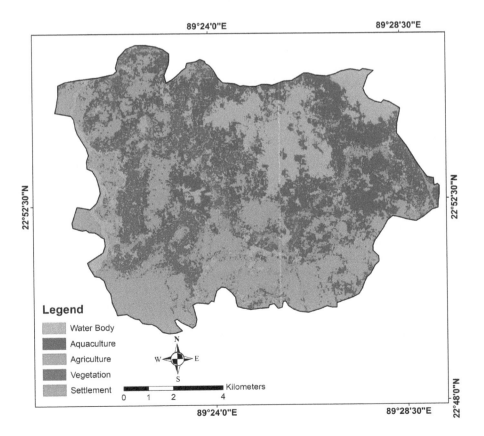

FIGURE 15.3 Land use land cover map of the study area for 2020.

TABLE 15.4
Results of the Areal Extent of LULCs and Their Ecosystem Services Value in the Study Area

LULC Classes	Area (ha)	% of Total	Ecosystem Service Coefficient (US$ ha^{-1} y^{-1})	Ecosystem Services Value (million US$)	Percentage Contribution (%)
Agriculture	5583.70	32.42	92	0.51	1.22
Aquaculture	4588.21	26.64	256	1.17	2.79
Settlement	935.21	5.43	0	0.00	0.00
Vegetation	1772.25	10.29	2007	3.56	8.44
Water body	4343.64	25.22	8498	36.91	87.55
Total	17223.01	100	10853	42.15	100

for 2020. Aquaculture and vegetation were the next most common types of land use, accounting for about 26.64% and 25.22%, respectively. Vegetation accounted for 10.29% of the total area. The settlement areas represented 5.43% and were the least common land use type in the Beel Dakatia region (Table 15.4). The accuracy of the land cover classification was measured using Google Earth images. The overall accuracy of the LULC map of 2020 was 85% with a kappa coefficient of 86%.

15.3.2 Estimation of Total ESV

Table 15.4 shows the areal extent of the land use type identified in the study area and the estimated ESV for the year 2020. The total ESV for 2020 was estimated at US$42.16 million (calculated using Equation 15.1) in the study region. Of the total, the water body contributed the largest percentage (87.56%) of the ESV. The other three, including agriculture, aquaculture, and vegetation, contributed roughly US$5.24 million, which is 12.45% of the gross ESV. However, due to the zero-value coefficient, settlement contributed no ESV.

15.3.3 Estimation of Ecosystem Function Value (ESV$_f$)

The ecosystem function values (calculated using Equation 15.2) and their rank based on total ESV are presented in Table 15.5. Of the 17 ecosystem functions, the most significant ecosystem services were water regulation, water supply, and waste treatment services. Their cumulative output contributed about 85.20% of the gross ESV. Among them, water regulation services incurred the highest value, accounting for 56.13% (US$23.66 million), followed by water supply (21.85%, US$9.21 million), waste treatment (7.21%, US$3.04 million), food production (4.06%, US$1.77 million), and nutrient cycling (3.87%, US$1.63 million) services. In contrast, climate regulation, soil formation, pollination, biological control, disturbance regulation, erosion control, genetic resources, and raw materials showed a substantially lower impact on total ESV, with a combined contribution of only 4.51%. Due to the zero coefficient value, gas regulation and habitat function of the study area did not contribute any value of services.

Of the four primary ecosystem services, regulating services contributed the highest amount to the total values (65.65%, US$27.67 million), followed by provisioning (27.40%, US$11.55 million), supporting (4.58%, US$1.73 million), and cultural services (2.85%, US$1.2 million) (Table 15.5).

In terms of the individual ESVs distributed in different LULCs, water body contributed the highest portion of the ESV (US$36.91 million) in the study area. It has a major effect on water regulation, waste treatment, water supply, and food production services. Conversely, with the highest land area, agricultural land use has shown the lowest contribution of all ecosystem functional services. On the other hand, aquaculture was influenced only by food production services. However, due

Valuing Ecosystem Services for Wetlands

TABLE 15.5
Estimated Results of the Individual Ecosystem Function Values

Ecosystem Function		ESV$_f$ (million US$)	%	Rank
Provisioning Services	Food production	1.71	4.06	4
	Water supply	9.21	21.85	2
	Raw material	0.56	1.33	7
	Genetic resources	0.07	0.17	12
Regulating Services	Water regulation	23.66	56.13	1
	Waste treatment	3.04	7.21	3
	Erosion control	0.43	1.02	8
	Climate regulation	0.4	0.95	9
	Biological control	0.13	0.31	11
	Gas regulation	0	0.00	16
	Disturbance regulation	0.01	0.02	15
Supporting Services	Nutrient cycling	1.63	3.87	5
	Pollination	0.08	0.19	13
	Soil formation	0.02	0.05	14
	Habitat	0	0.00	16
Cultural Services	Recreation	0.2	0.47	10

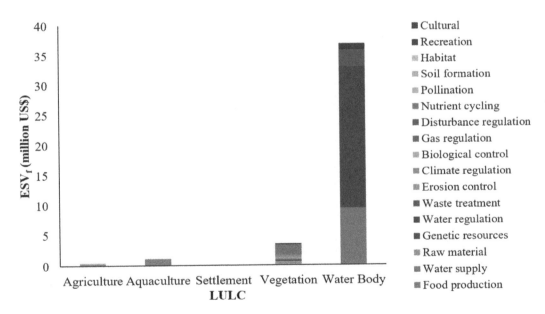

FIGURE 15.4 Distribution of ecosystem function value (ESV$_f$) for different land use land cover in the study area.

to the zero coefficient value, no ecosystem function has been affected by the land use comprising settlements in the study area (Figure 15.4).

Figure 15.5 shows the spatial pattern of the value of four primary ecosystem services in the study area. The value of provisioning service (PSV) was largely accrued in the northern part of the study region, attributable to water bodies and aquaculture. Of the four PSVs, the maximum ESV was produced by the water supply service of the area. The highest value has also been accumulated

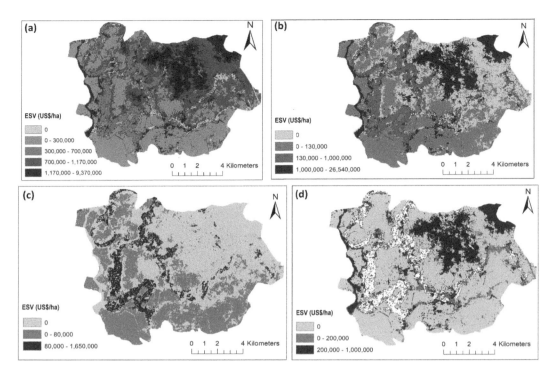

FIGURE 15.5 Spatial distribution of individual ecosystem function values, including (a) provisioning services, (b) regulating services, (c) supporting services, and (d) cultural services.

TABLE 15.6
Coefficient of the Sensitivity of ESVs of the Beel Dakatia, Bangladesh

LULC Class	Sensitivity Coefficient	Rank
Agriculture	0.01	4
Aquaculture	0.03	3
Settlement	0.00	5
Vegetation	0.08	2
Water body	0.88	1

in the water body in terms of regulating (RSV) and cultural services value (CSV). Water regulation and nutrient cycling were the highest contributors to RSV and CSV, respectively. The high value of supporting services was concentrated only in vegetation land use.

15.3.4 Significance of Land Use Types for ESV

Table 15.6 presents the sensitivity coefficient of ESVs (calculated using Equation 15.3) of different LULCs available in the study area. To determine the importance of each land use type, a ranking system was introduced based on their sensitivity factors. Water body ranked first in the study area and was identified as the most important land use type contributing to the total ESVs, followed by vegetation, aquaculture, agriculture, and settlement. This is attributed to the high coefficient value of water body. In sum, water body and vegetation are essential indicators for Beel Dakatia when considering total ESVs.

15.4 DISCUSSION

This study applied open source GIS and the value transfer method to determine the ESV of a freshwater wetland of southwest coastal Bangladesh. In addition to its economic importance, the study has identified and assessed the various land use practices of the studied wetland using standard remote sensing techniques. The ESV has been considered an important tool to inform decision-makers by emphasizing the economic benefits derived from sustainable management of natural ecosystems (Du et al., 2008; Hasan et al., 2020). Beel Dakatia is a freshwater floodplain wetland in southwest coastal Bangladesh that produced a total ESV of US$42.16 million for 2020 in this study. Similar findings were found in different ESV evaluation studies. For example, Huq et al. (2019) produced an economic value of $5.50 billion in the post-monsoon season of 2014 in Gopalgonj-Khulna Beel of southwest coastal Bangladesh. The Akkulam-Veli Wetland located in Kerala, India, generated an annual flow of US$ 575,852–680,352 for the year 2013, estimated using the benefits transfer approach (Ghermandi et al., 2016). Sun et al. (2018) estimated the ESV of coastal wetlands as $40,648 per ha per year by using a meta-analysis study on Liaoning province, China. These findings indicate that the ecosystem services of wetlands substantially contribute to the well-being of coastal communities annually. In preparing future management plans, we would suggest that the conservation of these ecosystems is a priority, taking into consideration the ecological, economic, and social importance of the coastal wetland identified in this study and the resources provided by such a wetland (Camacho-Valdez et al., 2013).

In this study, water body ranked as the third largest LULC in the area; however, it produced 87.56% of total ecosystem services. Since water body alone contributes a major portion of ESV, a subtle change in the area of that particular land use would have a substantial effect on the total ESV of this area. However, the water body area of this wetland has been increasingly occupied by modified land use types such as shrimp cultivation, which could alter the total ESV in the future. The local people have been constructing commercial shrimp aquaculture at the expense of open water fisheries, placing strong pressure on natural ecosystems (Kabir and Aftab, 2017). In addition, the faulty construction of polder and lack of flow increased siltation in the river bed, causing the river to dry up during the dry season, which resulted in the loss of aquatic ecosystem services (Ali and Syfullah, 2017). On the other hand, agriculture and aquaculture are the two dominant land use types in this area. Aquaculture land use demonstrated the highest value of food production services. However, despite low area coverage, vegetation contributes ecosystem services two times greater than the aggregate value of agriculture and aquaculture (see Table 15.4). This may be attributed to the higher ecological functions provided by the vegetation compared with the two other types. Furthermore, these two ecosystem types achieved a higher sensitivity factor value (see Table 15.6), which demonstrated their high importance in the total ESV. Therefore, the more natural the type of land cover, such as water body and vegetation, the higher the value of ecosystem services that can be produced.

The benefit transfer method has been extensively used in the literature and has proved to be more efficient in evaluating the ecosystem benefits in monetary units. However, many scholars have criticized this method for being used at coarse resolution, with uncertainties because of the complex, dynamic, and nonlinear nature of ecosystems (Turner et al., 2003; Hasan et al., 2020; Hou et al., 2020), which restricts the economic valuation and produces double-counting issues (Hasan et al., 2020). In this study, LULC class was used as a proxy for a biome that did not match perfectly with each class. For example, the land use class "aquaculture" was used as a proxy for the food production value of wetlands based on an estimate that could be further validated through more studies to explore the coefficient value of aquaculture. Furthermore, this study did not consider pollution (e.g., water pollution, waste pollution, and soil degradation), which may generate negative externalities and should be addressed in future study. The approach employed in this study estimated the ESV of each land use class by using an uncertain coefficient value, which may lead to greater uncertainty in estimating the total ESV. Furthermore, the ESV coefficient used in this study was supposed to remain constant over time; hence, the ESV provided by each LULC was assumed to be homogeneous throughout the study area. However, in reality, the value follows spatial heterogeneity (Zank et al., 2016). Therefore, combining high-resolution satellite images with field surveys can help to improve the reliability of the proxy-based methods (Hasan et al., 2020). The field survey provides fine-scale data that improve the accuracy of the LULC map, eliminates the biases in oversimplification of LULC estimation (Eigenbrod et al., 2010), and minimizes the uncertainty in the analysis (Hou et al., 2013).

15.5 CONCLUSION AND POLICY IMPLICATIONS FOR SUSTAINABLE CONSERVATION

The present study has revealed the importance of LULC for wetland ecosystem services supply. The application of open source software made the task more convenient. The SCP of QGIS is open source software that accelerated the land cover classification used in this study. The plugin integrates with QGIS data management architecture, which enables it to analyze the dynamic ecological models. The plugin has opened new opportunities for ecologists to deal with more complex models at various spatial scales.

The study suggests that a better understanding of ESV has great importance for promoting the sustainable use of wetland services. The results found that only 35.51% of the total wetland area was occupied by the natural ecosystem (water body and vegetation), which could be a threat to the existence of this wetland in its natural condition in the future. Therefore, proper land use planning should be adopted to allocate the wetland area in a sustainable manner. The study found that the net ESV of Beel Dakatia was US$42.16 million for 2020. More natural ecosystem types, such as water body together with vegetation, contributed almost 96% of the total ESV. The results reflect that these two types of land use have played a critical role in the supply of ecosystem services in this wetland. In terms of individual ecosystem services, water regulation contributed to the highest ESV, accompanied by water supply and waste treatment. Their accumulated contribution was about 85%, and all these services were provided by water body and vegetation LULCs. Therefore, the ESV analysis of this study emphasizes implementing sustainable conservation planning with special attention to the protection of water body and vegetation to reverse the unsustainable degradation of these ecosystem services. This could be achieved through planning design and conservation priority settings for local ecosystem services by incorporating spatial optimization and multi-criteria decision analysis. Beyond these findings, we recommend further research on the spatio-temporal change of wetlands ESV by integrating high-resolution remote sensing data and field surveys. Monitoring such changes to wetlands would allow policymakers to estimate the economic and environmental consequences of future development projects considered at the expense of the natural ecosystem.

REFERENCES

Ahmed, Irina, B. James Deaton, Rakhal Sarker, and Tasneem Virani. 2008. "Wetland Ownership and Management in a Common Property Resource Setting: A Case Study of Hakaluki Haor in Bangladesh." *Ecological Economics* 68(1–2): 429–436. doi: 10.1016/j.ecolecon.2008.04.016.

Akber, M. A., M. M. Patwary, M. A. Islam, and M. R. Rahman. 2018a. "Storm Protection Service of the Sundarbans Mangrove Forest, Bangladesh." *Natural Hazards* 94(1): 405–418. doi: 10.1007/s11069-018-3395-8.

Akber, Md Ali, Md Wahidur Rahman Khan, Md Atikul Islam, Md Munsur Rahman, and Mohammad Rezaur Rahman. 2018b. "Impact of Land Use Change on Ecosystem Services of Southwest Coastal Bangladesh." *Journal of Land Use Science* 13(3): 238–250. doi: 10.1080/1747423X.2018.1529832.

Ali, Shahjahan, and Khaled Syfullah. 2017. "Effect of Sea Level Rise Induced Permanent Inundation on the Livelihood of Polder Enclosed Beel Communities in Bangladesh: People's Perception." *Journal of Water and Climate Change* 8(2): 219–234. doi: 10.2166/wcc.2016.236.

Aschonitis, V. G., M. Gaglio, G. Castaldelli, E. A. Fano, V. G. Aschonitis, M. Gaglio, and G. Castaldelli. 2016. "Criticism on Elasticity-Sensitivity Coefficient for Assessing the Robustness and Sensitivity of Ecosystem Services Values View Project NOACQUA-Community Responses and Ecosystem Processes in Intermittent Streams View Project Criticism on Elasticity-Sensiti." *Ecosystem Services*, Elsevier 20(C): 66–68. doi: 10.1016/j.ecoser.2016.07.004.

Bai, Junhong, Baoshan Cui, Huicong Cao, Ainong Li, and Baiyu Zhang. 2013. "Wetland Degradation and Ecological Restoration." *The Scientific World Journal*. doi: 10.1155/2013/523632.

Barbier, E. B., M. Acreman, and D. Knowler. 1997. "Economic Valuation of Wetlands: A Guide for Policy Makers and Planners Valuation of Watershed Hydrological Services View Project A Restoration Synthesis View Project." IUCN Publications Services Unit, Switzerland, pp. 1–138, ISBN 2-940073-21-X.

Ben Zank, Kenneth J. Bagstad, Brian Voigt, and Ferdinando Villa. 2016. "Modeling the Effects of Urban Expansion on Natural Capital Stocks and Ecosystem Service Flows: A Case Study in the Puget Sound, Washington, USA." *Landscape and Urban Planning* 10: 523632. doi: 10.1016/j.landurbplan.2016.01.004.

Boyd, D. S., and G. M. Foody. 2011. "An Overview of Recent Remote Sensing and GIS Based Research in Ecological Informatics." *Ecological Informatics* 6(1): 25–36. doi: 10.1016/j.ecoinf.2010.07.007.

Camacho-Valdez, Vera, Arturo Ruiz-Luna, Andrea Ghermandi, and Paulo A. L. D. Nunes. 2013. "Valuation of Ecosystem Services Provided by Coastal Wetlands in Northwest Mexico." *Ocean and Coastal Management* 78 (June): 1–11. doi: 10.1016/j.ocecoaman.2013.02.017.

Chen, Daoyi, Shahriar Shams, César Carmona-Moreno, and Andrea Leone. 2010. "Assessment of Open Source GIS Software for Water Resources Management in Developing Countries." *Journal of Hydro-Environment Research* 4(3): 253–264. doi: 10.1016/j.jher.2010.04.017.

Congedo, Luca. 2017. "Semi-Automatic Classification Plugin Documentation Release 5.3.6.1 Luca Congedo." Available at: https://buildmedia.readthedocs.org/media/pdf/semiautomaticclassificationmanual-v3/latest/semiautomaticclassificationmanual-v3.pdf.

Coppin, P., I. Jonckheere, K. Nackaerts, B. Muys, and E. Lambin. 2004. "Digital Change Detection Methods in Ecosystem Monitoring: A Review." *International Journal of Remote Sensing*, 25(9): 1565–1596. doi: 10.1080/0143116031000101675.

Costanza, Robert, Ralph D'Arge, Rudolf De Groot, Stephen Farber, Monica Grasso, Bruce Hannon, Karin Limburg, et al. 1997. "The Value of the World's Ecosystem Services and Natural Capital." *Nature* 387: 253. doi: 10.1038/387253a0.

Costanza, Robert, Rudolf de Groot, Paul Sutton, Sander van der Ploeg, Sharolyn J. Anderson, Ida Kubiszewski, Stephen Farber, and R. Kerry Turner. 2014. "Changes in the Global Value of Ecosystem Services." *Global Environmental Change* 26: 152–158. doi:10.1016/j.gloenvcha.2014.04.002.

Du, Hui-shi, Zhi-ming Liu, and Nan Zeng. 2008. "Effects of Land Use Change on Ecosystem Services Value: A Case Study in the Western of Jilin Province." In *Geoinformatics 2008 and Joint Conference on GIS and Built Environment: Monitoring and Assessment of Natural Resources and Environments* 7145, 71451X.1-71451X.9; 2. doi: 10.1117/12.813051.

Eigenbrod, Felix, Paul R. Armsworth, Barbara J. Anderson, Andreas Heinemeyer, Simon Gillings, David B. Roy, Chris D. Thomas, and Kevin J. Gaston. 2010. "The Impact of Proxy-Based Methods on Mapping the Distribution of Ecosystem Services." *Journal of Applied Ecology* 47(2): 377–385. doi: 10.1111/j.1365-2664.2010.01777.x.

Finlayson, C. Max, Pierre Horwitz, and Philip Weinstein. 2015. *Wetlands and Human Health*. doi: 10.1007/978-94-017-9609-5.

Ghermandi, Andrea, Albert Moses Sheela, and Joseph Justus. 2016. "Integrating Similarity Analysis and Ecosystem Service Value Transfer: Results from a Tropical Coastal Wetland in India." *Ecosystem Services*. doi: 10.1016/j.ecoser.2016.09.014.

Hasan, Sarah, Wenzhong Shi, and Xiaolin Zhu. 2020. "Impact of Land Use Land Cover Changes on Ecosystem Service Value – A Case Study of Guangdong, Hong Kong, and Macao in South China." *PloS One* 8;15(4): e0231259. doi: 10.1371/journal.pone.0231259.

Hou, Lei, Faqi Wu, and Xinli Xie. 2020. "The Spatial Characteristics and Relationships between Landscape Pattern and Ecosystem Service Value along an Urban-Rural Gradient in Xi'an City, China." *Ecological Indicators* 108 (January): 105720. doi: 10.1016/j.ecolind.2019.105720.

Hou, Y., B. Burkhard, and F. Müller. 2013. "Uncertainties in Landscape Analysis and Ecosystem Service Assessment." *Journal of Environmental Management* 127: S117–S131. doi: 10.1016/j.jenvman.2012.12.002.

Huq, Nazmul, Antje Bruns, and Lars Ribbe. 2019. "Interactions between Freshwater Ecosystem Services and Land Cover Changes in Southern Bangladesh: A Perspective from Short-Term (Seasonal) and Long-Term (1973–2014) Scale." *Science of the Total Environment* 650 (February): 132–43. doi: 10.1016/j.scitotenv.2018.08.430.

Islam, Shafi Noor. 2010. "Threatened Wetlands and Ecologically Sensitive Ecosystems Management in Bangladesh." *Frontiers of Earth Science in China* 4 (4): 438–48. doi: 10.1007/s11707-010-0127-0.
Islam, Shafi Noor. 2016. "Deltaic Floodplains Development and Wetland Ecosystems Management in the Ganges–Brahmaputra–Meghna Rivers Delta in Bangladesh." *Sustainable Water Resources Management* 2(3): 237–56. doi: 10.1007/s40899-016-0047-6.
Islam, Shafi Noor, and A. Gnauck. 2007. "Effects of Salinity Intrusion in Mangrove Wetlands Ecosystems in the Sundarbans: An Alternative Approach for Sustainable Management." In *Proceedings of the International Conference W3M "Wetlands: Modelling, Monitoring, Management."* December 2007, pp. 244–248.
Kabir, Kazi Humayun, and Sharmin Aftab. 2017. "Exploring Management Strategies for Freshwater Wetland: Policy Options for Southwest Coastal Region in Bangladesh." *Asian Development Policy Review* 5(2):70–80. doi: 10.18488/journal.107.2017.52.70.80.
Khan, S. M, E. Haq, S. Huq, A. A. Rahman, S. M. A. Rashid, and H Ahmed. 1994. *Wetlands of Bangladesh.* Dhaka: Holiday Printers Limited, pp. 1–88.
Kreuter, Urs P., Heather G. Harris, Marty D. Matlock, and Ronald E. Lacey. 2001. "Change in Ecosystem Service Values in the San Antonio Area, Texas." *Ecological Economics* 39(3): 333–346. doi: 10.1016/S0921-8009(01)00250-6.
Leon, L. F., and C. George. 2008. "WaterBase: SWAT in an Open Source GIS." *The Open Hydrology Journal* 2: 1–6. doi: 10.2174/1874378100802010001.
Luque, Sandra, and Grazia Zulian. 2017. "Mapping Cultural Ecosystem Services." In (eds.) B. Burkhard, J. Maes, *Mapping Ecosystm Services*. Pensoft Publishers. doi: 10.3897/ab.e12837.
Mitra, Sudip, Reiner Wassmann, and Paul L G Vlek. 2003. "Global Inventory of Wetlands and Their Role in the Carbon Cycle." *ZEF Discussion Paper on Development Policy*, Bonn.
Msofe, Nangware Kajia, Lianxi Sheng, Zhenxin Li, and James Lyimo. 2020. "Impact of Land Use/Cover Change on Ecosystem Service Values in the Kilombero Valley Floodplain, Southeastern Tanzania." *Forests* 11 (1): 109. doi: 10.3390/f11010109.
Pandeya, B., W. Buytaert, Z. Zulkafli, T. Karpouzoglou, F. Mao, and D. M. Hannah. 2016. "A Comparative Analysis of Ecosystem Services Valuation Approaches for Application at the Local Scale and in Data Scarce Regions." *Ecosystem Services* 22(B): 250–259. doi: 10.1016/j.ecoser.2016.10.015.
Parvin, Gulsan Ara, Md. Hashan Ali, Kumiko Fujita, Md. Anwarul Abedin, Umma Habiba, and Rajib Shaw. 2017. "Land Use Change in Southwestern Coastal Bangladesh: Consequence to Food and Water Supply." In (eds.) M. Banda and R. Shaw, *Land Use Management in Disaster Risk Reduction*. Springer. doi: 10.1007/978-4-431-56442-3_20.
QGIS Development Team. 2013. "'QGIS Geographic Information System. Open Source Geospatial Foundation Project.'" 2013. https://qgis.org/en/site/.
Rahman, A. 1995. *Beel Dakatia: The Environmental Consequences of a Development Disaster*. Dhaka, Bangladesh: University Press.
Rebelo, L. M., M. P. McCartney, and C. M. Finlayson. 2010. "Wetlands of Sub-Saharan Africa: Distribution and Contribution of Agriculture to Livelihoods." *Wetlands Ecology and Management* 18(5): 557–572. doi: 10.1007/s11273-009-9142-x.
Schuyt, K., and Brander, L. 2004. Living Waters: The economic values of the world's Wetlands. *Environmental Studies* (p. 32). Amsterdam: WWF.
Siew, Mei Kuang, Mohd Rusli Yacob, Alias Radam, Abdullahi Adamu, and Emmy Farha Alias. 2015. "Estimating Willingness to Pay for Wetland Conservation: A Contingent Valuation Study of Paya Indah Wetland, Selangor Malaysia." *Procedia Environmental Sciences* 30: 268–272. doi: 10.1016/j.proenv.2015.10.048.
Steiniger, S., and G. J. Hay. 2009. "Free and Open Source Geographic Information Tools for Landscape Ecology." *Ecological Informatics* 4(4): 183–195. doi: 10.1016/j.ecoinf.2009.07.004.
Sun, Baodi, Lijuan Cui, Wei Li, Xiaoming Kang, Xu Pan, and Yinru Lei. 2018. "A Meta-Analysis of Coastal Wetland Ecosystem Services in Liaoning Province, China." *Estuarine, Coastal and Shelf Science* 200(January): 349–58. doi: 10.1016/j.ecss.2017.11.006.
Turner, R. Kerry, Jouni Paavola, Philip Cooper, Stephen Farber, Valma Jessamy, and Stavros Georgiou. 2003. "Valuing Nature: Lessons Learned and Future Research Directions." *Ecological Economics* 46(3): 493–510. doi: 10.1016/S0921-8009(03)00189-7.
Zhang, Fei, Ayinuer Yushanjiang, and Yunqing Jing. 2019. "Assessing and Predicting Changes of the Ecosystem Service Values Based on Land Use/Cover Change in Ebinur Lake Wetland National Nature Reserve, Xinjiang, China." *Science of the Total Environment* 656(March): 1133–44. doi: 10.1016/j.scitotenv.2018.11.444.
Zhao, Bin, Bo Li, Yang Zhong, Nobukazu Nakagoshi, and Jia Kuan Chen. 2005. "Estimation of Ecological Service Values of Wetlands in Shanghai, China." *Chinese Geographical Science* 15(2): 151–156. doi: 10.1007/s11769-005-0008-8.

16 Identification of Groundwater Prospect of Bilaspur District
A Multi-Criteria Decision Making Approach

Soumen Bramha, Gouri Sankar Bhunia, and S. R. Kamlesh

CONTENTS

16.1 Introduction 225
16.2 Study Area 227
16.3 Methods 227
    16.3.1 Pre-Field and Field Work Activities 228
    16.3.2 Post-Field Activities 228
    16.3.3 Rainfall Estimation 229
    16.3.4 Drainage Density 230
    16.3.5 Average Groundwater Level Depth 230
    16.3.6 Integration of Thematic Layers 230
    16.3.7 Analytic Hierarchy Process (AHP) 230
    16.3.8 Multi-Criteria Decision-Making Analysis to Find Groundwater Potential Zone 231
16.4 Results and Discussion 231
    16.4.1 Lithology 231
    16.4.2 Soil 231
    16.4.3 Drainage Density 232
    16.4.4 Rainfall 232
    16.4.5 Average Pre-Monsoon and Post-Monsoon Groundwater Depth 233
    16.4.6 Elevation 233
    16.4.7 Slope 235
    16.4.8 Land Use/Land Cover 235
    16.4.9 Vegetation Vigor 238
16.5 Conclusion 239
References 240

16.1 INTRODUCTION

In socio-ecological systems, groundwater is a vital source for drinking, agriculture, irrigation of horticulture, industry, wetland management, etc. Over the past few decades, groundwater levels have deteriorated significantly due to the population explosion, climate change, and land use changes (Mukherjee, 2016). Yeserterner (2008) confirmed that decreased rainfall is a leading cause of the depletion of groundwater, as it is a primary source of stream drainage and aquifer recharge. In addition, rapid urbanization affects the subsurface flow regime and the quality of groundwater by decreasing infiltration and recharge, decreasing evapotranspiration, potentially increasing groundwater extraction by industrial and commercial operations that do not inherently need high-quality water, and often through households tapping shallow aquifers for irrigation (Rinaudo et al., 2015). Furthermore, urbanization influences groundwater quality because economic, industrial, and retail expansion brings possible new pollution sources due to accidental spillages or long-term escape of chemical products, which are often contaminated with other sources of adulteration. Polluted soil harms land biodiversity and endangers health. Small-size trades such as tanneries, printers, laundries, and metal dispensers can largely operate in isolation and produce fluid seepages, for example, spent sanitizers, diluents, and emollients, that contaminate the adjacent soil. There can also be visible leakages from wastewater lagoons and sanitary sewer systems. Storm water, along with pathogenic bacteria and viruses, can carry notable loads from impermeable surfaces. In places where sanitary treatment is inadequate (e.g., cesspits, latrines, and septic tanks), or due to aging substructures where treatment methods are highly susceptible, pathogenic water quality problems may result (Hunt et al., 2010).

The sustainable management of groundwater resources has all the geographies of "wicked" or "messy" problems (Global Sustainable Development Report, 2019), identified by many characteristics: there is no comprehensive interpretation of the problem; the solution is not real or wrong, but either better or worse; stakeholders have various challenges based on local issues; restrictions and constraints. The UN Environmental Programme (UNEP, 2008) stated that stores of groundwater comprise over 90% of readily accessible freshwater on Earth. Though groundwater has been obscure and underappreciated, even if documented, groundwater dilapidation is reaching the threshold of concern. With advancements in exploration and pumping, the use of groundwater in several places around the world exceeds the use of surface water. Characterizing groundwater systems is difficult because of their hidden nature with the growing world population and increased agricultural productivity and industrial and economic growth. Groundwater is largely used for farming activities, specifically irrigation in most countries, except in the humid inter-tropical region (Zektser and Everett, 2004). The diminishing levels of groundwater also contribute to decreased quality of groundwater, particularly saltwater contamination in some coastal and estuarine regions. With the experience of acid sulfate soils, sunken groundwater levels have also supported the acidification of many wetlands in the region.

Integrated groundwater management (IGM) is a planned procedure that endorses the coordinated management of groundwater and associated properties so as to accomplish stable economic, social, wellbeing, and ecosystem consequences over space and time. Integrated assessment (IA) of water system is commonly used to explore organizational decision- and policy-making in the form of well-known benchmarks. IA was created for acid rain, climate change, soil erosion, sustainability of water, and air quality, conservation of forests and fisheries. and environmental health. By considering water resource management, Jakeman and Letcher (2003) summarize the key features and principles of IA as follows:

- A problem-focused operation needs guided and possibly project-based intervention.
- A collaborative, transparent context; improving communication.
- A procedure supplemented by stakeholder participation and enthusiastic espousal.
- Combination of research and strategy.
- Intricate association between natural and human environments.
- Awareness of spatial dependencies, inputs, and challenges.
- An iterative, adaptive method.
- An emphasis on key aspects.
- Acknowledgment of indispensable missing information for insertion.
- Team-shared properties, norms, and values, remedial equilibrium.
- Documentation, characterization, and decrease of significant uncertainties in forecasts.

Earlier authors have used the combination of remote sensing and geographical information system (GIS) to demarcate the groundwater potential zone (Choudhari et al., 2018; Deepa et al., 2016).

Moreover, Saaty's Analytical Hierarchy Process (AHP) uses the Multi-Criteria Decision Analysis (MCDA) technique to evaluate the groundwater potentiality and water resource engineering in various regions (Saha, 2017; Das et al., 2018). The analytical network process (ANP)-based system is a grid that represents relative measurements of the influence of elements that interact with respect to control criteria (Dagdeviren and Ihsan, 2007). A more common framework of the analytical hierarchy process (AHP) used in multi-criteria decision analysis is the analytic network process (ANP). ANP is more precise and easier than making decisions or AHP. Inside a set of elements, and among various sets of elements, it deals with dependency. As is always expected in the real world, the ANP prioritizes not only elements, but also classes or clusters of elements. Currently, several research works have described that MCDA offers an active tool for water resource management by totaling structure, auditability, transparency, and correctness of decisions (Mallick et al., 2014; Pinto et al., 2017). ANP and AHP are based on consumer expectations among the frameworks and include alternatives with utility weights, differing in types and quantities of pair comparisons and also in the way in which the utility weights are actually measured (Cheng and Li, 2004). With the uncertainty and dynamics of this changing environment, ideas cannot be simply recognized, as they are associated with other essential aspects that evidently also cannot be acknowledged. Therefore, with uncertainty and complexities, ANP is more efficient than AHP in the Bayesian framework (Tran et al., 2004). Hence, the present work employed AHP-coupled MCDA, remote sensing, and GIS techniques to integrate physical, climatic, hydrogeological, geomorphic, and land use characteristics in order to evaluate the groundwater potential zone and to put together a forthcoming guide map for groundwater exploration in an attempt to confirm the ideal and sustainable management and development of this living resource.

16.2 STUDY AREA

Bilaspur district is located in the northwest part of the Chattishgarh state, extended between 21°42′53.549″ and 22°36′20.889″ N latitude and between 81°43′6.798″ and 82°29′44.945″ E longitude (Figure 16.1). It covers an area of 3557.9 km². It is surrounded by Durg and Raipur districts to the south, Kawardha and Mantdla districts to the west, Koriya district to the north, and Korba and Champa districts to the east. The district is well connected by road (National highway no. 200) and railway and is 120 km away from the state capital, Raipur. The Seonath river flows from west to east and is located in the south of the district. The northern part of the district is characterized by a dendritic pattern and the southern part by a trellis drainage pattern. The drainage density radically lessens in the plains, signifying the earlier nature of the underlying formations (shale, limestone, and dolomite) than the developments in the northern part of the district (granites, gneisses, schists, and quartzites). The southern part of the district is a plain land with gentle slopes, also called Chattishgarh plain. The land is very fertile and is mostly used for agricultural purposes with a few irrigational facilities. The average annual rainfall of the district is around 1082 mm. July and August are the heaviest rainfall months. The annual temperature varies between 10 °C and 45 °C, and the hottest months are May and June.

16.3 METHODS

To delineate the groundwater potential zone, unified RS-GIS and AHP techniques were used. In a developing country, where satisfactory and good-quality data are deficient, this technique is very beneficial for its cost-effectiveness and user-friendliness. The thematic layers of lithology, geomorphology, elevation, slope, soil, rainfall, pre-monsoon and post-monsoon water level, land use/land cover (LULC), and vegetation characteristics were used for the demarcation of groundwater prospective zones (GPZs). AHP is very useful for multi-parametric evaluation and is most commonly used for environmental resources management under the multi-criteria decision approach (Malczewski and Rinner, 2015). In this technique, themes are identified with their rank and priorities, which supports assembling the criteria in hierarchical order through a pair-wise comparison

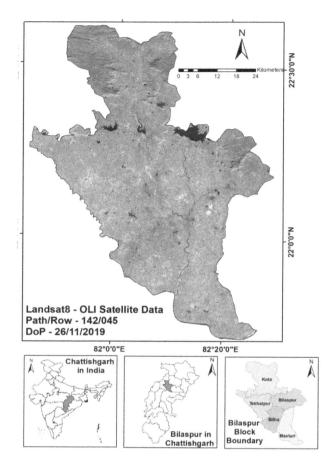

FIGURE 16.1 Location map of Bilaspur district in Chhattisgarh (India).

matrix. Consistency index (CI) and consistency ratio (CR) were computed based on the Saaty analytical process (Saaty, 1980).

16.3.1 PRE-FIELD AND FIELD WORK ACTIVITIES

The methodology designed in this study is comprised of three major stages: pre-field work, field work, and post-field work. In the pre-field work, a review of earlier research works, reconnaissance of the survey area and analogous research work on this topic were conducted. Suitable data were assembled, giving specific attention to the superiority of documents. Rainfall data were collected from the Indian Meteorological Department (IMD) station at various time periods (2010–2019). Pre- and post-field water depth data were collected from the Groundwater booklet. The Landsat8 Operational Land Imager (OLI) satellite image (Path/Row – 142/045; Date of Pass (DoP) – 26/11/2019) was obtained from the United States Geological Survey (USGS) Earth Explorer Community, composed of 11 spectral bands to delineate the landscape variables. ALOS digital elevation model (DEM) data with 30 m spatial resolution were used to identify the topographical variables and drainage characteristics of the study area.

16.3.2 POST-FIELD ACTIVITIES

Data collection evaluation of primary and secondary data derived from pre-field operations and field work is the core task of post-field work. Some of the important tasks are georeferencing of base map and satellite data using Universal Transverse Mercator (UTM) projection with World Geodetic Surface (WGS) 84 datum, north 44 zone, creation of the vector layer of the study area boundary, demarcation of the area of interest (AOI), extracting DEM, and subsetting the satellite image using the study area boundary. The detailed base maps of lithology, soil, drainage density, LULC, vegetation cover, elevation, slope, rainfall, and average groundwater depth were considered to determine the groundwater prospects for the study region. The thematic maps of lithology and soil were collected from published literature and websites. After that, polygon layers of lithology and soils were prepared in the GIS environment. The slope analysis was carried out through QGIS (version 2.4.0) 3-D analyst extension. For each cell, the slope tool computes the extreme rate of modification in value from that cell to its neighbor. The output slope raster can be calculated in degrees (Burrough and McDonell, 1998) using the following equation:

$$\text{Slope} = \tan\theta \sqrt{\left(\left[\frac{dz}{dx}\right]^2 + \left[\frac{dz}{dy}\right]^2\right)} \times 57.29578$$

A supervised image classification method with maximum likelihood algorithm is used to regulate the LULC characteristics (Akyürek et al., 2018) based on Bayes theorem. It uses the statistics of each class in each band and likelihood of specified pixel reliant upon LULC class. The normalized difference vegetation index (NDVI) is normally used in the evaluation of vegetation vigor and canopy coverage. The basic principle of NDVI is that the rates of reflection fluctuate for the near-infrared (band$_4$) band (700–1100 nm) and the red (band$_3$) band (600–700 nm) of an image, so these variances can yield an image that exemplifies the prominence of green plants and is least influenced by topography. The following equation is used to calculate NDVI:

$$\text{NDVI} = \frac{\text{NIR} - \text{RED}}{\text{NIR} + \text{RED}}$$

In general, the NDVI value varies from -1 to +1, but in reality, extreme negative values suggest water content. Values near zero mean bare earth, and the reference value of 0.5 reflects dense vegetation cover.

16.3.3 RAINFALL ESTIMATION

Rainfall data were collected from the IMD station of Chattishgarh State for the period between 2005 and 2015. The assessment of precipitation is a point observation and cannot be used for the region under discussion as a representative value. To obtain more consistent and descriptive results for the uniform depth of rainfall in the district, a deterministic interpolation technique is a prerequisite. Arithmetic mean and isohyetal polygon interpolation methods are considered to estimate the areal depth of rainfall, as they take into account the impact of physiographic factors, including slope, elevation, and proximity to the coast. Therefore, to calculate the mean monthly rainfall of the district, data from meteorological stations have been used to give the arithmetic mean:

$$P = \frac{P_1 + P_2 + P_3 \dots P_n}{N}$$

where P is the average intensity of rainfall of an area, P_1, P_2, P_3, and P_n are the rainfall records at stations 1, 2, 3 …, and N is the number of meteorological stations. These data are then geographically introduced using the inverse distance weighted (IDW) technique, which unites the hypotheses of immediacy to follow Thessian polygons with regular variations of the trend surface (Pinto et al., 2017).

16.3.4 DRAINAGE DENSITY

The drainage density (D_d) is the opposite function to porousness. It can be measured for the length of a stream channel by allocating the distance of the stream into an area analysis, which will create D_d value. A high value of D_d is favorable for runoff and consequently reveals a low groundwater potential zone.

16.3.5 AVERAGE GROUNDWATER LEVEL DEPTH

Choropleth maps for the groundwater level for pre-monsoon (May) and post-monsoon (November) have been prepared using data obtained from the Central Ground Water Board (CGWB) and Bilaspur. In addition, an IDW interpolation technique was applied to prepare the groundwater depth map. The mean error and root mean square error (RMSE) are considered as lower than 0.5 and 1.0, respectively.

16.3.6 INTEGRATION OF THEMATIC LAYERS

Maps of the nine thematic layers, such as annual average water level (e.g., pre-monsoon water level and post-monsoon water level), rainfall lithology, soil, drainage, slope, topography, land use, and vegetation coverage are transformed into the raster format using the spatial analyst tool of QGIS software v2.4.0 and are then considered for overlay analysis. The weighted overlay analysis (WOA) approach uses spatial analytical methods to generate a groundwater prospective region. For each individual parameter of each thematic layer map, weights are allocated to the unique function in the hydrogeological environment of the area, according to the performance of the multi-criteria decision-making (MCDM) (AHP) technique in WOA. On a scale of 1–9, the weights of the themes and their attributes are allocated depending on their impacts. In order to classify the groundwater potential index, all the themes and their sub-classes with normalized weights are inserted into the weighted overlay model in the QGIS spatial analyst program.

16.3.7 ANALYTIC HIERARCHY PROCESS (AHP)

In setting the objectives, AHP is the most regularly used multi-criteria approach due to its flexibility, as it balances the qualitative and quantitative decision characteristics. Formulating the decision problem in a system of tiered structure is the beginning of AHP, and it is an effective way to establish multifaceted systems. The nine points of the Saaty values of the parameters are applied to each map according to the importance of their effect on the adequate occurrence of groundwater (Saaty, 1980). The Saaty 9 point values are taken into account from group interviews, 15 recently published international research papers, and groundwater experts and officials from CGWB. The relative significance values are allocated using Saaty's 1–9 scale, where a score of "1" denotes equal impact between the two thematic maps, and a score of "9" designates the extreme impact of one thematic map relative to the other. It is competent both anatomically for representing a scheme and functionally for governing and hierarchal information systems.

The AHP determines the uncertainty in decisions based on eigenvalue and the CI (Saaty, 2004). CI is used to measure the degree of consistency using the following equation:

$$CI = \frac{\lambda_{max} - n}{n-1} = CI = \frac{9-9}{9-1} = 0$$

where λ_{max} is the maximum eigenvalue of the pair-wise comparison matrix, and n is the number of factors used in this analysis. The CR is calculated based on the following equation to govern the consistency analysis and scale decision, as follows:

$$CR = \frac{CI}{RCI}$$

where RCI is the random consistency index. The value of RCI for distinct values is defined and is equal to 1.45 ($n = 9$). If the CR value is less than or similar to 0.1, the accuracy is satisfactory, or if the CR value is equal to 0.00, the pair-wise comparison matrix assessment is completely compatible. If the CR is greater than 0.1, the comparison matrix must be recalculated to accurately evaluate the decision importance in relation to the dominant element affecting groundwater occurrence in the overall thematic layer index.

16.3.8 Multi-Criteria Decision-Making Analysis to Find Groundwater Potential Zone

To measure the relative values of the themes, the interaction between the nine thematic layers was generated using MCDM and is computed as well:

$$\text{MCDM} = \sum_{i=1}^{n}\sum_{j=1}^{m}\left[\alpha_i(\beta_{i,j} x_{i,j})\right]$$

where $\beta_{i,j}$ = weight of the jth class of ith theme obtained by AHP
α_i = weight of the ith theme obtained by ANP
n = total number of thematic layers
m = total number of classes in a thematic layer
$x_{i,j}$ = the pixel value of the jth class of the ith theme

The probable groundwater map is substantiated by well yield data from the sample area. Groundwater scenarios of the several regions are equated with the yield data.

16.4 RESULTS AND DISCUSSION

16.4.1 Lithology

This is another highly significant aspect controlling the appearance of groundwater abundance and reliability (Bhunia et al., 2012). As stated by earlier researchers (Ayazi et al., 2010; Hussein et al., 2017), lithology stimulates both the porosity and the permeability of aquifer rocks. In evaluating and regulating groundwater, each of the lithological units has equal importance. Lithologically, the study area is classified into various sub-groups, i.e., bastar gneisses, Chandi limestone-stromatolitic limestone, gunderdehi formation-shale, hirri formation-dolomite, maniyari formation-gypsiferous shale, pandaria formation-cavernous limestone and shale, tarenga formation-cherty shale and dolomite, and unclassified metamorphics BRS – Granite and gneiss (Figure 16.2a). The most apposite probable groundwater zones observed in the lithology class are illustrated in Figure 16.2a. Geologically, the study area is characterized by Chandi limestone-stromatolitic limestone, maniyari formation-gypsiferous shale, and gunderdehi formation-shale due to its effective absorption and water recharge capability.

The result is then put into sequence as follows: Chandi limestone-stromatolitic limestone > maniyari formation-gypsiferous shale > gunderdehi formation-shale > pandaria formation-cavernous limestone and shale > hirri formation-dolomite > tarenga formation-cherty shale and dolomite > bastar gneisses> unclassified metamorphics BRS – Granite and gneiss, respectively.

16.4.2 Soil

The physiognomies, categories, and dissemination of soil for a certain area are reliant upon geomorphology, relief, lithology, time, and other features. Soil characteristics determine the association

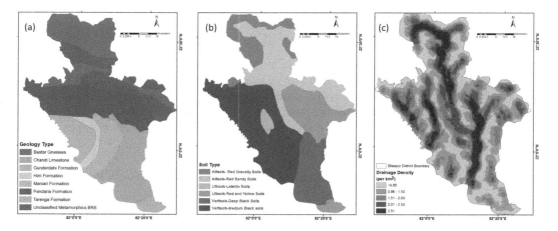

FIGURE 16.2 Thematic map of lithology (a), soil (b), and drainage density (c) of Bilaspur district.

between runoff and infiltration rate, which in turn, normalizes the degree of penetrability, which regulates the groundwater suitability (Das et al., 2019). Soil texture is a medium that regulates the groundwater susceptibility, which is imperative in defining the inherent susceptibility. Soil cover in the study area is very penetrable and is combined of reddish and sandy materials gathered from the weathering of granite and gneiss. The soils in the study area are categorized into (i) Alfisols – Red gravelly soils, (ii) Alfisols – Red sandy soil, (iii) Ultisols – Laterite soils, (iv) Ultisols – Red and yellow soils, (v) Vertisols – Deep black soils, and (vi) Vertisols – Medium black soils (Figure 16.2b). The results showed that red and yellow soil and deep black soils are found to be the most important in terms of area coverage. Depending on their infiltration rate, the distribution of soils has been determined. Sandy soil has a high hydraulic conductivity, and thus, greater priority has been assigned, whereas clayey soil has a lower infiltration rate, and low priority has therefore been allocated.

16.4.3 Drainage Density

The drainage density of the study area varies between 0.43 and 3.91 per km². The very high drainage density areas are observed in the northern, central, and small pockets of the western study area. On the other hand, an extremely low drainage level is located in the district, while the remaining areas, within the moderate and low drainage densities, are clustered in the south, east, and northeast of the study area (Figure 16.2c). Drainage density is an analogous permeation feature and is therefore a critical aspect for assessing the groundwater zone. High drainage density is conducive to runoff and thus implies a potential region of low groundwater prospect. The high values are allocated to the low drainage density zone, and vice versa. The coarse drainage texture reveals rock formations that are extremely malleable and ductile, while the fine drainage texture is much more prevalent in less fractured rocks (Waikar and Nilawar, 2014).

16.4.4 Rainfall

Rainfall measurement is a point observation and may not be considered as a illustrative value for the study area. It regulates the amount of water that would be available to infiltrate into the groundwater system. Hence, it is essential to find the operative uniform depth of rainfall of the investigated area to obtain more consistent and illustrative outcomes. The areal depth of rainfall in the catchment was assessed using the simple arithmetic mean. The average annual rainfall obtained from isohyetal interpolation is considered for this purpose. However, the rainfall is the main source of recharge. The rainfall map is grouped into five classes: <1000, 1001–1100, 1101–1200, 1201–1300,

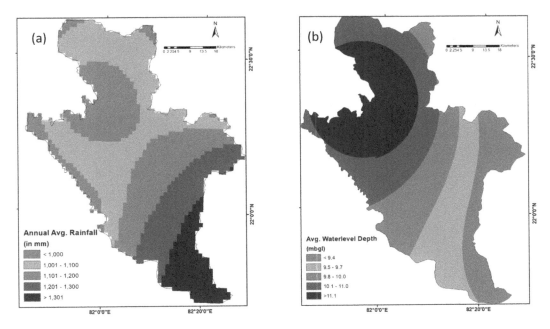

FIGURE 16.3 Spatial distribution of average annual rainfall (a) and average depth of water level (b) in Bilaspur district.

and >1301 mm/year (Figure 16.3a). It is observed that the southern and southeastern parts of the district receive the largest amount of rainfall, while the northern and extreme western parts of the district receive the lowest amount of rainfall. This small discrepancy of rainfall has little impact on the accessibility and occurrence of groundwater, as most of the water is lost as surface runoff because of the hard rock terrain. However, high rainfall is suitable for high groundwater probability, and is hereafter given higher precedence. It also regulates the amount of water that would be available to infiltrate into the groundwater system (Agarwal et al., 2013). In the natural water cycle, rainfall plays an important role in regulating groundwater suitability. Thereafter, during the study of groundwater prospective research, the areas with heavy rainfall are considered to have more weighted importance relative to areas with less annual rainfall (Table 16.1).

16.4.5 Average Pre-Monsoon and Post-Monsoon Groundwater Depth

Groundwater level data were collected during the pre-monsoon (May) and post-monsoon (November) seasons from CGWB, and the average water level depth was calculated. In the pre-monsoon season, the average groundwater depth generally varies between 6.285 and 13.155 meters below ground level (mbgl) (mean ± standard deviation 10.29 ± 2.20). In the post-monsoon season, the average groundwater level ranges from 3.73 to 9.95 mbgl, with a mean ± standard deviation 6.45 ± 2.18 (Figure 16.3b). In the northeast of the study area, having a high propensity for groundwater, and the south and southeast of the study site, the thematic maps of groundwater were classified into five groups, and the results showed that post-monsoon groundwater depth data are more appropriate for this investigation.

16.4.6 Elevation

Using QGIS software, the elevation map of the research area was prepared from ALOS DEM data with a spatial resolution of 30 m. The elevation map is categorized into six distinct groups as

TABLE 16.1
Pair-Wise Comparison Matrix for AHP

Thematic Map	Lithology	Rainfall	Drainage Density	Soil	Slope	Average Groundwater Depth	Vegetation Vigor	Elevation	LULC
Lithology	1.00	1.09	1.20	1.50	1.33	0.50	1.71	3.00	2.40
Rainfall	0.92	1.00	1.10	1.38	1.22	0.55	1.57	2.75	2.20
Drainage Density	0.83	0.91	1.00	1.25	1.11	0.60	1.43	2.50	2.00
Soil	0.67	0.73	0.80	1.00	0.89	0.75	1.14	2.00	1.60
Slope	0.75	0.82	0.90	0.89	1.00	0.67	1.29	2.25	1.80
Average Groundwater Depth	0.50	0.55	0.60	0.75	0.67	1.00	0.86	1.50	1.20
Vegetation Vigor	0.58	0.64	0.70	0.88	0.78	0.86	1.00	1.75	1.40
Elevation	0.33	0.36	0.40	0.50	0.44	1.50	0.57	1.00	0.80
LULC	0.42	0.45	0.50	0.63	0.56	1.20	0.71	0.80	1.00

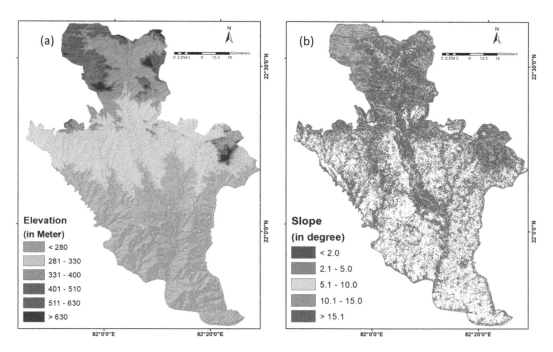

FIGURE 16.4 Spatial variation of topography (a) and slope (b) of Bilaspur district.

sub-criteria depending on the type and characteristics of the terrain: (i) <280 m, (ii) 281–330 m, (iii) 331–400m, (iv) 401–510 m, (v) 511–630 m, and (vi) >630 m (Figure 16.4a). In the elevation map, low elevated area is considered to be "very good" for groundwater potentiality, and vice versa (Table 16.2).

16.4.7 Slope

Slope is one of the most vital aspects governing groundwater occurrence. Flat regions are able to preserve moisture and allow groundwater to be recharged when associated with a steep slope environment where water passes easily downwards (Sisay, 2007). The slope of the district varies between 0° and 85°. Based on the slope, the study area is classified into five categories: (i) <2°, (ii) 2.1°–5.0°, (iii) 5.1°–10.0°, (iv) 10.1°–15.0°, and (v) 15.1° (Figure 16.4b). The slope becomes less and less steep towards the central and southern parts of the district. The slopes with a flat (<2°) and gentle areas (2°–5°) are more susceptible to the presence of groundwater in comparison to steep slopes. Gentle and flat slopes allow low runoff and are very good for groundwater recharge. Precipitous landscapes trigger fast runoff and do not accumulate water. The topographic situation provides an understanding of the overall path of groundwater movement and its influence on groundwater recharge and discharge (Tesfaye, 2010). Since, there is an inverse relationship between steep slope and groundwater potentiality.

16.4.8 Land Use/Land Cover

LULC plays a substantial role in the existence and determination of groundwater prospects. A supervised image classification technique was adopted to identify the types of LULC characteristics, and 13 classes were demarcated, namely, dense forest, degraded forest, open forest, mixed forest, industrial area, built-up area, crop land, agricultural fallow, barren land, dry fallow, sand,

TABLE 16.2
Assigned and Normalized Weights of the Different Classes of Each Theme

Thematic Layer	Sub-Category	Assigned Weight	Normalized Weight
Geology	Bastar Gneisses	2	0.06
	Chandi Limestone	8	0.22
	Gunderdehi Formation	6	0.17
	Hirri Formation	4	0.11
	Maniyari Formation	7	0.19
	Pandaria Formation	5	0.14
	Tarenga Formation	3	0.08
	Unclassified Metamorphic BRS	1	0.03
Soil	Alfisols – Red Gravelly soils	3	0.14
	Alfisols – Red Sandy soils	4	0.18
	Ultisols – Laterite soils	2	0.09
	Ultisols - Red and yellow soils	1	0.05
	Vertisols – Deep black soils	5	0.23
	Vertisols – Medium black soils	7	0.32
Drainage Density (per km^2)	<0.85	2	0.09
	0.86–1.50	3	0.14
	1.51–2.00	4	0.18
	2.01–2.50	6	0.27
	>2.51	7	0.32
Annual Average Rainfall (mm)	<1000	2	0.10
	1001–1100	3	0.15
	1101–1200	4	0.20
	1201–1300	5	0.25
	>1301	6	0.30
Average Water Level Depth (mbgl)	<9.4	6	0.30
	9.5–9.7	5	0.25
	9.8–10.0	4	0.20
	10.1–11.0	3	0.15
	>11.1	2	0.10
Elevation (m)	<280	6	0.29
	281–330	5	0.24
	331–400	4	0.19
	401–510	3	0.14
	511–630	2	0.10
	>631	1	0.05
Slope (°)	<2.0	5	0.33
	2.1–5.0	4	0.27
	5.1–10.0	3	0.20
	10.1–15.0	2	0.13
	>15.1	1	0.07
Land Use/Land Cover Class	Industrial area	1	0.02
	Open forest	4	0.08
	Mixed forest	7	0.14
	Dense forest	7	0.14
	Sand	3	0.06
	Built-up area	1	0.02
	Degraded forest	5	0.10

(Continued)

TABLE 16.2 (CONTINUED)
Assigned and Normalized Weights of the Different Classes of Each Theme

Thematic Layer	Sub-Category	Assigned Weight	Normalized Weight
	Surface water body	6	0.12
	Crop land	5	0.10
	Barren land	1	0.02
	Dry fallow	2	0.04
	Agricultural fallow	3	0.06
	River	4	0.08
Vegetation Vigor	Very high	5	0.29
	High	3	0.18
	Medium	2	0.12
	Low	1	0.06
	Water body	4	0.24

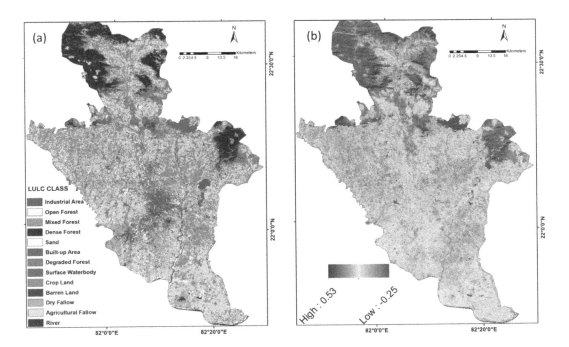

FIGURE 16.5 Spatial variation of land use/land cover (a) and vegetation vigor (b) of Bilaspur district.

surface water body, and river. The Landsat8 OLI satellite image of 2019 with 30 m spatial resolution was used as the source of data to derive the LULC map (Figure 16.5a). The LULC classification for weighted analysis is determined on the basis of the nature of LULC, the coverage and attributes of water infiltration, and the characteristics of water retention on the land surface (Bhunia, 2020). The form and characteristics of LULC influence groundwater recharge in the following order: dense/mixed forest > river/water bodies > degraded/open forest > cultivated land > agricultural fallow > sand/dry fallow > barren land > industrial/built-up (Table 16.2). Hence, forest and surface water bodies are more suitable for groundwater occurrence due to better infiltration and percolation activities.

16.4.9 Vegetation Vigor

The vegetation characteristics of the study area are analyzed using the NDVI index. A higher value of NDVI shows dense vegetation cover, and vice versa (Figure 16.5b). The results of the analysis show that the north and small pockets of the east of the district are covered by dense vegetation. A low density of vegetation cover is portrayed in the south and southwest of the study area. In the present study, high-density vegetation cover is given higher rank in terms of groundwater potency, and the scrub land and bare surface area is assigned as low groundwater potentiality (Table 16.2).

A pair-wise comparison matrix was calculated based on the AHP technique to recognize the precedence and rank of the criteria (Table 16.2). Partialities are assigned on a scale of 1–9 to each pair of themes. The pair-wise assessment is estimated from a matrix: $C=[C_{kp}]_{n*n}$; where C_{kp} is the precedence of the pair-wise comparison for the kth and pth theme. Finally, a vector of thematic weights $w=[w_1, w_2, ..., w_n]$ was obtained from the pair-wise comparison matrix. The ranks are assigned from the formula $C_w = \lambda_{max} w$; where λ_{max} is the highest eigenvalue of C (Malczewski and Rinner, 2015; Das et al., 2019).

Priority, ranks, and CR were calculated for the sub-categories of each theme as represented in Table 16.3. The CR values for features of lithology, rainfall, drainage density, soil, slope, average groundwater depth, vegetation vigor, elevation, and LULC are 15.91%, 14.68%, 13.47%, 11.09%, 12.00%, 8.82%, 9.93%, 6.85%, and 7.26%, respectively. The groundwater potential zone is evaluated by the MCDA technique and the ranks of the thematic layers from the pair-wise comparison, illustrated in Figure 16.6.

Individual weights for each theme were assigned based on the hierarchical ranking from the AHP analysis. The defined and normalized weights are illustrated in Table 16.2. Based on the weighted average value, the study area is classified into "very good" (>42.81), "good" (36.60–42.80), "moderate" (30.54–36.59), "low" (25.20–30.53), and "very low" (<25.19) groundwater potential zones. The results also indicated that 20.50% (505.64 km²) of the area has been classified as having "very good" groundwater potential and 27.65% (659.73 km²) area as having "good" groundwater potential, with more than 18.89% (667.15 km²) having moderate, 18.67% (976.84 km²) "low," and 14.31% (723.52 km²) "very low" groundwater potential (Table 16.4).

Finally, with the MCDA, the area's groundwater potential map was prepared and re-classified into five categories. Therefore, the suitability of the groundwater maps for the very good to the low potential classes is identified based on calculated values. In the north and east of the district, a very high groundwater potential zone is identified (Figure 16.6).

Very low groundwater potential is identified in the south of the district. The central and west of the district are identified as having medium groundwater potential. Favorable lithological

TABLE 16.3
Priority and Ranks of the Themes

Thematic Map	Priority (%)	Rank	Assigned Weight
Lithology	15.91	1	8
Rainfall	14.68	2	7
Drainage Density	13.47	3	6
Soil	11.09	5	4
Slope	12.00	4	5
Average Groundwater Depth	8.82	7	4
Vegetation Vigor	9.93	6	3
Elevation	6.85	9	2
LULC	7.26	8	1

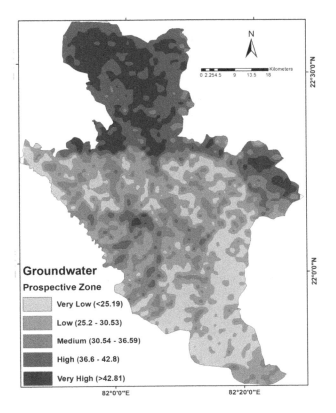

FIGURE 16.6 Prospective groundwater zone in Bilaspur district.

TABLE 16.4
Areal Distribution of Groundwater Potential Zone of Bilaspur District

Groundwater Potential Zone	Weighted Average	Area (km²)	Percent
Very Low	<25.19	723.52	20.50
Low	25.2–30.53	976.84	27.65
Medium	30.54–36.59	667.15	18.89
High	36.6–42.8	659.73	18.67
Very High	>42.81	505.64	14.31

characteristics, gentle slope, dense forest cover, coarse-grained soil, and a high rainfall zone aid in the preparation of a high potential zone.

16.5 CONCLUSION

Integrated geospatial technique is helpful for the evaluation and mapping of groundwater zones. There are several aspects, like lithology, soil, drainage density, annual average rainfall, average depth of water level, elevation, slope, LULC, and vegetation vigor, that influence groundwater recharge and discharge. Since these factors do not all have similar significance in governing groundwater, weighted values were determined using an AHP method. In the present research work, geospatial technology and MCDM techniques have been effectively used to demarcate a groundwater prospective zone (GPZ). Integrated remote sensing and GIS are widely used in identifying phreatic aquifers through the overlay analysis. The weights of the predisposing parameters have been assigned based on the AHP. The study reveals that 48.15% of the total area falls into the "very low"–"low" potential zone, which is mainly characterized by impervious lithological sceneries, high slope, lower annual rainfall, and impervious LULC characteristics. Another 18.89% of the study area falls into the moderate GPZ, and the remaining 32.98% comes under the "high to very high" GPZ. It is clear that porous lithological situations, gentle slopes, penetrable soil texture, dense vegetation cover, and low water levels aid in the generation of a highly probable zone. Hence, the proposed method is operative and essential for planning and managing groundwater in a particular watershed area. Since the central and south of the district are covered by agricultural land, this research will aid in expanding the irrigational facility and progressing the agricultural efficiency of the area.

REFERENCES

Abdul-Aziz Hussein, A.A., Govindu, V., Nigusse, A.G.M. 2017. Evaluation of groundwater potential using geospatial techniques. *Appl Water Sci* 7:2447–2461. doi: 10.1007/s13201-016-0433-0.

Agarwal, E., Agarwal, R., Garg, R.D., Garg, P.K. 2013. Delineation of groundwater potential zone. An AHP/ANP approach. *Journal of Earth System Science* 122(3):887–898.

Akyürek, D., Koç, Ö., Akbaba, E.M., Sunar, F. 2018. Land use/land cover change detection using multi–temporal satellite dataset: A case study in Istanbul New Airport. *Int Arch Photogr Rem Sens Spat Inform Sci* XLII-3/W4:17–22. GeoInformation For Disaster Management (Gi4DM), 18–21 March 2018, Istanbul, Turkey.

Ayazi, M.H., Pirasteh, S., Rizvi, S.M., Safari, H., Ramli, F.M., Pradhan, B., Rizvi, S.M. 2010. Using ERS-1 synthetic aperture radar for flood delineation, Bhuj Taluk, Kuchch District Gujarat, India. *Int Geoinform Res Dev J* 1:13–22.

Bhunia, G.S. 2020. An approach to demarcate groundwater recharge potential zone using geospatial technology. *Appl Water Sci* 10:138. https://doi.org/10.1007/s13201-020-01231-1.

Bhunia, G.S., Samanta, S., Pal, D.K., Pal, B. 2012. Assessment of groundwater potential zone in Paschim Medinipur District, West Bengal – a meso-scale study using GIS and remote sensing approach. *J Environ Earth Sci* 5(2):41–59.

Burrough, P.A., McDonell, R.A. 1998. *Principles of Geographical Information Systems.* New York: Oxford University Press, 190 pp.

Cheng, E.W.L., Li, H. 2004. Contractor selection using the analytic network process. *Constr Manage Econ* 22:1021–1032.

Choudhari, P.P., Nigam, G.K., Singh, S.K., Thakur, S. 2018. Morphometric based prioritization of watershed for groundwater potential of Mula river basin, Maharashtra, India. *Geol Ecol Landsc* 2(4):256–267.

Dagdeviren, M., Ihsan, Y. 2007. Personnel selection using analytic network process; Istanbul Ticaret. *Universitesi Fen Bilimleri Dergisi Yil* 6(11):99–118.

Das, B., Pal, S.C., Malik, S., Chakrabortty, R. 2019. Modeling groundwater potential zones of Puruliya district, West Bengal, India using remote sensing and GIS techniques. *Geol Ecol Landsc* 3(3):223–237. doi: 10.1080/24749508.2018.1555740.

Das, S., Pardeshi, S.D., Kulkarni, P.P., Doke, A. 2018. Extraction of lineaments from different azimuth angles using geospatial techniques: a case study of Pravara basin, Maharashtra, India. *Arab J Geosci* 11:160. https://doi.org/10.1007/s12517-018-3522-6.

Deepa, S., Venkateswaran, S., Ayyandurai, R., Kannan, R., Prabhu, M.V. 2016. Groundwater recharge potential zones mapping in upper Manimuktha Sub basin Vellar river Tamil Nadu India using GIS and remote sensing techniques. *Model Earth Sys Environ* 2(3):137.

Global Sustainable Development Report. 2019. The future is now science for achieving sustainable development. *United Nations publication issued by the Department of Economic and Social Affairs.* Available at: https://sustainabledevelopment.un.org/content/documents-/24797GSDR_report_2019.pdf.

Hunt, R.J., Borchardt, M.A., Richard, K.D., Spencer, S.K. 2010. Assessment of sewer source contamination of drinking water wells using tracers and human enteric viruses. *Environ Sci Technol* 44(20):7956–7963.

Jakeman, A.J., Letcher, R.A. 2003. Integrated assessment and modelling: Features, principles and examples for catchment management. *Environ. Model. Software* 18:491–501.

Malczewski, J., Rinner, C. 2015. *Multicriteria Decision Analysis in Geographic Information Science*. Springer, New York.

Mallick, J., Singh, C.K., Al-Wadi, H., Ahmed, M., Rahman, A., Shashtri, S., Mukherjee, S. 2014. Geospatial and geostatistical approach for groundwater potential zone delineation. *Hydrol Process* 29(3):395–418. doi: 10.1002/hyp.10153.

Mukherjee, D. 2016. A review on artificial groundwater recharge in India. *SSRG Int J Civil Eng (SSRG – IJCE)* 3(1):57–62.

Pinto, D., Shrestha, S., Babel, M.S., Ninsawat, S. 2017. Delineation of groundwater potential zones in the Comoro watershed, Timor Leste using GIS, remote sensing and analytic hierarchy process (AHP) technique. *Appl Water Sci* 7:503–519.

Rinaudo, J.D., Montginoul, M., Desprats, J.F. 2015. The development of private bore-wells as independent water supplies: challenges for water utilities in France and Australia. In: Rafton, Q., Chan, N., Daniell, K., Nauge, C., Rinaudo, J.-D. (eds) *Understanding and Managing Urban Water in Transition*. Dordrecht: Springer.

Saaty, T.L. 1980. *The Analytic Hierarchy Process: Planning, Priority Setting, Resource Allocation*. New York: McGraw-Hill.

Saaty, T.L. 2004. Fundamentals of the analytic network process – multiple networks with benefits, costs, opportunities and risks. *J. Systems Science and Systems Engineering* 13(3):348–379.

Saha, S. 2017. Groundwater potential mapping using analytical hierarchical process: A study on Md. Bazar Block of Birbhum District, West Bengal. *Spat Inform Res* 25(4):615–626.

Sisay, L. 2007. Application of remote sensing and GIS for groundwater potential zone mapping in Northern Ada'a plain (Modjo catchment). University/Publisher Addis Ababa University. Available at: http://etd.aau.edu.et/dspace/handle/123456789/386.

Tesfaye, T. 2010. Ground water potential evaluation based on integrated GIS and RS techniques in Bilate river catchment, South rift valley of Ethiopia. *Am Sci Res J Eng Technol Sci (ASRJETS)* ISSN (Print): 2313–4410, ISSN (Online), 2313–4402, Global Society of Scientific Research and Researchers. http://asrjetsjournal.org

Tran, L.T., Knight, C.G., O'Neill, R.V., Smith, E.R. 2004. Integrated environmental assessment of the Mid-Atlantic region with analytical network process. *Environ Monit Assess*, Kluwer Academic Publishers 94:263–277.

UNEP. 2008. *Vital Water Graphics – An Overview of the State of the World's Fresh and Marine Waters*, 2nd edn. Nairobi: UNEP. ISBN 92-807-2236-0.

Waikar, M.L., Nilawar, A.P. 2014. Identification of groundwater potential zone using remote sensing and GIS technique. *Int J Innov Res Sci Eng Technol* (*ISSN*: 2319-8753) 3:12163–12174.

Yesertener, C. 2008. Assessment of the declining groundwater levels in the Gnangara groundwater mound. Report for the Department of Water, Western Australia. *Hydrogeological Record Series Report No. HG14*, Government of Western Australia 2008, ISBN 978-1-921468-35-3.

Zektser, I.S., Everett, L.G. (eds) 2004. Groundwater resources of the world and their use, IHP-VI series on groundwater No 6. UNESCO, Paris.

17 Long-Term Drought Assessment and Prediction Driven by CORDEX-RCM

A Study on a Hydro-Meteorologically Significant Watershed of West Bengal

Shreyashi S. Mitra, Abhisek Santra, Akhilesh Kumar, and Shidharth Routh

CONTENTS

17.1 Introduction 244
17.2 Methodology 246
    17.2.1 Study Area 246
    17.2.2 Data Sources and Software 247
        17.2.2.1 Use of Open Source Software 248
    17.2.3 Methods 248
        17.2.3.1 Standardized Precipitation Evapotranspiration Index (SPEI) ...248
        17.2.3.2 Definition 250
        17.2.3.3 Drought Frequency 250
        17.2.3.4 Drought Duration 251
        17.2.3.5 Drought Magnitude and Drought Intensity 251
        17.2.3.6 Trend Analysis 251
17.3 Results and Discussion 251
    17.3.1 Nature of Drought 251
        17.3.1.1 Meteorological Drought 251
        17.3.1.2 Agricultural Drought 256
        17.3.1.3 Hydrological Drought 256
        17.3.1.4 Socio-Economic Drought 256
    17.3.2 Drought Frequency 257
    17.3.3 Total Drought Duration 260
    17.3.4 Drought Progression 262
        17.3.4.1 Temporal Progression of Drought 262
        17.3.4.2 Spatial Progression of Drought 263
    17.3.5 Drought Magnitude and Intensity 263
    17.3.6 Trend Analysis 265
17.4 Conclusion 270
Acknowledgments 271
References 271

17.1 INTRODUCTION

Water makes up almost 71% of Earth's surface, but the proportion of freshwater sources is just 3%, of which only 1% is available in the form of surface and groundwater. With extreme climate events becoming the norm, rainfall is becoming more erratic and unreliable. Furthermore, as the days keep getting hotter, evaporation also increases, leaving behind it a string of charred land. Despite 90% of the population living within 10 km of freshwater sources (Kummu et al., 2011), frequent droughts have pushed many big cities to the brink of Day Zero, with Cape Town in South Africa being a case in point in recent years. It is estimated that by 2040, 10 of the world's biggest cities will be experiencing shortages of freshwater sources, with demand expected to outrun supply by 40% by 2030. While human actions and population growth make a major contribution to this, climate change, especially reflected by phenomena like drought and subsequent water scarcity, is expected to be a major spoiler as well (Sengupta and Cai, 2019).

All this pressure on the existing sources of water has a bearing on the social, humanitarian, and economic well-being of people. Though a common natural occurrence across the globe, drought is one of the most complex and least understood climatic phenomena (Tian et al., 2018; Yang et al., 2016). There is a lot of confusion within the scientific and policy-making community about its characteristics. Much of this is due to the lack of precise and objective definition of various drought components: its onset, duration, and severity, and the impact on other natural processes (Mishra and Desai, 2005; Wilhite and Glantz, 1985). And this has a bearing on policy-makers' response to drought.

Unlike other natural hazards such as floods, cyclones, or earthquakes, the effects of drought often accumulate over time and may linger for years even after the conditions bringing about its onset have subsided. Because of this, drought is sometimes referred to as a creeping phenomenon. Furthermore, drought may mean different things to different people. As Mishra and Desai (2005) and Mishra and Singh (2010) noted, for a meteorologist, a drought is a deviation from normal precipitation, while for an agriculturist, it means a lack of adequate soil moisture for sustaining crops. Similarly, while a hydrologist might construe drought as a measure of fall in streamflow, surface, and groundwater levels, for a policy-maker, it may mean anything from this list or even famine or social disorder. Such divergences in objectivity and the definition of drought add to the uncertainty about the nature of drought.

In general, droughts can be broadly classified into four categories (Wilhite and Glantz, 1985). The first is meteorological drought, which is categorized by deficient precipitation relative to normal. This generally happens when rainfall misses its long-term climatological average for a given region or time (Santos, 1983). When this condition extends for more than 6 months, i.e., a season, the deficiency in rainfall leads to significant loss of soil moisture, and the drought is categorized as an agricultural drought. It is given this name because the absence of adequate soil moisture severely impedes crop production to the extent that its effects begin to become visible in overall agricultural productivity. If rainfall conditions don't improve even then, the drought extends beyond merely decreasing soil moisture to interfere with streamflow and water levels in rivers, lakes, and reservoirs too (Dracup et al., 1980; Zelenhasić and Salvai, 1987), morphing into a hydrological drought. Further precipitation deficiency turns the drought condition into a socio-economic crisis, impacting food supply, water availability, and the general economics of the region. This is labeled socio-economic drought, and the only way out is a cycle of above-average wet years. While most researchers and organizations, such as the National Centers for Environmental Information (NCEI) and the World Meteorological Organization (WMO), have stuck to the above four-pronged drought categorization (Arndt, 2020; Svoboda and Fuchs, 2016), some researchers, like Mishra and Singh (2010), have advocated adding groundwater drought as a separate fifth category.

Over the years, many drought indices have been developed to monitor drought characteristics. These include some very well-known indices, like the Palmer Drought Severity Index (PDSI), the Standardized Precipitation Index (SPI), and the most recent one in the form of the Standardized

Precipitation Evapotranspiration Index (SPEI). PDSI, which was a landmark in the development of drought indices, works on the concept of a water balance equation and takes into account precipitation, moisture supply, evaporation demand, and runoff at the surface level (Palmer, 1965). And while the development of a self-calibrated PDSI (sc-PDSI) solved many of the issues about spatial comparability and extreme event characterization (Wells et al., 2004), it still didn't solve the problems of fixed temporal scale and autoregressive characteristics whereby the PDSI index values were retrospectively impacted by conditions up to 4 years in the past (Guttman, 1998). This issue of fixed time scales was overcome by SPI, which took into account the difference in the period from the arrival of water input to its availability for use, underscoring the importance of time scales in water deficit and water accumulation (McKee et al., 1993). The flexibility to calculate SPI at different time scales made it more comparable in time and space (Hayes et al., 1999). But SPI still had one sore point. By taking only precipitation as an input, it assumed that rainfall variability is more pronounced than other climatological factors like temperature and potential evapotranspiration (PET). It also assumed other variables to be temporally stationary.

But recent research has underscored the importance of temperature-induced drought stress for net primary production and forest mortality (Adams et al., 2009; Hartmann et al., 2018; Linares and Camarero, 2012; McGuire et al., 2010; Taha et al., 2020), evapotranspiration and enhanced summer drought stress (Bodner et al., 2015; Kenawy et al., 2018; Rebetez et al., 2006), forest fires (Barriopedro et al., 2011; Dwomoh et al., 2019), agricultural production (Lobell et al., 2011; Ray et al., 2018), and related ecological impacts (Breshears et al., 2005; Chadd et al., 2017; Lund et al., 2018; Scasta and Rector, 2014), all of which are an essential part of drought impact studies. With temperature projected to rise further till at least the middle of the century, any drought model that leaves temperature aside isn't preferable for long-term future drought predictions. SPEI, based on precipitation and PET, combines PDSI's sensitivity to changes in evaporation demand with SPI's flexibility in time scales to create a simple and statistically robust drought index (Vicente-Serrano et al., 2010; Keyantash and Dracup, 2002). The use of temperature in drought statistics also assumes importance in the light of climate change, specifically global warming, and the impact it might have on future water scenarios.

Since it is a widely accepted drought index, there's no paucity of literature reviews to back it as an effective method of drought characterization. Spinoni et al. (2018) used SPI, SPEI, and the Reconnaissance Drought Index (RDI) to analyze drought scenarios across Europe using Representative Concentration Pathway (RCP) 4.5 and 8.5 projected data and found that drought frequency is increasing during spring and summer across the continent. Chen and Sun (2017) used SPEI and Coupled Model Intercomparison Project Phase 5 (CMIP5) climate data to characterize present and future drought changes over Eastern China and reported aggravated drought conditions under the projected scenarios. Yu et al. (2014) studied the nature of severe drought and its frequency across China and used SPEI to conclude that above-moderate droughts have become more serious over the years. A similar study by Nam et al. (2015) used a combination of SPI, SPEI, and sc-PDSI in South Korea and predicted an increase in drought severity and magnitude. Numerous other studies using SPEI and other drought indices have been used elsewhere to study the nature of drought and its behavior going into the future (Gao et al., 2017; Khan et al., 2017; Dibike et al., 2017; Smirnov et al., 2016). Most of the studies rely on open-source software (OSS) to carry out computation and result analyses. The availability of global- or regional-scale data in the private domain could optimally benefit the research community if OSS could be used to analyze it for various purposes. This substantially ensures the transfer and transmission of research findings. A fair number of works mentioned in the research have been carried out using OSS either solely or in combination with other commercial software.

In India too, drought studies have been at work for decades. Surendran et al. (2019) established that in the Indian subcontinent, the humid region has a negative trend of RDI during drought as compared with semiarid and arid regions, probably due to the impact of climate change. Sharma and Mujumdar (2017) studied the nature of concurrent meteorological droughts and heatwaves across the country and concluded that over the years, there has been an increase in their frequencies as well as their spatial extent. A similar study by Zhang et al. (2017) attempted to reconstruct drought events in major wheat-growing regions in India to demonstrate its implications for wheat production. In the process, the study found that meteorological droughts have become much more severe. While Mallya et al. (2016) and Thomas et al. (2015) concluded that drought severity and frequency had increased across India and in the Bundelkhand region in the last few decades, Dutta et al. (2015) carried out a drought assessment in Rajasthan and revealed related crop stress specifically in 2002. Naresh Kumar et al. (2012) reported an increase in the spatial extent of area under moderate drought in recent decades. Mishra and Desai (2005) conducted a drought study in the upper catchment of the Kangsabati river basin and reported more severe droughts with an increased areal extent in the twenty-first century.

But a major drawback of these studies has been that, except for a few studies, most of the research work on droughts in India, especially in the lower Gangetic plains, has focused primarily on SPI, which as noted earlier, doesn't take temperature into account, a crucial factor in water balance. Increased temperature leads to higher water demand, and any study that seeks to make drought projections for the future must factor it in (Bisht et al., 2019). The present study is an attempt in the aforementioned direction, with the prime objective being not only to compare the results of different drought indices and their suitability for the Kangsabati river basin but also to use a well-established method, SPEI, to identify possible drought incidences in the future by analyzing the nature of drought trends. Based on the available input, the study seeks to prepare a comprehensive drought register detailing the severity, nature, and duration of drought events in the basin for 118 years, i.e., from 1982 to 2100. Apart from this, to encourage reproducibility of results, the work focuses on using mostly OSS – R, Climate Data Operator (CDO), and MS Excel being the top choices. OSS provides the advantage of better support and collaboration (since anyone with knowledge can help); bypasses vendor lock-in, such that the available options can be tweaked and created to serve an entirely new purpose; facilitates extensive customization; is less buggy; and is widely used and available. While R was used for calculating SPEI, drought statistics, and raster plotting, CDO was used for working with NetCDF files of precipitation and temperature. It was used for remapping, subsetting, and carrying out basic calculations. MS Excel was used for preparing and organizing tables. Almost every paper mentioned in the present study has used this OSS for one or more portions of the work (Laaha et al., 2017; Vicente-Serrano et al., 2010; Bisht et al., 2019; Spinoni et al., 2014).

17.2 METHODOLOGY

17.2.1 Study Area

Drained by the confluence of two rivers, Kumari and Kangsabati, the basin is one of the most geographically diverse river basin systems in the eastern part of India (Figure 17.1). Spread over two states – West Bengal and Jharkhand – the elongated basin covers an area of 6012 km^2. Based on terrain, river flow, and drainage, the basin can be broadly classified into three distinct catchments (Santra Mitra et al., 2019).

The basin has a distinct climatic pattern, with seasonality impacting the climate in the upper catchment, and a huge reservoir and dense forest cover in the middle catchment, while relative proximity to the sea is the dominant factor in the lower reaches. Incidentally, while summer in the upper catchment is dry and hot, it becomes hot and humid as one moves towards the lower reaches. The basin receives most of its rainfall in 3 months or less, with rainfall being slightly more regular in areas closer to the sea (Kumar et al., 2019).

Predominantly agrarian, the basin is home to some of the most populated cities and industrial hotspots in the state. Given the lack of irrigation infrastructure, particularly in the upper catchment, the onset and cessation of monsoon have a major impact on the agricultural productivity and overall socio-economic condition of the people (Revadekar and Preethi, 2012).

Long-Term Drought Assessment and Prediction 247

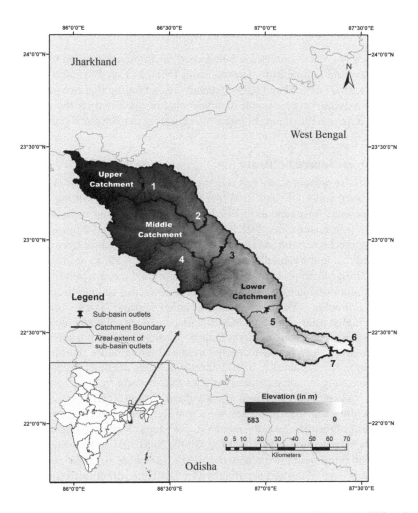

FIGURE 17.1 A study area map showing the terrain and the catchments of Kangsabati River basin.

17.2.2 DATA SOURCES AND SOFTWARE

The study uses two sets of climatic data sources. Since a minimum of 30–35 years is required and considered sufficient for any climatic influence to become visible, a period starting 35 years before 2017 was used. For the historical period, i.e., from 1982 to 2017, rainfall data at 0.25° × 0.25° grid size was sourced from the India Meteorological Department (IMD), while maximum and minimum temperature time series data were sourced from the National Oceanic and Atmospheric Administration (NOAA)'s Physical Science Laboratory (NOAA-PSL), which is available from 1979 onwards. The efficacy of IMD and NOAA datasets for drought studies have been established in numerous studies, both in India and across the globe (Beria et al., 2017; Smitha et al., 2018; Mishra and Liu, 2014; Bisht et al., 2019; Sharma and Mujumdar, 2017).

For the projected time, a multi-model ensemble (MME) of six driving models was used to extract precipitation and temperature time series data. An MME is usually considered a good option, since it provides reliable estimates by encompassing the range of uncertainties associated with each climate model (Weigel et al., 2010). To take the most probable climatic scenario into account, data of RCP 4.5 from the first ensemble member (r1i1p1) of each model were used. To maintain spatial consistency between projected and historical time series data, the RCP 4.5 dataset was remapped to

0.25° × 0.25° using CDO's bilinear interpolation technique mentioned in (Das et al., 2012; Akhter et al., 2017).

Furthermore, since two data sources are being used, a simple correlation test was also done on historical precipitation and temperature datasets from IMD, NOAA-PSL, and MME. There was a healthy correlation, with R^2 of 0.89 and 0.93, respectively, indicating that two sets of data series can be used. The use of existing observational data was also necessitated by the requirement to keep the use of projected data to a minimum in order to accurately map climatic conditions into drought statistics.

17.2.2.1 Use of Open Source Software

Except for MS Excel, the present study has relied entirely on OSS for most of the work. The most frequently used software was CDO and R. While CDO was used for working with the CORDEX-RCM dataset, R was used for calculating SPEI values, summarizing drought statistics, creating heatmaps, and generating raster plots. NetCDF data files downloaded from ESGF-LIU were first merged (using the mergetime command) to create a single file with data from 2020 to 2100. Then, the file was remapped (using the remapbil command) to 0.25° × 0.25° to maintain the spatial homogeneity with 1982–2017 gridded precipitation and temperature datasets from IMD and NOAA PSL. Once remapped, the area of interest (AOI) was subsetted (using sellonlatbox) out of it to reduce computational load, following which each value was either multiplied or subtracted with a constant to match the units. Finally, after creating a MME (using ensmean) of six driving models, point data were extracted for seven locations using the remapnn command and exported as text files using the outputtab command. The values for both precipitation and temperature (maximum and minimum) were then arranged monthly to calculate SPEI values in R using the SPEI package. The open source R was then used extensively to calculate drought stats like drought frequency, duration, magnitude, and intensity. The R package trend was used for calculating Mann Kendall statistics and Sen's slope.

To calculate drought stats, four packages were mainly used – dplyr, lubridate, openxlsx, and reshape2. The general algorithm broadly involved five steps. First, the dataset was divided into 20-year time periods by placing markers at 2020, 2040, 2060, and 2100 using cut and break from lubridate. After identifying years with negative SPEI values as drought years, the severe and extreme drought years were identified separately as well. Then, drought frequency was estimated using the group_by and summarize (by sum) functions from dplyr. Calculating drought duration is slightly complex and involves three steps – identifying the first month of each drought, calculating its length, and then combining these two pieces of information to create a drought duration file. lag and rle functions from dplyr were used for this process. Once the first month, duration, and frequency of a drought have been identified, its magnitude and intensity can be easily calculated using the summarize (by sum) function.

R software was also used for creating multiple heatmaps and raster plots. While heatmap.2 from the gplot package was used in conjunction with gridExtra, gridGraphics, grid, and lubridate to create multiple heatmaps on one plot, levelplot from rasterVis package was used for plotting geospatial raster maps. The code for calculating drought stats and creating multiple heatmaps can be found at: https://github.com/akhi9661/jrf-project-codes.

17.2.3 Methods

17.2.3.1 Standardized Precipitation Evapotranspiration Index (SPEI)

The basic underlying principle of SPEI rests on the concept of climatic water balance (CWB), which is defined as the difference between precipitation and PET. Based on its value, which can be either positive or negative, the CWB can indicate climate-induced surpluses or deficits in the water budget and its regional distribution. While the SPEI R package provides three methods to compute PET – Hargreaves (Rhee and Cho, 2016), Thornthwaite (Thornthwaite, 1948; Bonsal et al., 2017), and Penman-Monteith (Allen et al., 1998b) – the first method was chosen in this study.

While Penman-Monteith is recommended by the Food and Agriculture Organization (FAO) (Allen et al., 1998a), the sheer amount of data input it requires makes it undesirable for many places, especially those with data scarcity, like the present study area (Santra Mitra et al., 2015). Instead, Hargreaves was used to calculate PET and subsequently CWB and SPEI. It has been found to have a better response than Thornthwaite, which uses only average temperature (Bandyopadhyay et al., 2012).

Following the equations in the work of Hargreaves and Allen (2003) and Hargreaves (1994), the PET is calculated, after which CWB is estimated by subtracting precipitation from it. CWB is calculated for each month and is then aggregated at the required time scales.

Of the many distribution fittings available, a log-logistic probability distribution with unbiased probability-weighted-moments (ub-PWMs)–based distribution fitting parameters was used. Log-logistic shows a gradual decrease in the curve for low values; it has coherent probabilities for very low values; and the values never go below the origin parameter of the distribution. The three parameters α, β, and γ (called the scale, shape, and origin parameter, respectively) are estimated automatically by the R package using the plotting-position approach defined in a study by Hosking (1986). The use of an unbiased estimator in PWMs is usually recommended to maintain the spatial and temporal comparability of SPEI values across different time scales and regions (Vicente-Serrano et al., 2010).

The unbiased PWMs can be obtained using Equation 17.1:

$$w_s = \frac{1}{N}\sum_{i=1}^{N} \frac{\binom{N-i}{s} D_i}{\binom{N-1}{s}} \quad (17.1)$$

where D_i is the CWB for the month i, N is the number of data, and w_s is the weighted moment of order s. This is then used to calculate the log-logistic parameters. Similarly, the probability distribution function of D according to log-logistic can be calculated using Equation 17.2:

$$F(x) = \left[1 + \left(\frac{\alpha}{x-\gamma}\right)^\beta\right]^{-1} \quad (17.2)$$

The function $F(x)$ can then be used to obtain SPEI values following the equation in Abramowitz and Stegun (1965) (Equation 17.3):

$$SPEI = W - \frac{C_0 + C_1 W + C_2 W^2}{1 + d_1 W + d_2 W^2 + d_3 W^3} \quad (17.3)$$

where constants $C_0 = 2.515517$, $C_1 = 0.802853$, $C_2 = 0.010328$, $d_1 = 1.432788$, $d_2 = 0.189269$, $d_3 = 0.001308$, and $W = -2\ln(P)$, P being the probability of exceeding a determined D value. P depends on function $F(x)$ and is related to it as (Equation 17.4)

$$F(x) = \begin{cases} 1-P, & \text{if } P \leq 0.5 \\ P, & \text{if } P > 0.5 \end{cases} \quad (17.4)$$

When $P > 0.5$, the sign of SPEI is reversed. The log-logistic distribution gives an SPEI series with an average value of zero and a standard deviation of 1.

17.2.3.2 Definition

In the present study, drought is assumed when SPEI values go below zero. While SPEI >−1.0 means a mild drought, the drought is considered to be of moderate severity when SPEI is equal to or less than −1.0 but is still greater than −1.5. The drought is categorized as severe for SPEI less than or equal to −1.5. For all SPEI values below −2.0, the drought is categorized as extremely severe (Edwards, 1997; McKee et al., 1993).

Based on time scales, the WMO defines four types of drought: meteorological, agricultural, hydrological, and socio-economic (Svoboda and Fuchs, 2016). These are calculated at time scales of 3, 6–9, 12–15, and 24 months, respectively (Table 17.1). The categorization is based on the relative impact of water deficiency on different components of the water balance system.

17.2.3.3 Drought Frequency

The frequency of drought is defined as the cumulative sum of all the instances within a period where the SPEI value has changed its sign. To get a better idea of the drought liability during a study period, the number of droughts per 100 years was calculated using the equation described in a publication by Ghosh (2018) and provided here (Equation 17.5).

$$FD_{i,100} = \frac{N_i}{i \times n} \times 100 \tag{17.5}$$

where $FD_{i,100}$ is the drought frequency per 100 years, N_i is the number of months with droughts for time scale i (which can be 3, 6, 9, 15, or 24), and n is the number of years in the dataset. For the present study, n is equal to 20 years.

To calculate severe or extreme drought frequency, N_i can be replaced with the cumulative value of the number of instances (Ghosh, 2018) where the SPEI value has been lower than −1.5 and −2.0, respectively, as opposed to just being negative. To further simplify the visualization of severe and extreme drought frequencies, a spatial map with frequencies expressed as percentages was also generated.

TABLE 17.1
Drought Categorization Based on SPEI Time Scales and Their Significance for Water Balance

Time Scale	Significance
3-month SPEI	Reflects short- to medium-term moisture conditions; provides a seasonal estimation of water deficiency; used for meteorological drought analyses
6-month SPEI	Indicates seasonal to medium-term trends; water scarcity estimation across distinct seasons; used for agricultural drought analyses
9-month SPEI	Inter-seasonal drought trends; bridge between short-term seasonal droughts and long-term hydrological or multi-year droughts. Used for analyses of agricultural drought on streamflow and reservoirs
12-month SPEI	Medium- to long-term drought trends; used for analyses of hydrological droughts and their impact on streamflow, reservoirs, and even groundwater
15-month SPI	Multi-year precipitation and drought trends; used for analyses of hydrological and multi-year droughts on groundwater
24-month SPEI	Long-term drought trends; used for analyses of drought impacts on food supply, social relations, economy, and general well-being of the people

Sources: McKee et al., 1993; Svoboda et al., 2012

Long-Term Drought Assessment and Prediction

17.2.3.4 Drought Duration

Drought duration is defined as the cumulative sum of the number of months for which the SPEI values have been negative. In other words, it is the total number of months between a negative and a positive SPEI value. It must be pointed out here that in the present study, a drought duration has been ascribed to the preceding year and not the successive year (Spinoni et al., 2014). For instance, if a drought starts in 2060 but extends into and beyond 2061, then the drought will be counted as part of 2041–2060 and not split into 2041–2060 and 2061–2080.

17.2.3.5 Drought Magnitude and Drought Intensity

To quantify the extent of drought severity, the quantum of negative SPEI values has to be measured. As discussed in Adarsh et al. (2018), drought magnitude is the cumulative sum of all negative SPEI values over a drought period. It is the accumulated water deficit in the specified period. Drought intensity, on the other hand, is defined as the average drought magnitude across a drought period and is calculated as the ratio of the two.

17.2.3.6 Trend Analysis

To identify trends in drought magnitude or area under drought, a combination of Mann Kendall (Mann, 1945; Kendall and Gibbons, 1990) and Theil-Sen's slope (Theil, 1950; Sen, 1968) was used. Mann Kendall is a non-parametric test used to identify positive or negative trends in a time series. To quantify statistical significance in the trend, the p-value is usually taken as a reliable metric, and p-values of 0.1 and below are usually taken to imply statistical significance. Since autocorrelation in data can negatively impact the accuracy of Mann Kendall tests, the Modified Mann Kendall test that uses the variance correction approach proposed by Hamed and Rao (1998) was used. The R package "modified MK" was used to construct a time series using Sen's slope and lag-1 autocorrelation coefficient of the ranks of the data.

The Mann Kendall trend test has been widely used in numerous studies and is recommended for its robustness in identifying trends in time series. It was also used by Gao et al. (2017), Khan et al. (2017), Bisht et al. (2019), and Zhang et al. (2017) to identify trends in projected drought characteristics. Details of the specifics of Mann Kendall and Theil-Sen's slope can be found in a work by Kumar et al. (2019, 2020) and hence, have not been elaborated here.

17.3 RESULTS AND DISCUSSION

17.3.1 NATURE OF DROUGHT

To study the nature of drought patterns over 118 years, heatmaps spread across years and distributed by months were generated. This is useful to gain a deeper insight into the exact months that have been experiencing droughts and to carry out further studies to look for possible correlations they might have with existent climatic conditions like precipitation and temperature. It may be worthwhile to note here that drought, being a creeping phenomenon, doesn't happen in isolation and as a sudden reaction. Rather, it builds itself over the course of time. For that very reason, one drought type can easily be considered as an amalgamation of multiple instances of another drought type. For instance, two or three successive instances of meteorological drought can also be counted as one instance of agricultural drought. Similarly, two instances of successive agricultural drought can be counted as four successive meteorological droughts or one hydrologic drought. and so on.

17.3.1.1 Meteorological Drought

In the case of metrological drought, for the period 1982–2017, most of the severe and extreme drought events occurred in the twenty-first century (Figure 17.2a). While pre-monsoon months from January to May were the drought hotspot months for the first decade of the century, the focus has shifted to the second half of the year, i.e., from June to December, in the second decade, indicating

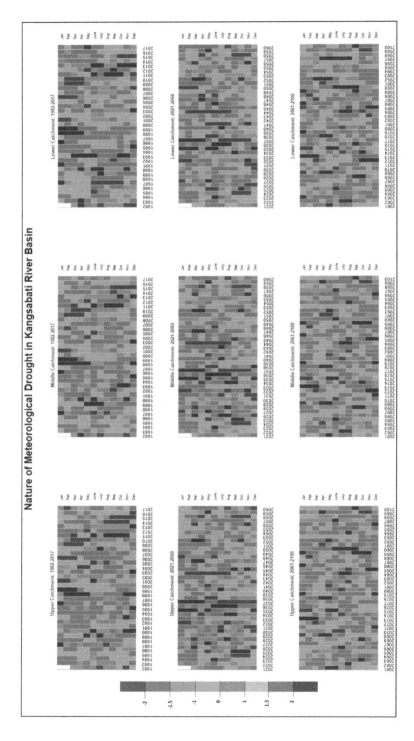

FIGURE 17.2 Heatmaps showing occurrences of (a) meteorological, (b) agricultural, (c) hydrological, and (d) socio-economic droughts during 1982–2100.

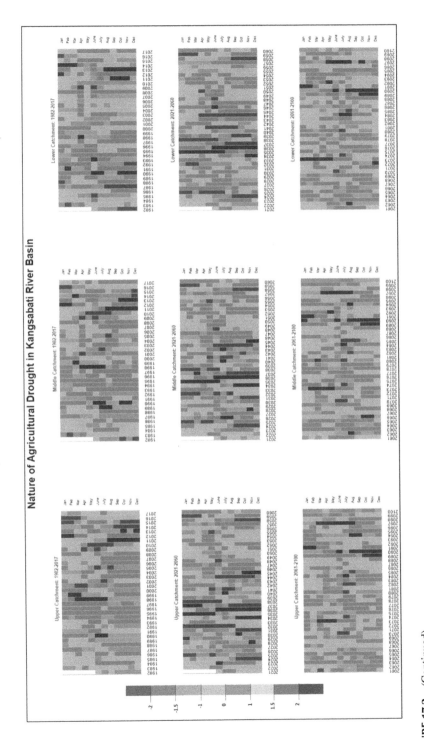

FIGURE 17.2 (Continued)

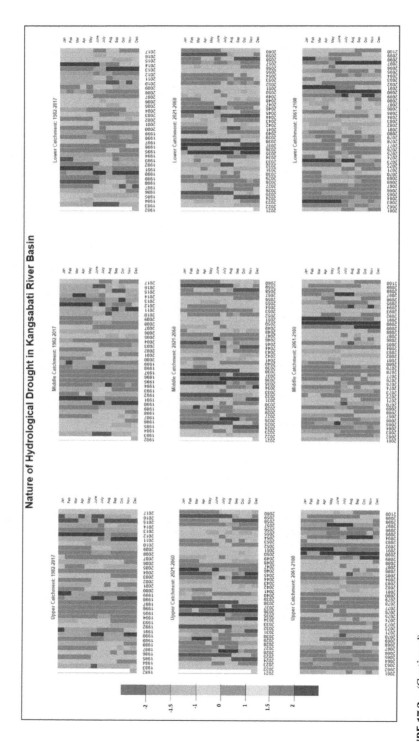

FIGURE 17.2 (Continued)

Long-Term Drought Assessment and Prediction

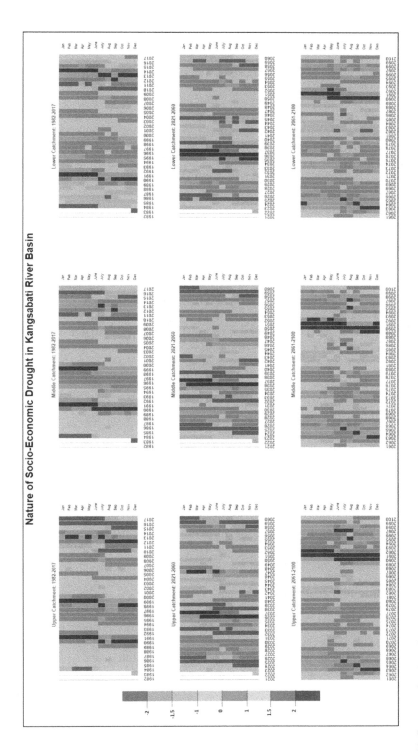

FIGURE 17.2 (Continued)

that all is not well with the monsoons in the years 2014–2016. This has been confirmed by Kumar et al. (2019), who observed that after the very high-precipitation years of 2011 and 2013, when the rainfall was almost 75% more than the long-term average, it dropped to less than 1000 mm for 2014, which is almost half of the average rainfall. From 2015 to 2017, the rainfall has remained more or less average to below average.

Of course, there have been some other scattered hotspots elsewhere as well, such as 1982–83 and 1988–89. The severity of the drought category is also symptomatic of the prevalent climatic conditions in the basin. For instance, in the upper catchment, where rainfall is more seasonal and summers are much hotter, the instances of severe and extreme drought occurrences are much higher than in the middle and significantly higher than in the lower catchment, where the rainfall is more uniformly distributed and the intra-seasonal temperature difference is less stark.

As for the predicted occurrence of meteorological drought, 2021–2040 is more prone to it than the succeeding 20-year time period. During 2021–2040, the instances of severe and extreme drought severity are higher in the upper catchment and decrease as one moves towards the lower reaches. The years from 2061–2100 are relatively less arduous and the idea of dry season events is restricted to generally extreme, in the best case scenario.

17.3.1.2 Agricultural Drought

Compared with meteorological drought, agricultural drought, though expected to be less frequent, is more severe in its outlook (Figure 17.2b). From 1982 to 2000, there are hardly any instances of extreme drought occurrence in the upper or middle catchment, while there have been two instances, 1982–83 and 1988–89, in the lower reaches. In the ensuing decades, i.e., from 2001 to 2060, each catchment is equally vulnerable to drought events, particularly severe and extreme drought events. The frequency and duration of such drought events have been accentuated in this period.

Starting in 2061, following the broad trends in other drought categories, the frequency, intensity, and duration of drought events decreased drastically, even more starkly than meteorological drought, indicating the reduction of climatic conditions that contribute to drought creep.

17.3.1.3 Hydrological Drought

In terms of hydrologic drought, the upper catchment has been particularly hard hit post 2010 (Figure 17.2c). It has carried this forward to successive decades until 2060, with its impact spreading equally to the other catchments. But it is the years after 2060 that show a slight divergence from the trends indicated by the aforementioned meteorological and agricultural drought. In the 2061–2100 period, while the incidence of extreme drought has decreased significantly, that of severe drought has picked up slightly. This indicates at least the possibility that agricultural drought has a higher chance of progressing into a hydrological drought, a fact that has been validated in the following sections, where the nature of temporal and spatial drought progression is discussed. Elsewise, the broad contours of drought categorization remain the same.

17.3.1.4 Socio-Economic Drought

As far as the nature of socio-economic or long-term drought is concerned, most of the drought events, extreme or otherwise, have occurred post 2010 in the upper catchment and before 1990 in the lower catchment, following the broad trend in the divergence of the nature of drought severity observed in the upper and lower reaches for meteorological, agricultural, and hydrological droughts. Post 2021, the nature of drought begins to resemble other drought types, albeit with increased instances of extreme drought (Figure 17.2d).

Post 2060, while the instances of the most severe drought events have increased, this can be attributed to the creeping nature of drought from agricultural to hydrologic and from hydrological to socio-economic, an observation discussed in detail under the results section, which looks at the nature of drought progression.

Long-Term Drought Assessment and Prediction

17.3.2 DROUGHT FREQUENCY

For the period from 2021 to 2100, analysis shows that while meteorological drought is still a very common phenomenon, with 2.5 to 1.8 instances every year, the trend is different for the different 20-year periods (Table 17.2). While it is most frequent in 2041-2060, the frequency of such droughts decreases in successive decades. Oddly enough, while the frequency of meteorological drought is higher during 2041–2060, the frequency of its being severe or extreme is higher during 2021–2040 and decreases progressively over the years, indicating a thaw in drought severity. For instance, while 18 severe and extreme drought events are expected per century for 2021–2040, the number is only around 12 and 3, respectively, during 2081–2100 (Table 17.2).

The basin-wise behavior also presents an interesting statistic. While the frequency of droughts increases from the upper to the lower catchment during 2021–2040 and 2081–2100, it decreases during the intervening period of 2041–2080. The same is true for the frequency of severe and extreme drought events. This aligns with the previous observations on the nature of drought severity, which pointed to a similar pattern of subtle changes in the climatic behavior of the basin and how it alters over 118 years.

Agricultural droughts, though less frequent than meteorological, showed a similar pattern. While there are 107 instances of agricultural drought per century in the upper catchment (109 in the lower) during 2041–2060, this number goes down to 75 in the upper and 88 in the lower catchment during 2081–2100. The frequency of severe and extreme droughts, too, is much lesser than for meteorological droughts. Akin to meteorological droughts, 2021–2040 has more instances of severe and extreme droughts than 2041–2060, despite having fewer drought occurrences compared with that period. But a noteworthy statistic is that while 9 to 11 extreme drought events per century are predicted for 2021–2040, the frequency of severe drought events is only around 8 to 9 events per century. Post 2060, no significant extreme drought events are observed in the lower catchment, while an extreme drought lasting a few months has been spotted in the upper and middle catchments.

The frequency of hydrological drought follows a similar pattern, though its occurrence, expectedly, is lower than that of agricultural droughts. The frequency of severe and extreme droughts has been predicted to reduce considerably as well, except for 2021–2040, which seems to be notably vulnerable to extreme drought events. Besides, no significant extreme drought events have been predicted in the basin for 2061–2080, while an extreme drought lasting for a few months might be observed in the ensuing decades.

Socio-economic drought events are thought to be more devastating due to their impact on food supply, water scarcity, and the general well-being of the population. These, too, are a relatively common phenomenon, with 33–36 instances during 2021–2040 and 22–27 instances during 2081–2100. Akin to trends for other drought categories, 2041–2060 is expected to have most instances of long-term drought events, albeit with no significant extreme drought events. On the other hand, 2061–2080 is fairly free of long-term droughts, with neither severe nor extreme droughts persisting for long periods.

A more intuitive and simpler picture of severe and extreme drought frequency, expressed as percentages of total drought frequency during 20 years, can be gleaned from the spatial distribution maps (Figure 17.3). The spatial maps point to some pretty gloomy scenarios. During 2021–2040, almost 13–21% of the basin is under severe socio-economic drought conditions. For other drought types, the percentage ranges from 6% to 11% for meteorological droughts, from 6% to 10% for agricultural droughts, and in a similar range for hydrological droughts, with the exception being that while the former two are higher in the middle and lower reaches, the last is higher in the upper catchment. The situation improves slightly during the last four decades: the percentages of severe droughts fall to less than 7% for meteorological and agricultural droughts and less than 3% for hydrologic and socio-economic droughts. Also, 2041–2060 shows improvement as far as severe drought events are concerned.

TABLE 17.2
Total Number of Droughts per Century for the Period 2021–2100

Period (Occurrences in 100 years)		Meteorological			Agricultural			Hydrological			Socio-Economic		
		Upper	Middle	Lower	Upper	Middle	Lower	Upper	Middle	Lower	Upper	Middle	Lower
2021–2040	DF	237.5	241.7	245.0	103.0	104.7	104.9	57.5	56.5	62.5	33.1	35.8	35.6
	SDF	17.5	21.7	25.0	8.0	8.0	9.1	4.8	4.1	4.9	5.6	7.1	5.6
	EDF	17.5	16.67	10.0	11.2	10.7	9.0	9.0	9.3	7.2	4.0	2.9	2.2
2041–2060	DF	260.0	245.0	261.7	106.7	103.7	108.8	61.5	59.3	61.7	38.0	37.3	39.3
	SDF	18.3	13.3	19.4	9.0	8.0	8.3	6.6	4.8	4.3	3.4	4.8	5.3
	EDF	10.8	11.67	16.1	5.2	6.0	8.1	1.6	2.6	4.5	–	–	–
2061–2080	DF	208.3	208.3	187.8	69.8	67.3	57.9	32.1	30.6	24.6	17.8	15.2	10.9
	SDF	14.2	11.67	11.7	3.2	3.0	2.0	0.4	0.4	–	–	–	–
	EDF	3.3	1.67	1.1	0.2	–	–	–	–	–	–	–	–
2081–2100	DF	194.2	200.0	220.6	74.7	83.7	87.6	41.2	49.3	49.6	22.6	26.9	26.2
	SDF	11.7	1–0	18.3	3.0	3.3	3.8	0.5	0.6	0.6	0.3	0.4	0.7
	EDF	2.5	3.33	0.6	–	0.7	–	0.1	0.2	0.1	–	–	–

DF: drought frequency; SDF: severe drought frequency; EDF: extreme drought frequency.

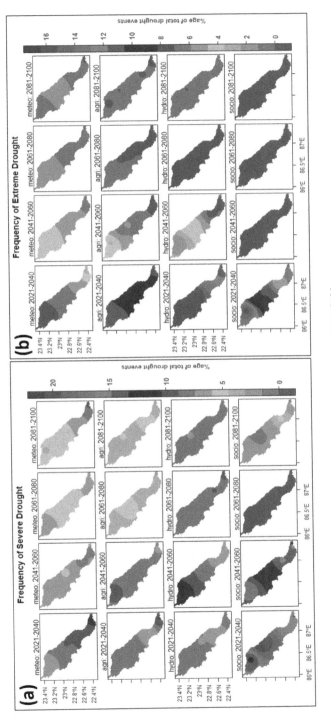

FIGURE 17.3 Map of projected (a) severe and (b) extreme drought frequencies during 2021–2100.

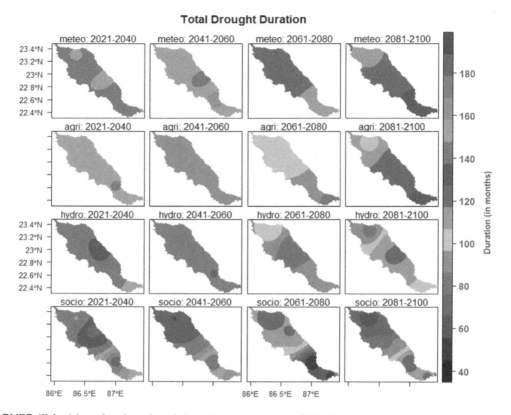

FIGURE 17.4 Map of projected total drought duration during 2021–2100.

The corresponding numbers in the case of extreme droughts are gloomier than those for severe droughts during 2021–2040 and improve greatly during the subsequent decades. For instance, while the proportion of extreme droughts as part of total hydrologic drought events during 2021–2040 is as high as 16%, it ranges from 8% to 15% in the upper catchment, from 6% to 8% in the middle catchment, and is less than 5% in the lower reaches for socio-economic droughts. Post 2060, the percentages of extreme droughts are less than 1%, with no extreme instances being the norm.

17.3.3 TOTAL DROUGHT DURATION

While drought frequency gives an idea of the drought instances likely to hit the basin, it doesn't indicate much about drought duration, which plays a crucial part in the overall water balance and is an important indicator of one drought type progressing into another. To overcome this blindsight, maps showing the distribution of total drought duration (Figure 17.4), severe drought duration, and extreme drought duration in months were generated (Figure 17.5a and Figure 17.5b).

The results of the total drought duration map reinforce what the drought frequency analysis was pointing to. They indicate that 2041–2060 is most devastating as far as drought possibilities are concerned. On average, the basin experienced 160 months of droughts in these 20 years, with the basin being under meteorological droughts for 140–160 months, under agricultural drought for around 165 months, a little longer under hydrological drought for around 175 months, and above 185 months under socio-economic drought in the upper catchment, around 170–175 months in the middle catchment, and around 140–165 months in the lower catchment (Figure 17.4).

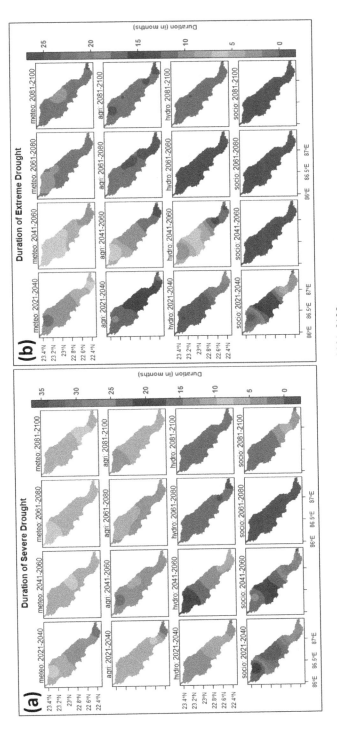

FIGURE 17.5 Map of projected (a) severe and (b) extreme drought duration during 2021–2100.

The preceding 20-year time period is slightly less gloomy, with an average drought duration of around 140–150 months for all drought types except socio-economic. For this drought type, the duration ranges from 120 to 150 months in the upper catchment, from 120 to 130 months in the middle catchment, and from 160 to 180 months in the lower catchment. The condition in the last four decades is a story in contrast and an indication of the cyclical nature of climate.

During 2061–2080, the upper catchment experiences more drought months than the middle and lower catchments, but the condition flips over diametrically during 2081–2100, when the lower catchment stays longer under drought in comparison to the middle and upper catchments. A stark example of this is socio-economic droughts. While the upper catchment experiences over 120 months of drought during 2061–2080, the number is lower than 60 months in the next 20-year period. Similarly, while fewer than 60 months experienced droughts in the lower reaches during 2061–2080, this rises to above 110 months in 2081–2100.

The illustration for severe and extreme drought duration is not greatly different (Figure 17.5). It, too, closely follows the frequency analysis curve, with 2021–2040 being the deadliest period and 2061–2100 remaining largely free from prolonged drought events. Consequently, it can be noticed that even though 2041–2060 is overrun with droughts, the duration of severe and extreme drought events is longer in the preceding two decades.

For instance, the total severe drought duration for socio-economic drought is as long as 35 months, with the lower catchment being more affected than the middle and upper catchments. In the case of extreme droughts, too, the duration ranges from as high as 25 months to 6 months, with the upper catchment being the most affected and the lower catchment the least. This is symptomatic of the rainfall conditions in the basin, a factor that was visible, like the types and frequency of droughts. But the situation is different for other drought types. An extremely severe meteorological, agricultural, and hydrologic drought is as likely a possibility as a severe drought of the same category.

17.3.4 Drought Progression

17.3.4.1 Temporal Progression of Drought

As stated before, drought is a creeping phenomenon, whose impacts linger for a long time. More often than not, one drought type extends into another, and this is visible in the drought progression trends too. If a meteorological drought occurs, there's a 62% chance in the upper catchment that it will progress to an agricultural drought, which itself has a 77% chance of creeping into a hydrologic drought. From there on, almost every drought incidence progresses into a long-term drought extending beyond 2 years. The progression of a hydrologic drought to a socio-economic drought in the upper catchment is close to 84%. And this makes sense (Table 17.3).

While a short-term drought usually happens when a couple of months miss their long-term rainfall averages (which isn't uncommon), an agricultural drought happens when an entire rainfall season is missing its mark. And once rainfall has missed an entire rainfall season, which in the Indian case is the monsoon months, there's very little chance that the remaining months will make up for the deficiency, and the drought situation can easily creep into a medium- to long-term hydrological

TABLE 17.3
Possibility of One Drought Type Progressing into Another

Drought Progression	Upper	Middle	Lower
		Chance in %	
If meteorological, then agricultural	61.68	62.56	65.41
If agricultural, then hydrological	77.21	75.33	78.05
If hydrological, then socio-economic	83.66	81.58	79.63

phenomenon. If this shortfall in rainfall lasts for more than two consecutive monsoon cycles, long-term socio-economic drought is never far off. The situation is only improved when an above-average rainfall season occurs, and this is evident in (Figure 17.6). It is observable that every severe- and extreme-category drought is followed by a period of fairly wet years. And there are more wet years as one moves from a short-term meteorological drought to long-term socio-economic drought.

In the lower reaches of the basin, the chances of a meteorological drought creeping into an agricultural drought, of an agricultural drought extending into a hydrologic drought, and of a hydrologic drought extending further into a socio-economic drought are 65%, 78%, and 80% respectively. The corresponding numbers in the middle catchment are 63%, 75%, and 82%, respectively. These numbers also indicate an interesting dynamic. While the probability that a meteorological drought will extend into an agricultural drought is lowest in the upper catchment area and progressively increases as we move towards the lower reaches, the situation progressively flips over as we begin to look at the progression of other drought types. In the case of a hydrologic drought extending to long-term socio-economic drought, the probability is lowest in the lower catchment and increases as we move north towards the upper reaches. In other words, the probability of drought progression is more extreme in the upper catchment and becomes progressively smoother as one moves towards the lower catchment regions.

17.3.4.2 Spatial Progression of Drought

Almost as a corollary to the temporal nature of drought progression, the spatial progression shows a similar trend too. When a rainfall-deficient situation persists, the spatial extent of the drought, too, begins to increase. (Figure 17.7) shows one such transition from June 2014 to December 2017. Since the SPEI values for different time scales had different maximum and minimum limits, to make a uniform spatial interpolation possible, the negative SPEI values were normalized to lie between 0 and 1. To do that, the values were first standardized using their mean and standard deviations and then divided by the maximum standardized value in each SPEI time scale to create a corresponding dataset between 0 and 1. The higher the normalized value, the lower the severity, and vice versa.

A short-lived precipitation deficit makes a dent at station 2 in the upper catchment area, which then gets progressively widened as we move from spei-3 to spei-24, i.e., from meteorological to socio-economic drought. The drought slowly progresses, as is evident from the percentage chances listed earlier, from meteorological drought to socio-economic drought with varying degrees of severity in different reaches of the basin. The severity is greater in the upper catchment areas and lessens as we move towards the lower reaches.

17.3.5 DROUGHT MAGNITUDE AND INTENSITY

While frequency and duration can give a pretty clear picture of the quantitative nature of the drought in a region, a reference to drought magnitude and mean intensity can paint an even better picture of the ground realities. It can provide a window to the extent of drought in a region.

A careful look at the drought magnitude reveals a picture in contrast as far as meteorological and agricultural drought is concerned (Table 17.4). In every 20 years, the magnitude of agricultural drought, including severe and extreme cases, is much greater than for meteorological drought, indicating that even though the duration and frequency of the latter are lower than for the former and comparable to hydrological drought, the extent of drought magnitude is very high, indicating starker water deficiencies on the 6- to 9-month time scales than others. This points to other potential challenges, one of them being possible stress on agricultural productivity and the related food security, social welfare, and other relevant socio-economic parameters.

The result for drought intensity brings into the picture another grim reality (Table 17.5). During 2021–2040, almost every drought type, except for socio-economic, has a mean intensity of −2.0 and below, indicating that the period is susceptible to extreme water deficiencies. The condition extends till the 2050s, albeit with even more intensity in the case of agricultural droughts. The conditions

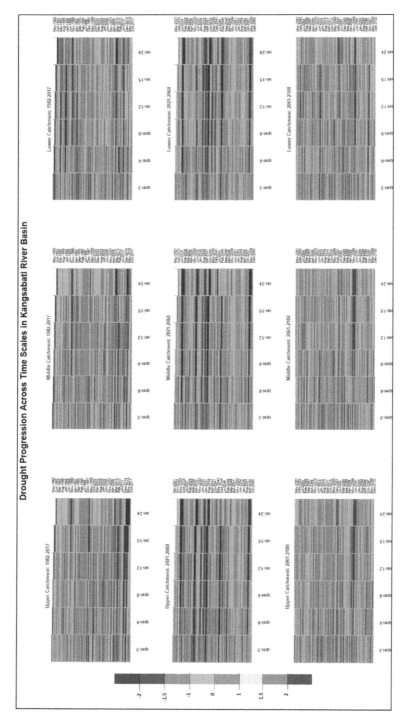

FIGURE 17.6 Heatmaps showing the temporal progression of drought in the Kangsabati river basin from 1982 to 2100.

Long-Term Drought Assessment and Prediction 265

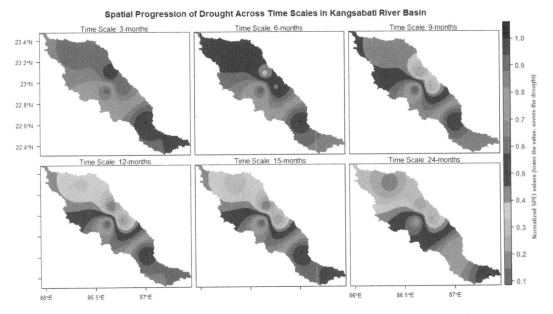

FIGURE 17.7 Map showing the spatial progression of drought for the period from June 2014 to December 2017.

improve a bit in the last four decades. Hydrological and socio-economic droughts, on the other hand, are milder in their intensity in comparison to the other two types. Except for 2021–2040, the intensity of socio-economic drought has remained above −1.0, indicating mostly milder droughts. Even for 2021–2040, it doesn't go beyond −1.5, implying mostly moderate droughts.

While the drought intensities of different drought types across the decades are informative enough, it's the basin-wise characteristics that are particularly interesting, in a way reinforcing what has been observed so far vis-à-vis frequency and duration analysis. In a way, they tie up with the cyclical nature of climate that we have been alluding to throughout. For instance, while drought intensities are higher in the upper catchment for meteorological droughts till 2080, the situation almost flips over in the next two decades, with the lower reaches being more impacted. Agricultural and hydrological droughts, on the other hand, have the exact opposite nature in comparison to meteorological. For these drought types, while lower reaches have more intense droughts, this lessens progressively as one moves towards the upper catchment. But post 2060, the situation flips here as well, with the upper reaches being more vulnerable.

Socio-economic droughts are quite different in their drought intensity response. For 2021–2040, while the middle catchment, oddly, has been slightly less intensely impacted, the upper and lower catchment experience slightly more intense droughts. On the other hand, 2041–2060 and 2081–2100 have a distinct catchment-specific response. During both periods, the lower reaches have been disproportionately more intensely impacted than the other catchments. On the other hand, 2061–2080 has remained relatively mild overall. Regardless, except for 2021–2040, the average intensity values never went below −1.0, indicating mostly mild droughts.

17.3.6 Trend Analysis

The Modified Mann Kendall Trend test for drought severity reveals some pretty significant trend characteristics, especially for agricultural and hydrological droughts. The months that are of particular concern include the winter months of January and February, the tail-end monsoon month of September, and the post-monsoon months of October, November, and December. The only exception is hydrological droughts, where almost every month seems to be under stress.

TABLE 17.4
Catchment-Specific Drought Magnitude for Different Drought Types during the Period 2021–2100

Drought Magnitude		Meteorological			Agricultural			Hydrological			Socio-Economic		
		Upper	Middle	Lower	Upper	Middle	Lower	Upper	Middle	Lower	Upper	Middle	Lower
2021–2040	MD	−42.21	−40.11	−41.95	−51.37	−46.55	−58.46	−41.94	−40.14	−44.09	−23.19	−15.44	−22.90
	SMD	−16.42	−18.38	−18.05	−20.65	−21.77	−29.43	−17.26	−15.31	−24.88	−13.61	−10.09	−13.45
	EMD	−13.51	−9.54	−10.51	−17.15	−15.41	−16.93	−16.74	−14.72	−10.37	−8.47	−4.35	−8.23
2041–2060	MD	−40.18	−39.98	−39.36	−62.11	−62.07	−65.62	−29.78	−30.76	−40.00	−9.69	−6.55	−16.06
	SMD	−16.73	−12.12	−13.67	−30.06	−29.98	−27.89	−16.15	−16.03	−17.98	−6.53	−3.55	−12.73
	EMD	−7.51	−11.76	−10.24	−12.79	−10.73	−19.06	−6.20	−6.26	−15.02	–	–	–
2061–2080	MD	−30.43	−31.22	−27.66	−34.21	−33.90	−31.87	−11.21	−11.91	−7.87	−3.77	−2.33	−2.75
	SMD	−8.90	−10.89	−8.31	−13.54	−12.18	−9.64	−1.54	−1.58	–	–	–	–
	EMD	−2.27	−2.22	−2.06	–	–	–	–	–	–	–	–	–
2081–2100	MD	−26.44	−27.65	−32.33	−32.75	−37.36	−32.08	−10.97	−13.42	−13.11	−3.14	−6.64	−7.21
	SMD	−8.51	−8.02	−13.39	−11.59	−10.79	−11.55	−4.56	−4.88	−4.85	−1.58	−3.17	−4.55
	EMD	−2.16	−4.38	–	–	−2.08	–	−2.01	−2.18	–	–	–	–

MD: drought magnitude; SMD: severe drought magnitude; EMD: extreme drought magnitude.

Long-Term Drought Assessment and Prediction

TABLE 17.5
Catchment-Specific Drought Intensity for Different Drought Types during the Period 2021–2100

Drought Intensity	Meteorological			Agricultural			Hydrological			Socio-Economic		
	Upper	Middle	Lower	Upper	Middle	Lower	Upper	Middle	Lower	Upper	Middle	Lower
2021–2040	−2.11	−2.01	−2.10	−2.57	−2.33	−2.92	−2.10	−2.01	−2.20	−1.16	−0.77	−1.15
2041–2060	−2.01	−2.00	−1.97	−3.11	−3.10	−3.28	−1.49	−1.54	−2.00	−0.48	−0.33	−0.80
2061–2080	−1.52	−1.56	−1.38	−1.71	−1.70	−1.59	−0.56	−0.60	−0.39	−0.19	−0.12	−0.14
2081–2100	−1.32	−1.38	−1.62	−1.64	−1.87	−1.60	−0.55	−0.67	−0.66	−0.16	−0.33	−0.36

While annual meteorological drought severities show a significant positive trend with $p = 0.05$ across the basin, only the upper catchment exhibits an upward trend of some significance ($p = 0.1$) for the same drought type during the winter months. But the monsoon season, oddly, shows a very high statistical significance in its trend, with $p < 0.0001$ (Table 17.6).

Agricultural droughts, on the other hand, are particularly significant in the winter and post-monsoon months. While January shows statistical significance with $p < 0.05$ in the upper and middle catchments, the lower reaches show an even stronger statistical significance with $p < 0.0001$. February, meanwhile, shows a non-significant trend in the upper catchment but strong statistical significance in the other catchments. At the tail-end of the monsoon, in September, while the upper catchment exhibits a strong statistical significance in its severity with $p < 0.0001$, the significance decreases to $p < 0.01$ in the middle catchment, and there is no significant trend in the lower reaches. October and November also exhibit a strong statistical significance in trend, while significance in December ranges from $p < 0.05$ in the upper and lower reaches to $p < 0.0001$ in the middle catchment. Similar significant trends with varying strengths can be found in seasonal Mann Kendall trends as well.

The scope and extent of significance in trends seem to have intensified for hydrologic droughts. With the exception of June, August, and September, almost every month has been exhibiting statistically significant trends in their severity. While the upper catchment in January, April, monsoons, and post monsoons, the lower catchment in March, October, December, and pre and post monsoons, and the middle catchment during October exhibit statistical significance with $p < 0.05$, the upper catchment in March, November, and during pre-monsoon, the middle catchment in November, December, and during winter, and the lower catchment during November exhibit statistical significance with $p < 0.01$. December in the upper catchment and post-monsoon in the middle catchment exhibit even stronger statistical significance with $p < 0.0001$.

These results indicate a few broad observations. First, in terms of the period, the twenty-first century has been particularly drought prone, and this is expected to continue until the mid-century, when temperatures under RCP-45 emission scenarios are expected to peak. Post 2060, the conditions leading to drought are expected to lessen, as is its occurrence. Second, over 118 years, the climatic conditions prevalent in the basin have also changed, as is visible from the subtle changes like drought severity in different catchments over the 20-year cycle. For instance, while the lower reaches were extremely impacted by droughts from 1982 to 2000, the situation flipped over in the ensuing decades. Furthermore, while the period from 1982 to 2000 was moderately conducive to drought events, the period from 2001 to 2060 is significantly favorable for the occurrence of drought, with deliverance coming only after 2060.

Furthermore, the drought trends indicate that while the factors causing droughts will mostly intensify until mid-century, the conditions will improve in the next few decades, only to get into a slump again in the last two decades. But this change won't be uniform. Climatic conditions in different reaches of the basin are likely to develop and behave differently, with seasonality shifting from one end of the basin to the other, almost like a cycle.

A few broad observations can be drawn from the results of drought frequency analysis as well. First, just like the nature of drought severity, the basin-wise analysis of drought frequency also points to subtle changes in the climatic conditions of the basin. While the frequency of drought events is higher during 2001–2060 as compared with 1982–2000, it decreases during 2061–2080, only to rise slightly again during 2081–2100, indicating the cyclical nature of climate. The recurrent nature of the meteorological drought has been indicated in another study covering the upper catchment of the Kangsabati river basin as well, though that study covered the period from 1965 to 2001 only (Mishra and Desai, 2005).

Second, over the years, while the frequency of drought increases from the advent of the twenty-first century and peaks during the 2050s, the proportion of severe and extreme severities has fallen considerably, indicating that while water deficiency will be the norm, it is unlikely to get completely out of hand. The droughts, for the most part, will be mild to moderate in nature.

Long-Term Drought Assessment and Prediction

TABLE 17.6
Mann Kendall Trends for Monthly Drought Severities

Time Series	Meteorological			Agricultural			Hydrological			Socio-Economic		
	Upper	Middle	Lower	Upper	Middle	Lower	Upper	Middle	Lower	Upper	Middle	Lower
January	0.000	−0.003	0.003	0.009**	0.010**	0.009##	0.009**	0.012	0.006	0.005	−0.031	0.016
February	−0.005	−0.008	−0.004	0.003	0.005**	0.005**	0.010	0.012	0.007	0.004	−0.029	0.022
March	0.004	0.004	−0.001	0.004	0.002	0.000	0.010#	0.012	0.008**	0.004	−0.031	0.021
April	0.002	0.003	0.003	0.000	0.000	0.006	0.012**	0.007	0.006	0.004	−0.028	0.025
May	0.001	−0.003	0.001	0.001	0.001	0.000	0.009*	0.007*	0.005*	0.004	−0.024	0.025
June	0.004	0.001	0.006	0.000	0.003	−0.001	0.003	0.001	0.001	0.004	−0.017	0.019
July	0.002	0.002	0.006	0.002	0.003	−0.001	0.002	0.003	0.005*	0.003	−0.012	0.022
August	0.001	0.001	−0.002	0.002	0.001	−0.001	0.001	0.005	0.004	0.005	−0.001	0.017
September	0.002	0.004	0.005	0.013##	0.013#	0.003	0.005	0.006	0.005	0.001	−0.014	0.025*
October	0.004	0.002	−0.001	0.012#	0.009	0.006*	0.009*	0.011**	0.010**	0.001	−0.034	0.017
November	0.000	−0.003	0.002	0.015#	0.008*	0.007	0.013#	0.014#	0.012#	0.004	−0.030	0.000
December	−0.003	−0.003	−0.002	0.010**	0.009##	0.009**	0.011##	0.015#	0.008**	0.004	−0.017	0.016
Annual	0.005**	0.004**	0.004**	0.004	0.005	0.008#	0.004	0.003	0.006#	0.005**	0.009	0.042
Winter	−0.005*	−0.006	−0.004	0.007	0.008**	0.007**	0.006	0.013#	0.006	0.005	−0.031	0.020
Pre-Monsoon	0.001	0.002	−0.002	−0.002	0.001	0.003	0.008#	0.008**	0.008**	0.004	−0.028	0.028
Monsoon	0.009##	0.005	−0.001	0.007*	0.006	−0.001	0.004**	0.005	0.004	0.003	−0.002	0.035
Post-Monsoon	0.000	0.001	−0.001	0.015#	0.008*	0.007*	0.011**	0.016##	0.010**	0.002	−0.016	0.016

Significance levels: ## at $p < 0.0001$, # at $p < 0.01$, ** at $p < 0.05$, and * at $p < 0.1$.

The trend in drought progression is also indicative of the climatic trends prevalent in the basin. As Kumar et al. (2019, 2020) observed, seasonality in precipitation increases in the basin as one moves from lower to upper catchment regions. Temperature, too, follows a similar trend. The summers are hotter in the upper parts of the basin in comparison to the lower reaches. And, as one moves south and towards the lower parts of the basin, the relative proximity to the sea means that these areas exhibit more uniformity in their rainfall occurrence. It's only when precipitation is too deficient that one drought type extends into another. This is perhaps the reason why the percentages are more spread out (62–84%) in the upper catchment but are closer in the lower catchment (65–80%). The uniformity in rainfall distribution and temperature gradient is what makes drought progression less likely in the lower reaches.

The spatial progression of drought also seems to align with similar studies conducted elsewhere in India (Mishra and Desai, 2005; Mishra and Singh, 2009; Sharma and Mujumdar, 2017; Naresh Kumar et al., 2012). Similarly, the prevalence of statistically significant Mann Kendall trend in meteorological droughts correlates with a study done by Bisht et al. (2019), who concluded that there was a strictly significant annual trend in meteorological drought severity for this part of the country.

17.4 CONCLUSION

The methodology adopted in the study encourages the use of OSS in data preparation, processing, analyses, etc. to reach the objective of finding out the drought situation of the Kangshabati river basin. The work is largely based on public domain data, too. Therefore, further comprehensive assessment of drought phenomena with a number of other socio-economic variables is possible without a commercial platform based on OSS only. Nevertheless, as far as the findings of the work are concerned, several facts have been established through the study. Water stress conditions in the form of droughts are bound to cause threats to the overall physical-economic balance of a region. Food security issues are often threatened by the concurrent existence of drought and agrarian dependencies, causing many difficulties for decision-makers in combating the problem. The area under study has been affected by these hydrologic extremes and their associated concerns, which makes this study relevant and useful for policy-makers. The findings also confirm the drought severity in this area on a periodic time scale. Therefore, it is imperative to carry out a drought appraisal against the backdrop of climate change in order to opt for the best-suited and strategic policy-making applicable in this area. The results of the SPEI-based drought study conducted here are spread over 118 years, from 1982 to 2100, in the Kangsabati river basin and have provided some interesting insights into the problem. The climate in the basin appears to follow a cyclical nature, which can be gauged from the behavior of drought severity, drought frequency, and drought duration every 20 years. Broadly speaking, while drought characteristics seem to be more visible during 1982–1983 and 1988–89, the next two decades are relatively peaceful. Post 2005, especially after 2010, the occurrences, severity, and duration of drought are on the rise, and this is expected to continue till the 2060s, after which it is relatively drought free. Post 2080, though, drought events are expected to rise again, albeit not in the same severity as before 2050. While the frequency and duration of drought events are higher during 2041–2060, the frequency of severe and extreme drought events is higher in the preceding decades of 2021–2040. Furthermore, while the severe drought durations are higher during 2041–2060, the extreme drought durations are higher during 2021–2040. The study has established that drought is a creeping phenomenon in the area under consideration. Drought occurrences showed a high possibility of transitioning from one drought type to another as well as increasing their spatial extent. Statistically, while two out of three meteorological droughts are expected to transition to an agricultural drought, four out of every five hydrological droughts are expected to progress into a long-term socio-economic drought. Also, while the possibility of a short-term drought progressing to a medium-term drought is higher in the upper catchment, the possibility of a medium-term drought progressing to a long-term drought is higher in the lower reaches. In terms of severity, agricultural drought is much more lethal than other drought types. This is unpleasant news for a basin whose economy depends mostly on agriculture. This also calls for investment in building more irrigation networks and strengthening the water supply chain. Going into the future, the years which are of particular concern are 2023–2025, 2031–2034, the entire 2040s, 2058–2060, and 2081–2084. While scattered drought events have been predicted for other years too, these periods are of particular concern given the severity and the duration of the droughts expected during these periods. In terms of season, winter and monsoon months appear to be particularly vulnerable, with the Mann Kendall test also showing mostly positive and statistically significant trends, particularly in the upper catchment, indicating a worsening of drought-inducing conditions i.e., either a rise in average temperature or a dip in rainfall during monsoons, of which the former seems to be the likely scenario. Annually, too, the trends in drought severity are exhibiting positive trends with pretty strong statistical significance.

ACKNOWLEDGMENTS

This work is a product of research financed by DST SERB, Govt. of India under the Core Research Grant scheme with the file number EMR/2016/006380 date: 28/08/2017. We acknowledge this agency for its financial assistance and other government and global agencies for providing us with the necessary data to conduct the research.

REFERENCES

Abramowitz, M. & Stegun, I. A. 1965. *Handbook of Mathematical Functions*. Dover Publications, New York, 361.

Adams, H. D., Guardiola-Claramonte, M., Barron-Gafford, G. A., Villegas, J. C., Breshears, D. D., Zou, C. B., Troch, P. A. & Huxman, T. E. 2009. The temperature sensitivity of drought-induced tree mortality portends increased regional die-off under global-change-type drought. *Proceedings of the National Academy of Sciences of the United States of America*, 106, 7063–7066.

Adarsh, S., Karthik, S., Shyma, M., Prem, G. D., Shirin Parveen, A. T. & Sruthi, N. 2018. Developing short term drought severity-duration-frequency curves for Kerala meteorological subdivision, India using bivariate copulas. *KSCE Journal of Civil Engineering*, 22, 962–973.

Akhter, J., Das, L. & Deb, A. 2017. CMIP5 ensemble-based spatial rainfall projection over homogeneous zones of India. *Climate Dynamics*, 49, 1885–1916.

Allen, R., Pereira, L., Raes, D. & Smith, M. 1998a. Chapter 2 – FAO penman-monteith equation. *Crop Evapotranspiration – Guidelines for Computing Crop Water Requirements – FAO Irrigation and Drainage Paper*, 56, 1–326.

Allen, R. G., Pereira, L. S., Raes, D. & Smith, M. 1998b. *Crop Evapotranspiration – Guidelines for Computing Crop Water Requirements – FAO Irrigation and Drainage Paper 56*. FAO, Rome, 300, D05109.

Arndt, E. 2020. *Did You Know? Monitoring References*. National Centers for Environmental Information (NCEI), USA.

Bandyopadhyay, A., Bhadra, A., Swarnakar, R., Raghuwanshi, N. & Singh, R. 2012. Estimation of reference evapotranspiration using a user-friendly decision support system: DSS_ET. *Agricultural and Forest Meteorology*, 154, 19–29.

Barriopedro, D., Fischer, E. M., Luterbacher, J., Trigo, R. M. & García-Herrera, R. 2011. The hot summer of 2010: Redrawing the temperature record map of Europe. *Science*, 332, 220–224.

Beria, H., Nanda, T., Bisht, D. S. & Chatterjee, C. 2017. Does the GPM mission improve the systematic error component in satellite rainfall estimates over TRMM? An evaluation at a pan-India scale. *Hydrology and Earth System Sciences*, 21, 6117–6134.

Bisht, D. S., Sridhar, V., Mishra, A., Chatterjee, C. & Raghuwanshi, N. S. 2019. Drought characterization over India under the projected climate scenario. *International Journal of Climatology*, 39, 1889–1911.

Bodner, G., Nakhforoosh, A. & Kaul, H.-P. 2015. Management of crop water under drought: A review. *Agronomy for Sustainable Development*, 35, 401–442.

Bonsal, B. R., Cuell, C., Wheaton, E., Sauchyn, D. J. & Barrow, E. 2017. An assessment of historical and projected future hydro-climatic variability and extremes over southern watersheds in the Canadian Prairies. *International Journal of Climatology*, 37, 3934–3948.

Breshears, D. D., Cobb, N. S., Rich, P. M., Price, K. P., Allen, C. D., Balice, R. G., Romme, W. H., Kastens, J. H., Floyd, M. L., Belnap, J., Anderson, J. J., Myers, O. B. & Meyer, C. W. 2005. Regional vegetation die-off in response to global-change-type drought. *Proceedings of the National Academy of Sciences of the United States of America*, 102, 15144–15148.

Chadd, R. P., England, J. A., Constable, D., Dunbar, M. J., Extence, C. A., Leeming, D. J., Murray-Bligh, J. A. & Wood, P. J. 2017. An index to track the ecological effects of drought development and recovery on riverine invertebrate communities. *Ecological Indicators*, 82, 344–356.

Chen, H. & Sun, J. 2017. Characterizing present and future drought changes over eastern China. *International Journal of Climatology*, 37, 138–156.

Das, L., Annan, J., Hargreaves, J. & Emori, S. 2012. Improvements over three generations of climate model simulations for eastern India. *Climate Research*, 51, 201–216.

Dibike, Y., Prowse, T., Bonsal, B. & O'neil, H. 2017. Implications of future climate on water availability in the western Canadian river basins. *International Journal of Climatology*, 37, 3247–3263.

Dracup, J. A., Lee, K. S. & Paulson Jr, E. G. 1980. On the statistical characteristics of drought events. *Water Resources Research*, 16, 289–296.

Dutta, D., Kundu, A., Patel, N. R., Saha, S. K. & Siddiqui, A. R. 2015. Assessment of agricultural drought in Rajasthan (India) using remote sensing derived vegetation condition index (VCI) and standardized precipitation index (SPI). *The Egyptian Journal of Remote Sensing and Space Science*, 18, 53–63.

Dwomoh, F. K., Wimberly, M. C., Cochrane, M. A., Numata, I. 2019. Forest degradation promotes fire during drought in moist tropical forests of Ghana. *Forest Ecology and Management*, 440, 158–168.

Edwards, D. C. 1997. *Characteristics of 20th Century Drought in the United States at Multiple Time Scales*. Air Force Inst of Tech Wright-Patterson AFB OH, USA.

Gao, X., Zhao, Q., Zhao, X., Wu, P., Pan, W., Gao, X. & Sun, M. 2017. Temporal and spatial evolution of the standardized precipitation evapotranspiration index (SPEI) in the Loess Plateau under climate change from 2001 to 2050. *Science of the Total Environment*, 595, 191–200.

Ghosh, K. G. 2018. Analysis of rainfall trends and its spatial patterns during the last century over the gangetic West Bengal, eastern India. *Journal of Geovisualization and Spatial Analysis*, 2, 15.

Guttman, N. B. 1998. Comparing the palmer drought index and the standardized precipitation index 1. *JAWRA Journal of the American Water Resources Association* 34, 113–121.

Hamed, K. H. & Rao, A. R. 1998. A modified Mann-Kendall trend test for autocorrelated data. *Journal of Hydrology*, 204, 182–196.

Hargreaves, G. & Allen, R. 2003. History and evaluation of hargreaves evapotranspiration equation. *Journal of Irrigation and Drainage Engineering-ASCE – J Irrig Drain Eng-ASCE*, 129, 53–63.

Hargreaves, G. H. 1994. Defining and using reference evapotranspiration. *Journal of Irrigation and Drainage Engineering*, 120, 1132–1139.

Hartmann, H., Moura, C. F., Anderegg, W. R. L., Ruehr, N. K., Salmon, Y., Allen, C. D., Arndt, S. K., Breshears, D. D., Davi, H., Galbraith, D., Ruthrof, K. X., Wunder, J., Adams, H. D., Bloeman, J., Cailleret, M., Cobb, R., Gessler, A., Grams, T. E. E., Jansen, S., Kautz, M., Lloret, F. & O'Brien, M. 2018. Research frontiers for improving our understanding of drought-induced tree and forest mortality. *New Phytologist* 218, 15–28.

Hayes, M. J., Svoboda, M. D., Wiihite, D. A. & Vanyarkho, O. V. 1999. Monitoring the 1996 drought using the standardized precipitation index. *Bulletin of the American Meteorological Society*, 80, 429–438.

Hosking, J. R. 1986. *The theory of probability weighted moments*, IBM Research Division, TJ Watson Research Center, New York, USA.

Kenawy, E.-R., Saad-Allah, K. & Hosny, A. 2018. Mitigation of drought stress on three summer crop species using the superabsorbent composite Gelatin-g-p(AA-co-AM)/RH. *Communications in Soil Science and Plant Analysis*, 49, 2828–2842.

Kendall, M. G. & Gibbons, J. D. 1990. *Rank correlation methods*. Ed. Edward Arnold, London, UK.

Keyantash, J. & Dracup, J. A. 2002. The quantification of drought: An evaluation of drought indices. *Bulletin of the American Meteorological Society*, 83, 1167–1180.

Khan, M. I., Liu, D., Fu, Q., Saddique, Q., Faiz, M. A., Li, T., Qamar, M. U., Cui, S. & Cheng, C. 2017. Projected changes of future extreme drought events under numerous drought indices in the Heilongjiang Province of China. *Water Resources Management*, 31, 3921–3937.

Kumar, A., Santra Mitra, S., Santra, A. & Routh, S. 2020. Precipitation and runoff trend analysis using Mann Kendall and Sen's slope estimator for Kangsabati River Basin in West Bengal. *International Conference on Sustainable Water Resources Management Under Changed Climate*. Jadavpur University, 13–15 March 2020, Kolkata, India.

Kumar, A., Santra Mitra, S., Santra, A., Routh, S. & Sinha, S. 2019. Evaluation of long-term precipitation trends and its seasonality over a catchment in lower gangetic basin of West Bengal. *ISH – HYDRO 2019 International Conference*, 18–20 December 2019 Osmania University, Hyderabad.

Kummu, M., De Moel, H., Ward, P. J. & Varis, O. 2011. How close do we live to water? A global analysis of population distance to freshwater bodies. *PloS One*, 6, e20578.

Laaha, G., Gauster, T., Tallaksen, L. M., Vidal, J.-P., Stahl, K., Prudhomme, C., Heudorfer, B., Vlnas, R., Ionita, M., Van Lanen, H. A. J., Adler, M.-J., Caillouet, L., Delus, C., Fendekova, M., Gailliez, S., Hannaford, J., Kingston, D., Van Loon, A. F., Mediero, L., Osuch, M., Romanowicz, R., Sauquet, E., Stagge, J. H & Wong, W. K. 2017. The European 2015 drought from a hydrological perspective. *Hydrology and Earth System Sciences*, 21, 3001–3024.

Linares, J. C. & Camarero, J. J. 2012. From pattern to process: Linking intrinsic water-use efficiency to drought-induced forest decline. *Global Change Biology*, 18, 1000–1015.

Lobell, D. B., Schlenker, W. & Costa-Roberts, J. 2011. Climate trends and global crop production since 1980. *Science*, 333, 616–620.

Lund, J., Medellin-Azuara, J., Durand, J. & Stone, K. 2018. Lessons from California's 2012–2016 drought. *Journal of Water Resources Planning and Management*, 144, 04018067.

Mallya, G., Mishra, V., Niyogi, D., Tripathi, S. & Govindaraju, R. S. 2016. Trends and variability of droughts over the Indian monsoon region. *Weather and Climate Extremes*, 12, 43–68.

Mann, H. B. 1945. Nonparametric tests against trend. *Econometrica: Journal of the Econometric Society*, 13, 245–259.

Mcguire, A. D., Ruess, R. W., Lloyd, A., Yarie, J., Clein, J. S. & Juday, G. P. 2010. Vulnerability of white spruce tree growth in interior Alaska in response to climate variability: Dendrochronological, demographic, and experimental perspectives. *Canadian Journal of Forest Research*, 40, 1197–1209.

Mckee, T. B., Doesken, N. J. & Kleist, J. 1993. The relationship of drought frequency and duration to time scales. *Proceedings of the 8th Conference on Applied Climatology*. Boston, 179–183.

Mishra, A. & Desai, V. 2005. Spatial and temporal drought analysis in the Kansabati river basin, India. *International Journal of River Basin Management*, 3, 31–41.

Mishra, A. & Liu, S. C. 2014. Changes in precipitation pattern and risk of drought over India in the context of global warming. *Journal of Geophysical Research: Atmospheres*, 119, 7833–7841.

Mishra, A. & Singh, V. P. 2009. Analysis of drought severity-area-frequency curves using a general circulation model and scenario uncertainty. *Journal of Geophysical Research, Atmospheres*, 114, 1–18.

Mishra, A. K. & Singh, V. P. 2010. A review of drought concepts. *Journal of Hydrology*, 391, 202–216.

Nam, W.-H., Hayes, M. J., Svoboda, M. D., Tadesse, T. & Wilhite, D. A. 2015. Drought hazard assessment in the context of climate change for South Korea. *Agricultural Water Management*, 160, 106–117.

Naresh Kumar, M., Murthy, C., Sesha Sai, M. & Roy, P. 2012. Spatiotemporal analysis of meteorological drought variability in the Indian region using standardized precipitation index. *Meteorological Applications*, 19, 256–264.

NOAA-Psl PSL Data. CPC unified gauge-based analysis of daily precipitation over CONUS. NOAA Physical Sciences Laboratory, Boulder, CO.

Palmer, W. C. 1965. *Meteorological Drought*. US Department of Commerce, Research Paper No. 45, Weather Bureau, Washington DC.

Ray, R. L., Fares, A., & Risch, E. 2018. Effects of drought on crop production and cropping areas in Texas. *Agricultural & Environmental Science*, 3, 170037.

Rebetez, M., Mayer, H., Dupont, O., Schindler, D., Gartner, K., Kropp, J. P. & Menzel, A. 2006. Heat and drought 2003 in Europe: A climate synthesis. *Annals of Forest Science*, 63, 569–577.

Revadekar, J. & Preethi, B. 2012. Statistical analysis of the relationship between summer monsoon precipitation extremes and foodgrain yield over India. *International Journal of Climatology*, 32, 419–429.

Rhee, J. & Cho, J. 2016. Future changes in drought characteristics: Regional analysis for South Korea under CMIP5 projections. *Journal of Hydrometeorology*, 17, 437–451.

Santos, M. A. 1983. Regional droughts: A stochastic characterization. *Journal of Hydrology*, 66, 183–211.

Santra Mitra, S., Santra, A. & Kumar, A. 2019. Catchment specific evaluation of Aphrodite's and TRMM derived gridded precipitation data products for predicting runoff in a semi gauged watershed of Tropical India. *Geocarto International*, 1–16. https://doi.org/10.1080/10106049.2019.1641563.

Santra Mitra, S., Wright, J., Santra, A. & Ghosh, A. R. 2015. An integrated water balance model for assessing water scarcity in a data-sparse interfluve in eastern India. *Hydrological Sciences Journal*, 60, 1813–1827.

Scasta, J. D. & Rector, B. S. 2014. Drought and ecological site interaction on plant composition of a semi-arid rangeland. *Arid Land Research and Management*, 28, 197–215.

Sen, P. K. 1968. Estimates of the regression coefficient based on Kendall's tau. *Journal of the American Statistical Association*, 63, 1379–1389.

Sengupta, S. & Cai, W. 2019. A quarter of humanity faces looming water crises. *The New York Times*. www.nytimes.com/interactive/2019/08/06/climate/world-water-stress.html.

Sharma, S. & Mujumdar, P. 2017. Increasing frequency and spatial extent of concurrent meteorological droughts and heatwaves in India. *Scientific Reports*, 7, 1–9.

Smirnov, O., Zhang, M., Xiao, T., Orbell, J., Lobben, A. & Gordon, J. 2016. The relative importance of climate change and population growth for exposure to future extreme droughts. *Climatic Change*, 138, 41–53.

Smitha, P., Narasimhan, B., Sudheer, K. & Annamalai, H. 2018. An improved bias correction method of daily rainfall data using a sliding window technique for climate change impact assessment. *Journal of Hydrology*, 556, 100–118.

Spinoni, J., Naumann, G., Carrao, H., Barbosa, P. & Vogt, J. 2014. World drought frequency, duration, and severity for 1951–2010. *International Journal of Climatology*, 34, 2792–2804.

Spinoni, J., Vogt, J. V., Naumann, G., Barbosa, P. & Dosio, A. 2018. Will drought events become more frequent and severe in Europe? *International Journal of Climatology*, 38, 1718–1736.

Surendran, U. et al. 2019. Analysis of drought from humid, semi-arid and arid regions of India using DrinC Model with different drought indices. *Water Resources Management*, 33, 1521–1540.

Svoboda, M. & Fuchs, B. 2016. *Handbook of Drought Indicators and Indices*. World Meteorological Organization (WMO) and Global Water Partnership (GWP). Integrated Drought Management Programme (IDMP), Integrated Drought Management Tools and Guidelines Series 2. Geneva, Switzerland.

Svoboda, M., Hayes, M. & Wood, D. 2012. *Standardized Precipitation Index User Guide*. World Meteorological Organization, Geneva, Switzerland.

Taha, R. S., Alharby, H. F., Bamagoos, A. A., Medani, R. A. & Rady, M. M. 2020. Elevating tolerance of drought stress in *Ocimum basilicum* using pollen grains extract: A natural biostimulant by regulation of plant performance and antioxidant defense system. *South African Journal of Botany*, 128, 42–53.

Theil, H. 1950. A rank-invariant method of linear and polynomial regression analysis, 3; confidence regions for the parameters of polynomial regression equations. *Indagationes Mathematicae*, 1, 467–482.

Thomas, T., Nayak, P. & Ghosh, N. C. 2015. Spatiotemporal analysis of drought characteristics in the Bundelkhand region of central India using the standardized precipitation index. *Journal of Hydrologic Engineering*, 20, 05015004.

Thornthwaite, C. W. 1948. An approach toward a rational classification of climate. *Geographical Review*, 38, 55–94.

Tian, L., Yuan, S. & Quiring, S. M. 2018. Evaluation of six indices for monitoring agricultural drought in the south-central United States. *Agricultural and Forest Meteorology*, 249, 107–119.

Vicente-Serrano, S. M., Beguería, S. & López-Moreno, J. I. 2010. A multiscalar drought index sensitive to global warming: The standardized precipitation evapotranspiration index. *Journal of Climate*, 23, 1696–1718.

Weigel, A. P., Knutti, R., Liniger, M. A. & Appenzeller, C. 2010. Risks of model weighting in multimodel climate projections. *Journal of Climate*, 23, 4175–4191.

Wells, N., Goddard, S. & Hayes, M. J. 2004. A self-calibrating Palmer drought severity index. *Journal of Climate*, 17, 2335–2351.

Wilhite, D. A. & Glantz, M. H. 1985. Understanding: The drought phenomenon: The role of definitions. *Water International*, 10, 111–120.

Yang, M., Yan, D., Yu, Y. & Yang, Z. 2016. SPEI-based spatiotemporal analysis of drought in Haihe River Basin from 1961 to 2010. *Advances in Meteorology*, 2016, 1–10.

Yu, M., Li, Q., Hayes, M. J., Svoboda, M. D. & Heim, R. R. 2014. Are droughts becoming more frequent or severe in China based on the standardized precipitation evapotranspiration index: 1951–2010? *International Journal of Climatology*, 34, 545–558.

Zelenhasić, E. & Salvai, A. 1987. A method of streamflow drought analysis. *Water Resources Research*, 23, 156–168.

Zhang, X., Obringer, R., Wei, C., Chen, N. & Niyogi, D. 2017. Droughts in India from 1981 to 2013 and implications to wheat production. *Scientific Reports*, 7, 44552.

18 Use of Open Source Software to Assess Spatio-Temporal Variation of Agricultural Drought at Regional Scale

Debarati Bera and Dipanwita Dutta

CONTENTS

18.1 Introduction	275
18.2 Study Area	277
18.3 Materials and Methods	278
18.3.1 Data Used	278
18.3.2 Vegetation Condition Index (VCI)	280
18.3.3 Standardized Precipitation Index (SPI)	280
18.3.4 Rainfall Anomaly Index (RAI)	282
18.3.5 Yield Anomaly Index (YAI)	282
18.4 Results and Discussion	282
18.4.1 Spatio-Temporal Pattern of Vegetational Drought Using VCI	282
18.4.2 Spatio-Temporal Pattern of Meteorological Drought using SPI	283
18.4.3 Rainfall Anomaly Index (RAI)	284
18.4.4 Crop Production and Yield Anomaly Index (YAI)	285
18.4.5 Yield Anomaly Index and Standardized Precipitation Index	286
18.4.6 Standard Precipitation Index and Vegetation Condition Index	289
18.4.7 Yield Anomaly Index and Vegetation Condition Index	291
18.5 Conclusion	291
References	292

18.1 INTRODUCTION

Drought is one of the most vulnerable and widespread natural hazards, occurring in almost every climate region around the world (Hao and Singh, 2015). Due to their multifaceted impacts, droughts have been considered as the "most far-reaching of all-natural disasters" on earth by the United Nations (UN, 2014). This creeping hazard spreads slowly with time; however, its devastating effects are severe, and it may reduce the gross domestic product (GDP) of a country significantly by damaging its agriculture, inland fisheries, and allied sectors. According to the World Economic Forum, it is one of the "most expensive weather-related disasters in the world" (www.weforum.org). This phenomenon leads to catastrophic consequences for global food security, economic losses, and the ecological environment (Jiao et al., 2019; Zhong et al., 2019). According to the International Decade for Natural Disaster Reduction, drought comprises 22% of the damages caused by all disasters and affects 33% of the population of the world annually (Keshavarz et al., 2013). Prolonged dry spells reduce the ecological uptake of CO_2, resulting in an increase in the concentration of this greenhouse gas in the atmosphere (Ciais et al., 2005). Drought is indirectly related to climate change and global warming, the major factors fueling the rising frequency of this hazard all over the world (Dai, 2013; Zhong et al., 2019). It is worthy of note that the frequency of drought has increased in the past couple of decades (Sheffield and Wood, 2008), and it will be more intense and severe in the near future as a consequence of global warming (Dai, 2011a, 2013). About 50% of the landmass of India is drought prone (Kamble et al., 2010), and the vulnerable area will increase due to rising temperatures and numbers of hot days. In the last century, India has experienced a significant rising trend in mean temperature (0.484 C 100 yr^{-1}), indicating an increase in drought vulnerability (Kumar et al., 1994; IPCC, 2013). Recurring drought events over several parts of the country pose a great threat to ecosystems, livelihood, and food security. Therefore, understanding the spatio-temporal variability of drought phenomena is crucial for proper management and implementing mitigation strategies to reduce the increasing vulnerability to drought (Mishra and Singh, 2011).

In general, a drought event can be categorized into four classes, viz., meteorological, agricultural, hydrological, and socio-economic drought, depending upon its character and the nature of its impact (American Meteorological Society, 1997; Zhong et al., 2019). A drought event commences primarily due to shortage of rainfall (meteorological drought), which concurrently affects surface and groundwater storage (hydrological drought), and as a cumulative consequence, the crop production of the area is affected (agricultural drought). The deleterious effects of drought may severely damage the socio-economic conditions of an area by threatening livelihood and property. Unlike the conventional methods of drought assessment based on sparsely distributed in situ datasets, geospatial techniques have become widely accepted for monitoring and modeling drought events with good precision (Park et al., 2017). The satellite datasets have become essential in drought studies because of several advantages, i.e., reliability, seamless coverage, synoptic view, and uninterrupted availability of data at varying temporal and spatial scales (Kogan, 1995; Gu et al., 2007; Dutta et al., 2015).

With the advancement of geospatial technology, a substantial number of satellite-derived drought indices evolved. The Normalized Difference Vegetation Index (NDVI) introduced by Rouse et al. (1974) has been successfully used for monitoring vegetation stress (Peters et al., 2002), crop phenology (Padhee and Dutta, 2019), and vegetation health, and modeling yield (Dutta et al., 2013). Previous studies also explored the potential of NDVI anomalies to assess drought-related crop stress (Dutta, 2010; Li et al., 2014; Palchaudhuri and Biswas, 2020). However, the Vegetation Condition Index (VCI) derived from long-term NDVI (Kogan, 1990) proved to be an efficient indicator of agricultural drought due to its ability to segregate the weather-related stress on vegetation (Gouveia et al., 2009; Wu et al., 2015). Patel and Yadav (2015) used VCI to develop a spatial vegetation drought index (SVDI) to monitor drought stress in the Bundelkhand region of central India. The proficiency of VCI has been ascertained in a study dealing with drought assessment over Sahelian countries conducted by Noureldeen et al. (2020). They found a significant correlation between VCI anomaly and crop yield anomaly. The standardized precipitation index (SPI) developed by McKee et al. (1993) has been widely accepted for assessing meteorological droughts over various regions of the world (Vicente-Serrano et al., 2006; Li et al., 2014; Sahoo et al., 2015; Xu et al., 2015). Previous research also established a significant interrelationship between meteorological and agricultural drought through VCI and SPI (Quiring and Ganesh, 2010; Dutta et al., 2015).

The western part of West Bengal is affected by frequent drought phenomena with the failure of crop production (Dash et al., 2019; Cornish et al., 2015). Palchaudhuri and Biswas (2020) assessed VCI images derived from long-term Moderate Resolution Imaging Spectroradiometer (MODIS) NDVI and identified severe drought events in a few blocks of Purulia district. An analysis of SPI-based meteorological drought trends also reveals increasing dry spells in the drought-prone districts of West Bengal, i.e., Bankura, Purulia, and Paschim Medinipur (Bhunia et al., 2019). In spite of the evidence of severe droughts in the western part of West Bengal, the spatio-temporal variability of agricultural drought remains largely unexplored by researchers. A comprehensive study of drought inventory through long-term Earth observation datasets, which is crucial for the implementation of sustainable drought mitigation policy, is still lacking in this area.

Against this background, the present study examines the spatio-temporal variability of drought events through long-term satellite-based datasets. It also tries to understand the interrelationship between agricultural and meteorological droughts identified through satellite-derived and in situ observations. The prime focus of this study is to monitor the spatial and temporal pattern of agricultural drought during the Kharif (rainfed) season and to identify its linkage with meteorological drought. It requires long-term seasonal datasets for exploring the spatio-temporal variability of drought. However, accurate computation of such voluminous datasets is complex and requires efficient software for managing the database. The premium commercial software used for geospatial analysis is not accessible to the public due to high cost and upgradation fees. There are several tools and software packages under the GNU (GNU's not Unix) project, which are freely available to all at no cost and with no hidden conditions. Among them, R, MatLab®, SPSS, Microsoft Excel, SAS (Statistical Analysis Software), etc. are widely used for statistical analysis. According to the Free Software Foundation, software can be categorized as free if it gives the user freedom to run, copy, distribute, study, change, and improve the software (Wilson and Tchantchaleishvili, 2013). Open source software is becoming more popular not only for its cost-saving benefit but also because it makes provision for further development and modification on demand. Thus, it promotes exploring new research skills by fostering innovative thinking. This study was performed using the R studio (version 3.5.1), which is completely free and open source, with an integrated development environment (IDE) and a better user-friendly interface. There are many freely available packages, and the user can create code according to demand.

18.2 STUDY AREA

Four underdeveloped districts of the western plateau of West Bengal, i.e., Bankura, Purulia, Purba, and Paschim Medinipur, were chosen for this study. Geographically, the area is situated between 21.94°N and 22.60° N latitudes and between 85.75° and 87.78° E longitudes (Figure 18.1). The study site, which is an extended part of Chhotanagpur Plateau, covers an area of about 27,222 km². The western part comprises highly undulating gullied topography with lateritic soil, whereas the southeastern part (Purba and Paschim Medinipur districts) is mainly dominated by plain land.

Darakeshwar, Kangsabati, Rupnarayan, and Subarnarekha are the major rivers flowing from west to east, following the general slope of the area. It is worth mentioning that 50% of the water flows as runoff into the sea due to the undulating topography. The area is covered by different subtypes of laterite and alluvium soils. A substantial portion of the area is characterized by residual soils, formed by the process of weathering of bedrocks. Here, the dominant soil categories have low moisture-holding capacity, which is not suitable for agricultural practice. This area has a monsoonal climate, characterized by wet summer and dry winter seasons. The mean annual rainfall varies between 1100 and 1500 mm and is heavily skewed to late summer (monsoon) (June to September). The summer season is characterized by oppressive heat and humid days. The maximum temperature ranges from 28 to 40 °C, while the minimum temperature ranges from 9 to 12 °C.

The land use land cover (LULC) of the region was prepared from Landsat8 (OLI) datasets using supervised classification techniques (Figure 18.2). It shows agricultural practices as the dominant economic activity (Table 18.1) over the area. Some parts of the area are covered by distinct forest patches. Apart from the agricultural activities, the livelihood of a large number of people is dependent upon forests and fisheries. Since most of the areas are predominantly rainfed, uncertainty in monsoonal rainfall acts as a major controlling factor in the productivity of Kharif crops. The study area is mainly dominated by two crop growing seasons, Kharif (July to October) and Rabi (October to March) (Panigrahy et al., 2005). This part of West Bengal is relatively drier and highly vulnerable to agricultural drought. A substantial portion of the total population in this area belongs to economically underdeveloped and tribal communities, making the area more susceptible to socioeconomic drought. Moreover, low coping strategies in this underprivileged region have exacerbated the drought risk (Saha, 2015).

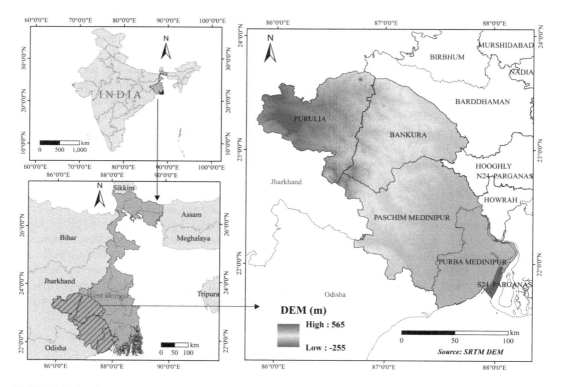

FIGURE 18.1 Study area.

18.3 MATERIALS AND METHODS

18.3.1 Data Used

MODIS (Terra) 16 day composite NDVI data (MOD13Q1) were collected (https://ladsweb.modaps.eosdis.nasa.gov) to analyze the agricultural drought conditions for the 2000–2019 period. This product is freely available with 250 m resolution from the year 2000. The datasets have been successfully used by many researchers in different domains, i.e., in vegetational dynamics (Kern et al., 2020), surface phenology (Padhee and Dutta, 2019), crop identification (da Silva et al., 2020) and vegetational stress (Testa et al., 2014; Palchaudhuri and Biswas, 2019; Phan et al., 2020).

CHIRPS (Climate Hazard Group Infrared Precipitation with station) daily rainfall data products were freely downloaded (ftp://ftp.chg.ucsb.edu/pub/org/chg/product/chirps-2.0/global_daily/tifp05) for the last 20 years (2000–2019) to assess the spatio-temporal dynamics of meteorological droughts. The CHIRPS datasets are quasi-global rainfall products, spanning from 50°S to50°N and available since 1981. They were developed by the U.S. Geological Survey (USGS) in collaboration with the Earth Resources Observation and Science (EROS) Center. With a spatial resolution of 0.05°, this satellite-based merged rainfall product also integrates the in situ rainfall observations. For the remote areas and where weather stations are inadequate, these satellite-based rainfall data are highly useful. Several researchers have successfully used CHIRPS data for early warning and monitoring of seasonal droughts (Katsanos et al., 2016; Lai et al., 2019). Apart from this, the daily gridded (0.25°) precipitation data provided by the India Meteorological Department (IMD), Pune were collected for the years 2000–2016. These ground-based rainfall data were used for validating our outcome and examining the potential of satellite-based (CHIRPS) rainfall data for monitoring drought variability in the study region.

Year-wise crop yield data were collected from the District Statistical Handbooks. The yield of the principal crop, i.e., rice, which grows in large parts of the Bankura, Purulia, Purba, and Paschim Medinipur districts in Kharif (rainfed) season, was considered for this study.

Variation of Agricultural Drought

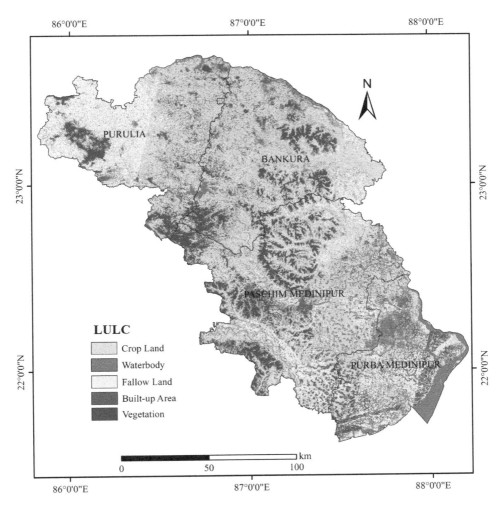

FIGURE 18.2 Land use and land cover map (supervise classification of Landsat 8 [OLI]).

TABLE 18.1
Area under Different Land Use and Land Cover

Land Use Land Cover Type	Area (%)
Crop Land	56
Natural Vegetation	20
Fallow Land	16
Waterbody	4
Built-up Area	4

R is an open source programming language and freeware for statistical computing. It is not GUI based but has one of the largest systems of packages and worldwide community support, which makes its use advantageous and cost-effective. It was created by Ross Ihaka and Robert Gentleman at the University of Auckland, New Zealand, and was developed by the R Development Core Team (R Core Team, 2020). The initial version was released in 1995, and the latest version of R is 4.0.3 ("Bunny-Wunnies Freak Out"). The RStudio is an IDE for R. The first public beta version (v0.92) was announced in February 2011. The R studio (RStudio Team, 2020) provides a modern and user-friendly interface of R. It enables better viewing and writing codes for different operations. There are more than 10,000 packages available for R, and these libraries cover a vast domain of analysis for different fields. The packages used in the study are available as ncdf4, raster, ggplot2, rgdal, sp, spatialEco, and RasterVis. The collected CHIRPS rainfall data (.nc format) were preprocessed using the ncdf4 package (Pierce, 2019). The present study integrates the spatial science, meteorology, and geostatistics domains. Here, 16-day composite SPI was estimated through the R studio platform following gamma distribution (discussed in Section 18.3.3). The raster (Hijmans, 2020), sp (Bivand et al., 2013), and rgdal (Bivand et al., 2020) packages were used for preprocessing MODIS NDVI data and calculating VCI. Apart from this, the spatial correlation (pixel-wise) between VCI and SPI (resampled) images of the previous 20 years was performed using the spatialEco package (Evans, 2020). RasterVis (Perpiñán and Hijmans, 2020), a popular visualization package, was used in this study to generate a multi-raster map in a single frame. All the statistical calculations and data visualization, along with map and diagram generation, were performed with the help of the ggplot2 (Wickham, 2016) package.

18.3.2 Vegetation Condition Index (VCI)

NDVI is a globally acceptable robust measurement of vegetation condition (Tian et al., 2019; Rhee and Yang, 2018) and quite a good reflector of water and climatic conditions. By using the range of NDVI, Kogan (1990) developed VCI, which is an efficient indicator for assessing the severity of agricultural drought. It can be computed as follows (Equation 18.1):

$$VCI = \frac{NDVI_i - NDVI_{Min}}{NDVI_{Max} - NDVI_{Min}} \times 100 \qquad (18.1)$$

VCI has the potential to monitor the onset and extent of agricultural drought and also identify its temporal duration (Dutta et al., 2015). Previous studies revealed that VCI has a significant inter-relationship with crop yield and can be used for early warning of agricultural drought (Sahoo et al., 2015).

This study examined the vegetation condition in the second fortnights of July and August over the last 20 years, in relation to drought-related crop stress during Kharif season. The VCI values are measured as percentages and vary from 0 to 100. Values lower than 35% indicate severe drought, 35% to 50% indicates moderate drought condition, and above 50% means the normal state (Kogan, 1995). The study area was cropped from the stacked 16-day composite MODIS NDVI of the Julian days 209 and 241. Thereafter, VCI was computed for each pixel following Equation 18.1.

18.3.3 Standardized Precipitation Index (SPI)

SPI, propounded by McKee et al. (1993), is the most useful meteorological index to identify the dry and wet periods in different accumulation periods (3 to 24 months). It is calculated using long-term rainfall data by applying the statistical distribution of gamma function and standard normal curves. The SPI value varies from +3 to −3, where positive values indicate higher precipitation than the long-term average, and negative values refer to deficit of rainfall. In order to categorize the level of drought intensity, SPI values are classified as shown in Table 18.2.

SPI is derived from the normalization of rainfall value, so it represents both dry and wet periods. However, it is incapable of assessing climate change and related phenomena, as it considers only the normalized value of a single parameter. It is helpful for measuring precipitation deficiency at different time scales and the severity of meteorological drought. Short-term precipitation anomalies are good for examining the soil moisture properties, whereas long-term anomalies may have

TABLE 18.2
SPI Classification and Drought Categories

SPI Values	Drought Category
>2	Extremely wet
1.50 to 1.99	Very wet
1.0 to 1.49	Moderately wet
0.99 to −0.99	Near normal
−1 to −1.49	Moderately dry
−1.50 to −1.99	Very dry
<−2.00	Extremely dry

McKee et al., (1993)

significant effects on groundwater, streamflow, and water reservoirs. Synchronizing with the fortnightly MODIS VCI data, the CHIRPS-based daily rainfall images were processed into 16-day composite rainfall data. The historical data of the study period were stacked together for the second fortnight of July and the second fortnight of August. Then, a gamma distribution function was applied to calculate the grid-based SPI over the study area (Equation 18.2).

$$g(x) = \frac{1}{\beta^\alpha \Gamma(\alpha)} x^{\alpha-1} e^{-x/\beta} \text{ for } x > 0 \quad (18.2)$$

where x is the precipitation amount, $\alpha > 0$ is the shape parameter, $\beta > 0$ is the scale parameter, and $\Gamma(\alpha)$ is the gamma function. For calculation of the α and β values, Equations 18.3 and 18.4 were followed.

$$\hat{\alpha} = \frac{1}{4A}\left(1 + \sqrt{1 + \frac{4A}{3}}\right) \quad (18.3)$$

and

$$\hat{\beta} = \frac{\bar{x}}{\hat{\alpha}} \quad (18.4)$$

where $\bar{x}$ is the mean rainfall and A is the area given by Equation 18.5:

$$A = \ln(\bar{x}) - n^{-1} \sum \ln(x) \quad (18.5)$$

Cumulative probability, G(x), of the total amount of precipitation for a unit time scale was calculated by Equation 18.6:

$$G(x) = \frac{1}{\hat{\beta}^{\hat{\alpha}}(\hat{\alpha})} \int_0^x e^{-x/\beta} dx \quad (18.6)$$

When precipitation is zero (x = 0), then gamma distribution is undefined. For x = 0, q = P(x=0), where P(x=0) is the probability of zero rainfall. Hence, the cumulative probability becomes (Equation 19,7):

$$H(x) = q + (1-q)G(x) \quad (18.7)$$

18.3.4 Rainfall Anomaly Index (RAI)

RAI, developed by Van Rooy (1965), indicates the positive and negative anomalies of precipitation value. It is a simple yet efficient index for rainfall variability, drought monitoring, and climate change that can be calculated on a weekly, monthly, or annual time scale. It reflects the deviation of rainfall from the long-term average and is formulated as follows (Equation 18.8):

$$RAI_i = \frac{R_i - \mu}{\sigma} \quad (18.8)$$

where R_i is rainfall of the ith period, μ represents long-term mean rainfall, and σ is the standard deviation of rainfall during the period.

RAI was calculated for the second fortnights of July and August using the 16 years' (2000 to 2016) in situ gridded rainfall data provided by IMD. In tune with the fortnightly MODIS dataset, the daily gridded data was converted into a 16-day composite value. The temporal mean and standard deviation were calculated for each grid over the mentioned period. Here, RAI was used to validate the potential of the satellite-based meteorological drought index.

18.3.5 Yield Anomaly Index (YAI)

Yield Anomaly Index is a very simple technique whereby the deviation in crop production is calculated from its historical mean (Equation 18.9). It reflects the variability of actual yield and specifically, the drought-related crop stress, which can be used as reference data for evaluating satellite-derived drought indices.

$$YAI_i = \frac{Y_i - \mu}{\sigma} \quad (18.9)$$

where YAI_i is Yield Anomaly Index, Y_i = yield of ith year, μ = long-term average yield, and σ = standard deviation.

District-wise crop yield data for 15 years (2000 to 2014) were collected from the District Statistical Handbooks. The average yield and standard deviation of the study period were computed and used for estimating the YAI of each year.

The metholody followed in this study is shown in Figure 18.3.

18.4 RESULTS AND DISCUSSION

18.4.1 Spatio-Temporal Pattern of Vegetational Drought Using VCI

Rao et al. (2015) demonstrated five different growth stages (sowing, seedling, tillering, reproductive, and grain filling) of the Kharif crop (paddy) for this study region. The tillering (mid-July to mid-August) and reproductive (mid-August to mid-September) phases are decisive stages for healthy vegetation growth. Previous literature has also revealed that the VCI is more significant in monitoring agricultural drought during maximum vegetation growth (Quiring and Ganesh, 2010; Vicente-Serrano, 2007). In order to assess the agricultural drought, 16-day composite VCI was calculated for the second fortnights of July and August of each year in the study period. The result shows a significant temporal and spatial variation of the VCI pattern in the study area (Figures 18.4 and 18.5). It was found that the average VCI value was distinctly lower during July 2005 in Purba Medinipur (17%), followed by Paschim Medinipur (25%) and Bankura (32%), indicating extreme drought conditions in those districts. In the second fortnight of July 2000, there was a severe drought in the Paschim Medinipur district (VCI ~33%). Moderate drought conditions (VCI<50%) were also

Variation of Agricultural Drought

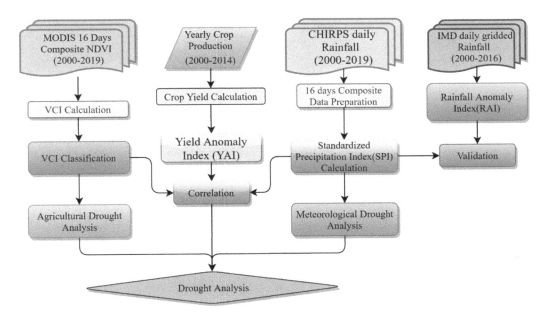

FIGURE 18.3 Methodological flow chart.

shown in the rest of the area. In July, moderate vegetational stress frequently occurred, mostly in the Purulia and Bankura districts.

The vegetation condition in the second fortnight of August represents the early reproductive stage of the Kharif season crop in this area; hence, vegetation stress during this period is critical for crop grain production. The average VCI values of most of the years were more than 60% for all districts, which indicates normal conditions for crop growth. However, moderate to severe drought conditions (VCI <50%) prevailed in Purulia in 2003. The average VCI also shows drought stress (VCI <50%) in Purba Medinipur district in the years 2007, 2011, and 2019. VCI values lower than 50% indicate a significant decline in the yield of Kharif crops and may be useful for early warning of drought.

Vegetation condition is highly affected by water availability, and the variation in stress is mainly regulated by uneven rainfall patterns. Other controlling factors are physiography and soil type, which determine the availability of water content in the soil for healthy growth of vegetation. Human interference, such as irrigational systems and application of fertilizer, pesticides, etc., also regulates the condition of vegetation health. Apart from rainfall, the complex interrelationship of environmental components determines the spatio-temporal variability of agricultural drought in an area.

18.4.2 Spatio-Temporal Pattern of Meteorological Drought using SPI

The 16-day composite SPI for the peak monsoonal months, July and August, was employed to analyze the occurrences of meteorological drought in the 2000–2019 period. Based on the gamma function, SPI uses long-term rainfall to measure the severity of meteorological drought. Here, the spatio-temporal variation of meteorological drought was identified through SPI by using long-term satellite-based CHIRPS rainfall data (Figures 18.6 and 18.7).

The SPI values were further categorized based on the classification scheme proposed by McKee et al. (1993) to identify the drought-affected years in each district. It reveals the evidence of extreme

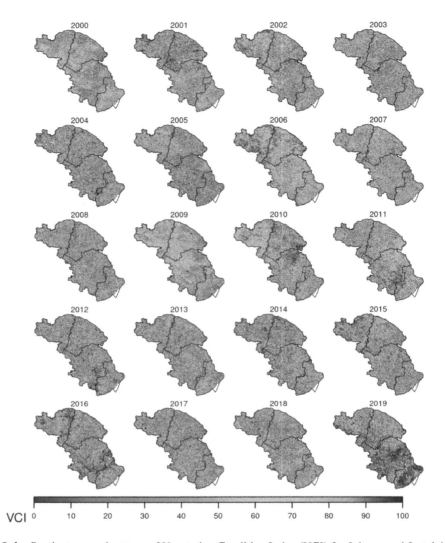

FIGURE 18.4 Spatio-temporal pattern of Vegetation Condition Index (VCI) for July second fortnight.

meteorological drought (SPI <−2) in Purba Medinipur and severe drought (SPI −1.5 to −2) in Paschim Medinipur during July 2001. The study demonstrates the occurrence of a mild drought that affected the entire region in the year 2010. Apart from this, mild droughts were observed in Purulia and Paschim Medinipur districts in 2011 and 2016, respectively. Extreme wet conditions with SPI >2 were found in July 2007 and 2017 over the entire region.

SPI of August shows that extreme dry conditions prevailed over the area in 2000. Mild dry conditions were also observed in Purulia (2005), Bankura (2005, 2010), and Paschim Medinipur districts (2012). In contrast, severe wet conditions were also identified in August 2013 over Purba and Paschim Medinipur. The fortnightly pattern of SPI provides evidence of very short-term drought, which may play a crucial role in agricultural drought and consequent reduction in yield of Kharif crops.

18.4.3 Rainfall Anomaly Index (RAI)

RAI was calculated using the gridded rainfall data of IMD for the period from 2000 to 2016. The year 2007 was identified as an extreme wet year for both July and August, which strongly agrees

Variation of Agricultural Drought

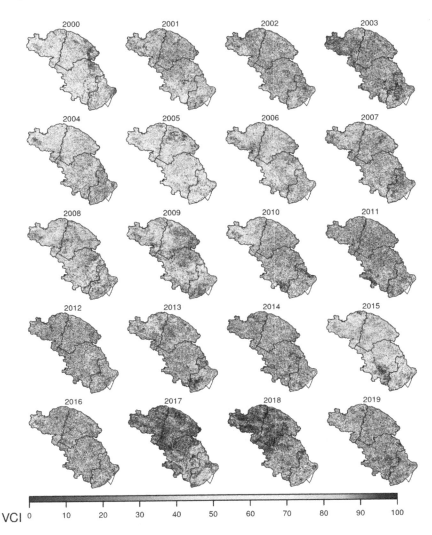

FIGURE 18.5 Spatio-temporal pattern of Vegetation Condition Index (VCI) for August second fortnight.

with the SPI values. Moderate to low wet spells (RAI −0.5 to −1) were evident in several years over the area (Figure 18.8).

In order to evaluate the outcome of satellite-derived meteorological drought analysis, the interrelationship between RAI and SPI based on IMD and CHIRPS rainfall, respectively, was estimated (Figure 18.9). A strong, statistically significant ($p < 0.05$) correlation ($r^2 = 0.80$) was found between RAI and SPI values for the period (2000–2016), indicating the potential of satellite-based rainfall data for assessing meteorological drought (Figure 18.9).

18.4.4 CROP PRODUCTION AND YIELD ANOMALY INDEX (YAI)

The production of rice, the major crop of the Kharif season in the area, was assessed for validating the outcome of satellite-based drought indices used in the study. The district-wise acreage and production of rice (2000–2014) were used to calculate the yield and YAI in the study area.

The district-wise rice yield (thousand tonnes per km^2) of Kharif season is represented in Figure 18.10, which reveals a significant temporal variation. The yield of rice was highest in

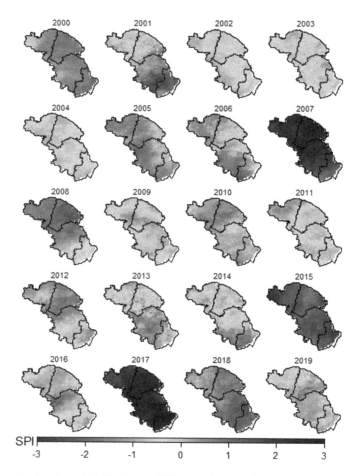

FIGURE 18.6 Standardized precipitation index (SPI) for July second fortnight.

Bankura, followed by Paschim Medinipur, and lowest in Purba Medinipur district in most of the years. Both Purba and Paschim Medinipur districts exhibited high yield (YAI >0.5) in the years 2004, 2008, 2011, 2012, and 2014 (Figure 18.11), while the yield of rice was high (YAI > 0.5) in 2001, 2006, 2007, and 2011 in Bankura and Purulia districts. It is worth noting that both these districts experienced significant reduction in yield (YAI <−1) in the years 2000 and 2010. In contrast, Purba Medinipur district exhibits the lowest YAI (YAI <−1.5) in 2005 and 2007 (Table 18.3). Both Paschim (YAI <−0.5) and Purba Medinipur (YAI <−1) districts reveal moderate to high yield loss in 2013.

18.4.5 Yield Anomaly Index and Standardized Precipitation Index

Agricultural drought is generally triggered by irregularity in the amount of rainfall received by an area. Shortage of rainfall and dry spells during the sowing period significantly affect crop health and reduce the yield of rainfed crops. To examine the relationship between monsoonal (July and August) rainfall and crop yield during the Kharif season (mainly rice), we analyzed the SPI and YAI of the area (Figure 9.12).

The study region has experienced dual extreme weather events, drought and flood. Purulia and Bankura districts are prone to drought, whereas heavy rainfall in Purba Medinipur and part of Paschim Medinipur caused frequent floods (Mittal et al., 2013). It is worth noting that the area

Variation of Agricultural Drought

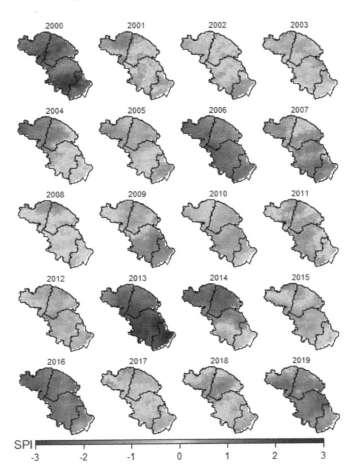

FIGURE 18.7 Standardized precipitation index (SPI) for August second fortnight.

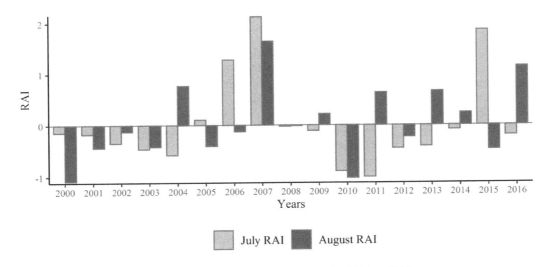

FIGURE 18.8 July and August Rainfall Anomaly Index (RAI) for 2000 to 2016.

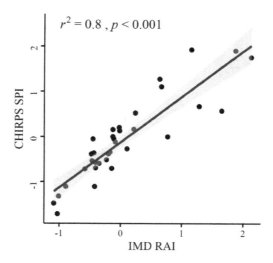

FIGURE 18.9 Correlation between CHIRPS SPI and IMD RAI.

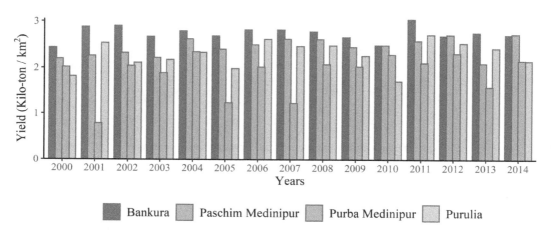

FIGURE 18.10 District-wise rice yield from 2000 to 2014.

is characterized by heterogeneous physiography, and low altitude in this part of the study area is the major reason for recurring flood events during monsoon. The study identified meteorological drought events (SPI <−1) in August 2000, 2003, 2005, and 2010 (Table 18.4) which agrees with the YAI computed from the ground-based yield dataset. The crop yields of Purulia and Bankura districts were distinctly reduced (YAI <−1) in those years. Purba Medinipur district witnessed near average yield (Table 18.3), as it has plenty of soil moisture due to low lands. The year 2007 experienced heavy downpours, causing crop damage in Purba Medinipur (Tables 18.3 and 18.4). The study reveals (Figure 18.12) that August SPI had a more significant influence on crop production than July SPI for the entire region. Statistically significant ($p < 0.05$) correlations between SPI and YAI were observed in Bankura ($R^2 = 0.37$), followed by Purulia ($R^2 = 0.34$) district. However, the correlation coefficient was significantly lower in Paschim Medinipur ($R^2 \sim 0$), and a negative trend was found in the flood-prone Purba Medinipur district. The study indicates good interrelationships between crop yield and rainfall in the western part of the area, comprising the Purulia and Bankura districts, rather than in the Purba and Paschim Medinipur districts.

Variation of Agricultural Drought

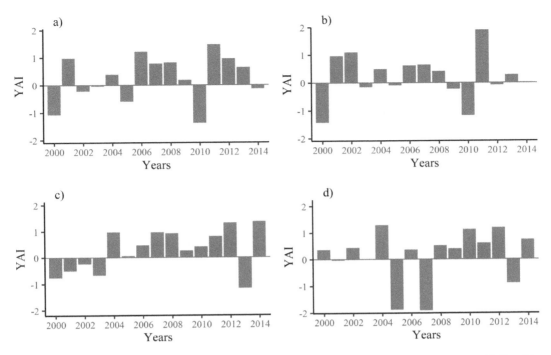

FIGURE 18.11 Yield Anomaly Index (YAI) for 2000 to 2014. (a) Purulia, (b) Bankura, (c) Paschim Medinipur, (d) Purba Medinipur.

TABLE 18.3
District-Wise Yield Anomaly Index (YAI) Values

Year	Purulia	Bankura	Paschim Medinipur	Purba Medinipur
2000	−1.60	−1.96	−1.30	0.22
2003	−0.37	−0.45	−1.19	−0.16
2005	−1.02	−0.38	−0.27	−2.05
2007	0.61	0.51	0.86	−2.08
2010	−1.96	−1.66	0.16	0.98
2011	1.45	2.01	0.66	0.46

18.4.6 STANDARD PRECIPITATION INDEX AND VEGETATION CONDITION INDEX

VCI measures the health condition of plants and the growing stage of vegetation, which is highly regulated by the availability of soil moisture. This study explores the interrelationship between meteorological and agricultural droughts in the area. A pixel-wise correlation between VCI and SPI was established for the study period, which indicates a good positive correlation. Generally, Kharif rice is transplanted in July, and panicle initiation takes place in August over the area (Rao et al., 2015). Lower VCI, as observed in the month of July, can be explained by reduced background scatter effects of rice crop in the month (Table 18.5). The study reveals a moderate coefficient of determination ($r^2 \sim 0.4$) in the study area (Figure 9.13). Short-term meteorological drought may not influence VCI, as vegetation growth is mainly controlled by the soil moisture available for the plants.

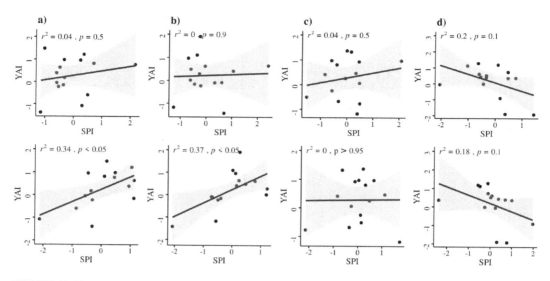

FIGURE 18.12 Relation between SPI and YAI. (a) Purulia, (b) Bankura, (c) Paschim Medinipur, (d) Purba Medinipur.

TABLE 18.4
District-Wise Average SPI Value for Second Fortnight of July and August

Year	Purulia		Bankura		Paschim Medinipur		Purba Medinipur	
	July	August	July	August	July	August	July	August
2000	0.30	−2.13	0.70	−2.07	0.49	−2.14	0.76	−2.38
2003	−0.58	−0.56	−0.46	−0.40	−0.32	−0.35	−0.4458	−0.26
2005	0.50	−0.87	0.33	−0.70	0.48	−0.28	0.3737	0.30
2007	2.16	0.44	2.41	0.23	2.07	0.66	1.4334	0.78
2010	−1.15	−0.32	−1.26	−0.57	−0.84	−0.84	−1.0364	−0.48
2011	−1.02	0.17	−0.25	0.26	−0.39	0.31	−0.6084	0.07
2017	0.30	−0.30	2.34	−0.39	2.50	−0.10	2.8349	−0.40

TABLE 18.5
District-Wise Average VCI Value for Second Fortnight of July and August

Year	Purulia		Bankura		Paschim Medinipur		Purba Medinipur	
	July	August	July	August	July	August	July	August
2000	39.91	55.81	36.48	54.70	33.65	58.63	44.29	51.56
2003	54.16	47.55	48.86	51.37	54.69	60.79	50.82	73.63
2005	49.70	63.84	32.73	66.00	25.16	73.66	17.07	64.70
2007	51.83	75.64	53.28	68.46	49.33	63.88	53.82	45.93
2010	68.33	57.54	76.74	56.48	70.43	67.97	54.73	68.80
2011	38.07	66.66	48.18	68.02	71.14	62.06	60.12	46.400
2017	44.00	87.98	40.63	89.03	44.10	83.25	44.44	76.31

Variation of Agricultural Drought

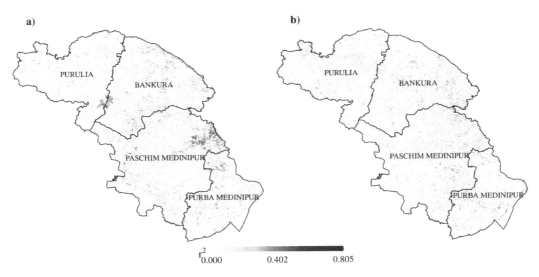

FIGURE 18.13 Correlation between VCI and SPI. (a) July second fortnight. (b) August second fortnight.

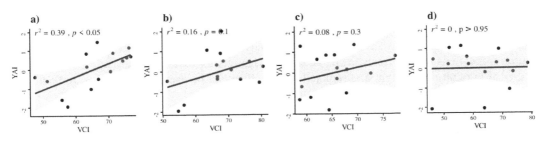

FIGURE 18.14 Relation between average VCI and YAI for August second fortnight. (a) Purulia, (b) Bankura,(c) Paschim Medinipur, (d) Purba Medinipur.

18.4.7 Yield Anomaly Index and Vegetation Condition Index

In order to identify the interrelationship between satellite-derived drought estimates and actual crop yield anomaly, we employed Pearson Correlation between VCI and YAI. After masking the VCI images with other LULCs, i.e., settlement, forest, waterbody, etc., the average VCI of crop land was estimated for each district (Table 18.5). The study reveals a moderate to strong positive correlation over Bankura ($r = 0.4$) and Purulia ($r = 0.62$) districts (Figure 18.14). However, it was not significant in the eastern part of the study area covering Purba and Paschim Medinipur districts. This indicates that the vegetation condition of the growing season has least influence on the yield of this area. This area is affected by recurring flood events, which may be a reason for the weak correlation between VCI and YAI. We have considered vegetation condition in the second fortnight of August, and most of the flood events occur during the end of the monsoon, when soils remain saturated with water. Therefore, variation in yield anomaly in this area cannot be determined by the vegetation condition of early or mid-Kharif season. In contrast, the drought-prone western part of this area shows a good interrelationship, and the growing season VCI plays a crucial role in determining the yield of Kharif rice in the area.

18.5 CONCLUSION

This study explored the potential of open source software to analyze the spatio-temporal dynamics of agricultural drought at a regional scale. The result shows a varying pattern of drought events over the western part of West Bengal in the 2000–2019 period. The study used VCI derived from MODIS NDVI for assessing agricultural droughts over the area. It reveals severe drought in 2005 over Bankura (VCI ~32%), Paschim (VCI ~25%), and Purba Medinipur (VCI ~17%) districts. Moderate drought stress (VCI <50%) was observed frequently in various parts of the area. In order to examine the interrelationship between agricultural and meteorological droughts, SPI and RAI, respectively derived from CHIRPS and station-based gridded IMD rainfall, were compared with VCI images. It is worth mentioning that SPI estimated from satellite-based merged rainfall data was found to be strongly correlated ($r^2 \sim 0.8$, $p < 0.01$) with the RAI based on in situ gridded IMD data. However, the study reveals a moderate coefficient of determination ($r^2 \sim 0.4$, $p < 0.05$) between VCI and SPI in the study area. The meteorological drought events were not found to be coherent with the dynamics of agricultural drought stress in most of the years. This can be explained by the undulating topography, high runoff, and poor water-holding capacity of soil, which largely controls the vegetation condition. A substantial part of Purba Medinipur and a few pockets of Paschim Medinipur experience frequent floods and consequent damage to the Kharif crop. The interrelationship between average VCI of August and YAI does not show any significant correlation in Purba or Paschim Medinipur, whereas it reveals moderate to strong positive correlation in Bankura ($r = 0.4$) and Purulia ($r = 0.62$) districts. Further study can be conducted including the soil moisture anomaly, an important controlling parameter of drought-related crop stress in the area. The entire study was performed using R studio, an open source package having excellent potential over commercial paid software in terms of cost and freedom to analyze large datasets at any dimension. Furthermore, R studio has many advantages in dealing with different raster data formats (.nc, .tif), especially for meteorological data handling. It also helps with advanced statistical analysis. This study proves the potential of open source software as well as freely available satellite datasets for assessing the spatio-temporal variability of agricultural drought at a regional scale, which can be useful for planning and adopting sustainable mitigation strategies to reduce drought vulnerability.

REFERENCES

Alexander, L., Allen, S. and Bindoff, N. L. (2013) Working group I contribution to the IPCC fifth assessment report climate change 2013: The physical science basis summary for policymakers. Cambridge University Press.

American Meteorological Society (1997) Meteorological drought-policy statement, *Bulletin of the American Meteorological Society*, 78, 847–849.

Bhunia, P., Das, P. and Maiti, R. (2019) Meteorological drought study through SPI in three drought prone districts of West Bengal, India, *Earth Systems and Environment*. Springer International Publishing, 4(1), pp. 43–55. doi: 10.1007/s41748-019-00137-6.

Bivand, R., Keitt, T. and Rowlingson, B. (2020) RGDAL: Bindings for the "Geospatial" data abstraction library. Available at: https://cran.r-project.org/package=rgdal.

Bivand, R. S., Pebesma, E. and Gomez-Rubio, V. (2013) *Applied spatial data analysis with {R}*, Springer. 2nd ed. pp. xviii, 405. ISBN: 978-1-4614-7617-7.

Ciais, P. et al. (2005) Europe-wide reduction in primary productivity caused by the heat and drought in 2003, *Nature*, 437(7058), pp. 529–533. doi: 10.1038/nature03972.

Cornish, P. S. et al. (2015) Improving crop production for food security and improved livelihoods on the East India Plateau II. Crop options, alternative cropping systems and capacity building, *Agricultural Systems*. Elsevier Ltd, 137, pp. 180–190. doi: 10.1016/j.agsy.2015.02.011.

Dai, A. (2011) Characteristics and trends in various forms of the Palmer Drought Severity Index during 1900–2008, *Journal of Geophysical Research*, 116(D12), p. D12115. doi: 10.1029/2010JD015541.

Dai, A. (2013) Increasing drought under global warming in observations and models, *Nature Climate Change*. Nature Publishing Group, 3(1), pp. 52–58. doi: 10.1038/nclimate1633.

Das, R. et al. (2019) Trends and vulnerability assessment of meteorological and agricultural drought conditions over Indian region using time-series (1982–2015) satellite data, *International Archives of the Photogrammetry, Remote Sensing and Spatial Information Sciences – ISPRS Archives*, 42(3/W6), pp. 453–459. doi: 10.5194/isprs-archives-XLII-3-W6-453-2019.

Dutta, D. (2010) *A comparative approach for predicting agricultural drought using PS-n crop growth model, NOAA-AVHRR NDVI and ARIMA model.* M.Sc. Thesis, University of Twente, Faculty of Geo-Information and Earth Observation (ITC), Netherlands.

Dutta, D., Kundu, A. and Patel, N. R. (2013) Predicting agricultural drought in eastern Rajasthan of India using NDVI and standardized precipitation index, *Geocarto International*, 28(3), pp. 192–209. doi: 10.1080/10106049.2012.679975.

Dutta, D. et al. (2015) Assessment of agricultural drought in Rajasthan (India) using remote sensing derived Vegetation Condition Index (VCI) and Standardized Precipitation Index (SPI), *Egyptian Journal of Remote Sensing and Space Sciences*, 18(1), pp. 53–63. doi: 10.1016/j.ejrs.2015.03.006.

Evans, J. S. (2020) spatialEco. Available at: https://github.com/jeffreyevans/spatialEco.

Gouveia, C., Trigo, R. M. and DaCamara, C. C. (2009) Drought and vegetation stress monitoring in Portugal using satellite data, *Natural Hazards and Earth System Sciences*, 9(1), pp. 185–195. doi: 10.5194/nhess-9-185-2009.

Gu, Y. et al. (2007) A five-year analysis of MODIS NDVI and NDWI for grassland drought assessment over the central Great Plains of the United States, *Geophysical Research Letters*, 34(6), p. L06407. doi: 10.1029/2006GL029127.

Hao, Z. and Singh, V. P. (2015) Drought characterization from a multivariate perspective: A review, *Journal of Hydrology*, Elsevier B.V., 527(May), pp. 668–678. doi: 10.1016/j.jhydrol.2015.05.031.

Hijmans, R. J. (2020) raster: Geographic data analysis and modeling. Available at: https://cran.r-project.org/package=raster.

Jiao, W. et al. (2019) A new multi-sensor integrated index for drought monitoring, *Agricultural and Forest Meteorology*, Elsevier, 268(July 2018), pp. 74–85. doi: 10.1016/j.agrformet.2019.01.008.

Kamble, M. V., Ghosh, K., Rajeevan, M. and Samui, R. P. (2010) Drought monitoring over India through normalized difference vegetation index (NDVI). *Mausam*, 61(4), pp. 537–546.

Katsanos, D., Retalis, A. and Michaelides, S. (2016) Validation of a high-resolution precipitation database (CHIRPS) over Cyprus for a 30-year period, *Atmospheric Research*, Elsevier B.V., 169, pp. 459–464. doi: 10.1016/j.atmosres.2015.05.015.

Kern, A., Marjanović, H. and Barcza, Z. (2020) Spring vegetation green-up dynamics in Central Europe based on 20-year long MODIS NDVI data, *Agricultural and Forest Meteorology*, Elsevier, 287(December 2019), p. 107969. doi: 10.1016/j.agrformet.2020.107969.

Keshavarz, M., Karami, E. and Vanclay, F. (2013) The social experience of drought in rural Iran, *Land Use Policy*, Elsevier Ltd, 30(1), pp. 120–129. doi: 10.1016/j.landusepol.2012.03.003.

Kogan, F. N. (1990) Remote sensing of weather impacts on vegetation in non-homogeneous areas, *International Journal of Remote Sensing*, 11(8), pp. 1405–1419. doi: 10.1080/01431169008955102.

Kogan, F. N. (1995) Application of vegetation index and brightness temperature for drought detection, *Advances in Space Research*, 15(11), pp. 91–100. doi: 10.1016/0273-1177(95)00079-T.

Kumar, K. R., Kumar, K. K. and Pant, G. B. (1994) Diurnal asymmetry of surface temperature trends over India, *Geophysical Research Letters*, 21(8), pp. 677–680. doi: 10.1029/94GL00007.

Lai, C. et al. (2019) Monitoring hydrological drought using long-term satellite-based precipitation data, *Science of the Total Environment*, Elsevier B.V., 649, pp. 1198–1208. doi: 10.1016/j.scitotenv.2018.08.245.

Li, R., Tsunekawa, A. and Tsubo, M. (2014) Index-based assessment of agricultural drought in a semi-arid region of Inner Mongolia, China, *Journal of Arid Land*, 6(1), pp. 3–15. doi: 10.1007/s40333-013-0193-8.

McKee, T. B., Doesken, N. J. and Kleist, J. (1993) The relationship of drought frequency and distribution to time scale, in *Eighth Conference on Applied Climatology, American Meteorological Society*, Anaheim, CA.

Mishra, A. K. and Singh, V. P. (2011) Drought modeling – A review, *Journal of Hydrology*, Elsevier B.V., 403(1–2), pp. 157–175. doi: 10.1016/j.jhydrol.2011.03.049.

Mittal, N., Mishra, A. and Singh, R. (2013) Combining climatological and participatory approaches for assessing changes in extreme climatic indices at regional scale, *Climatic Change*, 119(3–4), pp. 603–615. doi: 10.1007/s10584-013-0760-1.

Noureldeen, N. et al. (2020) Spatiotemporal drought assessment over Sahelian countries from 1985 to 2015, *Journal of Meteorological Research*, 34(4), pp. 760–774. doi: 10.1007/s13351-020-9178-7.

Padhee, S. K. and Dutta, S. (2019) Spatio-temporal reconstruction of MODIS NDVI by regional land surface phenology and harmonic analysis of time-series, *GIScience and Remote Sensing*, Taylor & Francis, 56(8), pp. 1261–1288. doi: 10.1080/15481603.2019.1646977.

Palchaudhuri, M. and Biswas, S. (2020) Application of LISS III and MODIS-derived vegetation indices for assessment of micro-level agricultural drought, *The Egyptian Journal of Remote Sensing and Space Sciences* 23(2) pp. 221–229. doi: 10.1016/j.ejrs.2019.12.004.

Panigrahy, S., Manjunath, K. R. and Ray, S. S. (2005) Deriving cropping system performance indices using remote sensing data and GIS, *International Journal of Remote Sensing*, 26(12), pp. 2595–2606. doi: 10.1080/01431160500114698.

Park, Seonyoung et al. (2017) Drought monitoring using high resolution soil moisture through multi-sensor satellite data fusion over the Korean peninsula, *Agricultural and Forest Meteorology*, Elsevier B.V., 237–238, pp. 257–269. doi: 10.1016/j.agrformet.2017.02.022.

Patel, N. R. and Yadav, K. (2015) Monitoring spatio-temporal pattern of drought stress using integrated drought index over Bundelkhand region, India, *Natural Hazards*, 77(2), pp. 663–677. doi: 10.1007/s11069-015-1614-0.

Perpiñán, O. and Hijmans, R. (2020) rasterVis. Available at: http://oscarperpinan.github.io/rastervis/.

Peters, A. J. et al. (2002) Drought monitoring with NDVI-based Standardized Vegetation Index, *Photogrammetric Engineering and Remote Sensing*, 68(1), pp. 71–75.

Phan, P. et al. (2020) Using multi-temporal MODIS NDVI data to monitor tea status and forecast yield: A case study at Tanuyen, Laichau, Vietnam, *Remote Sensing*, 12(11), 1814. doi: 10.3390/rs12111814.

Pierce, D. (2019) ncdf4: Interface to Unidata netCDF (version 4 or earlier) format data files. Available at: https://cran.r-project.org/package=ncdf4.

Quiring, S. M. and Ganesh, S. (2010) Evaluating the utility of the vegetation condition index (VCI) for monitoring meteorological drought in Texas, *Agricultural and Forest Meteorology*, 150(3), pp. 330–339. doi: 10.1016/j.agrformet.2009.11.015.

R Core Team (2020) R: A language and environment for statistical computing. Vienna, Austria. Available at: www.r-project.org/.

Rao, V. U. M., Subba Rao, A. V. M., Sarath Chandran, M. A., Kaur, P., Vijaya Kumar, P., Bapuji Rao, B., Khandgond, I. R., and Srinivasa Rao, C. H. (2015) District Level Crop Weather Calendars of Major Crops in India. Central Researh Institute for Dryland Agriculture, Hyderabad, 500 059, p. 40.

Rhee, J. and Yang, H. (2018) Drought prediction for areas with sparse monitoring networks: A case study for Fiji, *Water (Switzerland)*, 10(6), 788. doi: 10.3390/w10060788.

Rouse, J. W., Hass, R. H., Schell, J. A., Deering, D. W., and Harlan, J. C. (1974) *Monitoring the vernal advancement and retrogradation (green wave effect) of natural vegetation*. Final Report, RSC 1978–4, Texas A & M University, College Station, TX.

RStudio Team (2020) RStudio: Integrated development environment for R. Boston, MA. Available at: www.rstudio.com/.

Saha, P. (2015) Identifying the causes of water scarcity in Purulia, West Bengal, India -A geographical perspective, *IOSR Journal of Environmental Science Ver. I*, 9(8), pp. 2319–2399. doi: 10.9790/2402-09814151.

Sahoo, R. N. et al. (2015) Drought assessment in the Dhar and Mewat Districts of India using meteorological, hydrological and remote-sensing derived indices, *Natural Hazards*, 77(2), pp. 733–751. doi: 10.1007/s11069-015-1623-z.

Sheffield, J. and Wood, E. F. (2008) Projected changes in drought occurrence under future global warming from multi-model, multi-scenario, IPCC AR4 simulations, *Climate Dynamics*, 31(1), pp. 79–105. doi: 10.1007/s00382-007-0340-z.

da Silva, M. R. et al. (2020) Wheat planted area detection from the MODIS NDVI time series classification using the nearest neighbour method calculated by the Euclidean distance and cosine similarity measures, *Geocarto International*, 35(13), pp. 1400–1414. doi: 10.1080/10106049.2019.1581266.

Testa, S., Mondino, E. C. B. and Pedroli, C. (2014) Correcting MODIS 16-day composite NDVI time-series with actual acquisition dates, *European Journal of Remote Sensing*, 47(1), pp. 285–305. doi: 10.5721/EuJRS20144718.

Tian, S. et al. (2019) Forecasting dryland vegetation condition months in advance through satellite data assimilation, *Nature Communications*, Springer US, 10(1), pp. 1–7. doi: 10.1038/s41467-019-08403-x.

United Nations (2014) Land and drought, UNCCD, New York [WWW Document]. Available at: www.unccd.int/issues/land-and-drought (accessed 10.18.18).

Van Rooy, M. P. (1965) A rainfall anomaly index independent of time and space. NOTOS, *Weather Bureau of South Africa*, 14, 43–48.

Vicente-Serrano, S. M. (2006) Spatial and temporal analysis of droughts in the Iberian Peninsula (1910–2000), *Hydrological Sciences Journal*, 51(1), pp. 83–97. doi: 10.1623/hysj.51.1.83.

Vicente-Serrano, S. M. (2007) Evaluating the impact of drought using remote sensing in a Mediterranean, semi-arid region, *Natural Hazards*, 40(1), pp. 173–208. doi: 10.1007/s11069-006-0009-7.

Wickham, H. (2016) *ggplot2: Elegant graphics for data analysis*. Springer-Verlag, New York. Available at: https://ggplot2.tidyverse.org.

Wilson, M. and Tchantchaleishvili, V. (2013) The importance of free and open source software and open standards in modern scientific publishing, *Publications*, 1(2), pp. 49–55. doi: 10.3390/publications1020049.

Wu, D., Qu, J. J. and Hao, X. (2015) Agricultural drought monitoring using MODIS-based drought indices over the USA Corn Belt, *International Journal of Remote Sensing*, 36(21), pp. 5403–5425. doi: 10.1080/01431161.2015.1093190.

Xu, K. et al. (2015) Spatio-temporal variation of drought in China during 1961–2012: A climatic perspective, *Journal of Hydrology*, Elsevier B.V., 526, pp. 253–264. doi: 10.1016/j.jhydrol.2014.09.047.

Zhong, R. et al. (2019) Drought monitoring utility of satellite-based precipitation products across mainland China, *Journal of Hydrology*, Elsevier, 568(November 2018), pp. 343–359. doi: 10.1016/j.jhydrol.2018.10.072.

19 Snow Cover Monitoring Using Topographical Parameters for Beas River Catchment Area

Chetna Soni, Arpana Chaudhary, and Chilka Sharma

CONTENTS

19.1 Introduction 297
19.2 Materials and Methods 299
   19.2.1 Study Area 299
   19.2.2 Data Used..299
   19.2.3 Methodology...299
19.3 Results and Discussion 301
19.4 Conclusions...307
References..307

19.1 INTRODUCTION

Snow is a natural source of water for snowbound mountainous river basins. Indus, Shyok, Chenab, Stalaj, Brahmaputra, Raavi, Beas, Jhelum, and Kishanganga are the major perennial rivers that originate from the Himalayan region. Snow cover monitoring greatly influences the sustainability of freshwater supplies, agriculture, and hydropower generation for millions of people living in downstream regions (Kour et al., 2016; Sharma et al., 2014; Thayyen and Gergan, 2010).

The Himalayas extend to both east and west in the form of an arc with a length of about 2400 km. The width of the Himalayan range is 400 km at the western end and 150 km at the eastern end. The geologically young Himalayan Mountains run in three series, which are parallel to each other. The ranges are known as the Great Himalayas (Upper Himalayas or Himadri), with an average altitude of 6000 m, the Middle Himalayas, with an elevation range of 3700–4500 m, and the Outer Himalayas (Shivalik) having an altitude range of 900–1100 m. Mount Everest (8848 m), Kanchenjunga (8598 m), and Nanda Devi (7817 m) are in the Great Himalayan range. The altitude variation of the eastern Himalayas is greater than that of the western Himalayan range. Major streams such as Sutlej, Ganga, Indus, Jhelum, Ravi, and Brahmaputra originate from the Himalayan range only. Since the mountains are young, the rivers are perennial and carry a large amount of sediment along with water. The velocity and sediment-carrying capacity of the rivers are reduced as these rivers enter the plains (Singh et al., 1997).

The impacts of climate change on variability at the local and regional levels in high-altitude regions like the Himalayas can be studied by observing the snow cover pattern and extent (Changchun et al., 2008; Kour et al., 2016; Paudel and Andersen, 2011; Shen et al., 2014; Wang and Li, 2006; Ye et al., 1999). For a few decades, multi-sensor satellite remote sensing data have been used to estimate snow cover area (SCA) (Bergeron et al., 2014; Jain et al., 2009, 2010; Krajčí et al., 2016; Ronco et al., 2016; Salcedo and Cogliati, 2014). In the catchment level of south Mediterranean areas, fractional snow cover (FSC) maps were generated using snow cover duration (SCD) and SCA as seasonal indicators for the period 2000–2013 (Marchane et al., 2015).

Significant studies have been performed on snow cover variability, and the research outcomes revealed either an increasing or a decreasing trend in snow cover (Brown, 2000; Frei and Robinson, 1999; Kour et al., 2016; Maskey et al., 2011). An increase in snow cover pattern has been witnessed in the Karakorum region (Hewitt, 2005), the Tibetan plateau (Ye et al., 1999), and the upper reaches of the Tarim River basin (Weng and Hu, 2008). On the other hand, a decreasing trend has been reported globally, such as in western Eurasia (Brown, 2000), the northern hemisphere (Frei and Robinson, 1999), Tien Shan (Aizen et al., 1997), the Kaligandaki River basin (Mishra et al., 2014), and the Hind Kush Himalayan region (Gurung et al., 2011; Immerzeel et al., 2009; Menon et al., 2010; Pu et al., 2007). Investigations suggest that the pattern of snow cover is influenced not only by regional topography but also by climatic conditions such as temperature and rainfall (Kaur et al., 2012; Kour et al., 2015a, b; Mishra et al., 2014). Snow has been considered as a subtle indicator of climate change by numerous investigators (Foster et al., 1983; Kour et al., 2016; Wang and Li, 2006).

Multi-sensor datasets such as Moderate Resolution Imaging Spectroradiometer (MODIS) Terra/Aqua, Landsat 5, 7, and 8 (Thematic Mapper [TM], Enhanced Thematic Mapper Plus [ETM+], and Operational Land Imager [OLI]), AwiFS, terrestrial photography, and Synthetic Aperture Radar (SAR) sensors have been incorporated to improve SCA on sloping terrain (Pimentel, 2016; Thakur et al., 2013). MODIS 8-day standard snow cover datasets have been used by many researchers studying climate change using temporal snow cover in the Himalayan region (Gurung et al., 2011; Kour et al., 2016; Krishna and Sharma, 2013; Sharma et al., 2014). A large number of investigations have shown a reduction in snow cover in the Himalayas. Annual snow cover shows a linear pattern due to inter-annual variation in circulation pattern, reported to be $-1.25 \pm 1.13\%$, which is consistent with the decline during the 1990–2001 time period in the Hind Kush Himalayas (Gurung et al., 2011). A decreasing trend for annual and seasonal variation has been reported for SCA/D in the northwestern Himalayas, which covers eight major river basins: Indus, Jhelum, Beas, Sutlej, Chenab, Raavi, Shyok, and Kishanganga (Sharma et al., 2014). A relationship between mean SCA and mean air temperature has been observed throughout 2000–2012 during the winter season for the Chenab basin (Kour et al., 2016). Climate change impacts have been detected on the Gangotri glacier for the ablation period of 2006–2010 using MOD10A2 and MOD11A2 products (Krishna and Sharma, 2013).

The heterogeneous pattern of snow cover in the undulating topography of the Himalayan region is largely controlled by climate conditions (temperature and rainfall) and regional topographic characteristics (Anderton et al., 2004; Jain et al., 2009; Jost et al., 2007; Kaur et al., 2012; Pomeroy et al., 2013; Pomeroy et al., 1998; Sharma et al., 2014). Land features, along with topographical factors, influence the snow accumulation and ablation pattern. Mountain ranges such as the Himalayas interrupt the winds, which can redistribute the snowdrifts (Jain et al., 2009). Elevation, slope, and aspect dynamically contribute to snow melting with strong radiation conditions. The spatial pattern of snow accumulation and the ablation rate vary as a function of elevation, slope, aspect, and vegetation at a local scale (Dunn and Colohan, 1999; Paudel and Andersen, 2011). The long-term erraticism in snow cover supported by climatological and topographical data is limited in the Himalayan region. Satellite data with high temporal resolution are more suitable for snow cover distribution analysis. The TERRA/AQUA sensors of the MODIS satellite, having the highest continuity in terms of spatio-temporal (1-day revisit) and data quality aspects, provide identical prospects of studying snow distribution daily. Snow cover shows a negative correlation with wind speed, net radiation, and temperature, which validates the consistency of MODIS snow cover datasets (Maskey et al., 2011).

The chapter utilizes open-source software (OSS), whose main advantage is that it is free. OSS is developed by the contribution of experts and users worldwide. Source code can be modified, shared, and distributed by the user, who is therefore free to experiment, innovate, and gain experience (Badea and Badea, 2016). OSS is compatible with different operating systems. Various OSS has been designed using different programming languages (Bocher and Erwan, 2008) OSS provides all the common geographic information system (GIS) tools for viewing, creating, editing, and analyzing geospatial data as well as managing databases. OSS helps in developing desktop,

Snow Cover Monitoring

web, and mobile-based applications. OSS supports different data formats and provides web services (Mohammed, 2014). New versions are released periodically with improvements on past versions. Online tutorials, documentation, user guides, and an online user community are available for OSS users. Plugins are available for different functions; a user can add a new function through a plugin built in a language supported by OSS according to their need (Bocher and Erwan, 2008). This chapter describes a study conducted in one of the major tributaries of the Indus River, i.e., the Beas River basin. The MOD10A2 snow cover data product of MODIS Terra was utilized. SCA variations were observed due to topographic parameters using the geo-statistical approach with a complex variable mean for 19 years (2001 to June 2020). The snow cover distribution was examined using elevation, slope, and aspect.

19.2 MATERIALS AND METHODS

19.2.1 STUDY AREA

In northern India, the Beas River is an important river, which is also a part of the Indus River system. The Beas River originates from the Rothang Pass in the Himalayan region in central Himachal Pradesh, India, at an elevation of 3900 m. It flows for some 470 kilometers to the Sutlej River, Punjab. The drainage basin length is 20,303 km². The catchment area is largely comprised of rash gradients with bare rocks. High peaks are present on both sides of the river valley. The altitude of the catchment varies from 934 to 6582 m. The Beas catchment with the Thalut discharge station as an outlet has been chosen for this study. The extent of the river catchment is from 76.935 to 77.867°E and from 31.507 to 32.414°N, and it covers an area of 4176 km². A significant portion of the river is covered by snow during the winter season. Snow cover is observed to be maximum in January–February, and minimum in July–August. Snowmelts are the major sources of the Beas River in the summer season. This region has a sub-tropical climate and tends to be temperate on the hilltops. Spring (March–May), summer (June–August), autumn (September–November), and winter (December-February), are the four seasons in the Beas River basin (Singh et al., 1997). Snow, glaciers, forest, and some barren and rocky land with steep slopes comprise the major land cover of the area. The rest of the area is utilized for agricultural purposes, especially apple farming and terrace cultivation, settlements, and river channels. The study area is shown in Figure 19.1.

19.2.2 DATA USED

The MOD10A2 data product for 2001 to 2016 has been taken from *reverb.echo.nasa*. The MODIS Terra snow cover 8-day L3 Global 500 m Grid (MOD10A2) was used to estimate the SCA of the study area. The snow cover product MOD10A2 version 5 (Hall and Riggs, 2007) is available from February 24, 2000 to the present. The spatial resolution of the MOD10A2 snow cover product is 500 × 500 m², and the temporal resolution is 8 days, starting on the first day of each year and extending to a few days of the following year (Maskey et al., 2011). Onboard NASA's Terra satellite, an imagining instrument, i.e., the Advanced Spaceborne Thermal Emission and Reflection Radiometer (ASTER), is placed. Stereoscopic images can be acquired from ASTER, which has nadir and backward-looking near-infrared (NIR) channels 3A and 3B at 15 m spatial resolution (Abrams, n.d.). Cloudy pixels have been removed using the cloud mask function. Abnormalities and outliers have been removed using a statistical selection algorithm. For the creation of elevation zones, slope, aspect maps, and snow line, the ASTER digital elevation model (DEM) 30 m has been utilized.

19.2.3 METHODOLOGY

The MODIS snow detection process is a completed auto-driven algorithm for snow detection. The algorithm utilizes the reflectance of five bands (bands 1–2 for Normalized Difference Vegetation

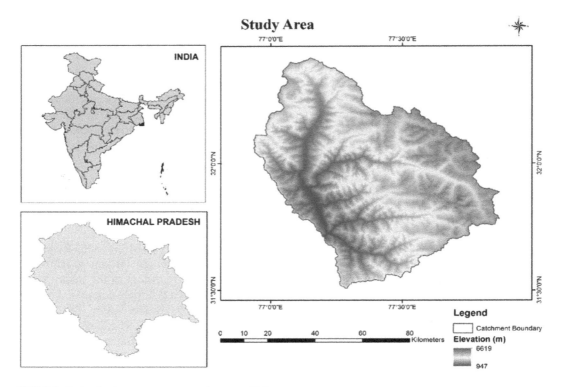

FIGURE 19.1 Study area – Beas catchment to Talut.

Index [NDVI] and bands 4–6 for Normalized Difference Snow Index [NDSI]). The reflectance of (0.545–0.565 µm) band-4 and (1.628–1.652 µm) band-6 wavelengths is utilized to calculate NDSI (Hall et al., 1995; Jain et al., 2009; Riggs and Hall, 2002). The NDSI process is widely used for snow cover analysis, as it is helpful for detecting snow cover under mountain shadows due to the capability of the satellite to detect diffuse radiation (Kulkarni et al., 2006). A discrimination between snow and cloud can be observed by NDSI, and the cloud has high reflectance in the visible and mid-infrared (MIR) wavelengths. Snow cover products are available using MODIS Terra and Aqua sensors. Snow mapping algorithms have been developed with an inclination towards Terra sensors. The MODIS Terra sensor is fully functional compared with the Aqua sensor (30%). MOD10A2 has been used to track the chronology of snow and to generate maximum snow cover extent. The maximum snow cover extent shows the presence of snow for more than a day. Snow cover product has been generated using an 8-day cycle because of the Terra platform's ground track period. The effect of clouds is a critical issue when generating snow cover products. The logic of the snow cover mapping algorithm minimizes the extent of cloud cover. Several studies (Gafurov and Bárdossy, 2009; Gurung et al., 2011; Parajka and Blöschl, 2008; Paudel and Andersen, 2011) have attempted different techniques, such as spatial filtering, linking Terra and Aqua MODIS data, temporal filtering, and 8-days maximum snow cover for the removal of cloud-obscured pixels from snow cover products. MOD10A2 images have been projected to the Geographic Latitude/Longitude projection system. WGS84 has been chosen as a datum plane. DEM to image registration has been carried out using QGIS Software. The study area (Beas basin up to Talut) has been extracted from a large area of MOD10A2 snow cover products. Snow cover maps based on altitude zones, slope zones, and aspect have been generated. The methodology chart is shown in Figure 19.2.

Snow Cover Monitoring

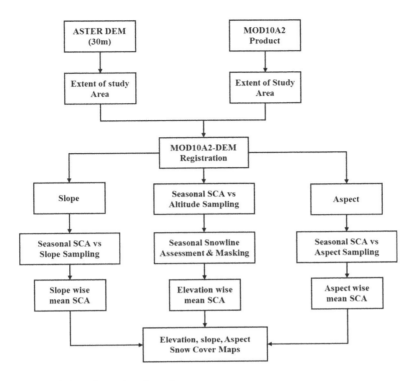

FIGURE 19.2 Methodology for estimation of snow cover area.

19.3 RESULTS AND DISCUSSION

The outcome presents spatio-temporal SCA analysis by considering the topographical parameters (elevation, slope, and aspect) in the rugged terrain of the Himalayan region. The spatio-temporal analysis was carried out using the MOD10A2 snow cover product for the duration of 2001 to 2016. Parallels and elevation control air and surface temperature (Jain et al., 2009). With every increment in altitude, the temperature decreases, leading to greater accumulation of snow in higher-elevation zones. The altitude and orientation of slopes are responsible for regulating the rainfall, climate, and weather of the Himalayan region (Jain et al., 2009). It is possible to separate large-scale variability created by topographic parameters from variability due to parameters like forest cover. Elevation, slope, aspect, and forest cover show a strong relationship with snow accumulation and snowmelt within the watershed (Jain et al., 2009; Jost et al., 2007). The existence of various microhabitats is modulated to a large degree by solar energy input, which is influenced by slope, orientation, and local topography (Antoine et al., 1998; Hörsch, 2003; Keller et al., 2005). In the present study, the total area has been divided into 10 elevation zones with a 500 m elevation difference. The altitude range was kept at an interval of 500 m to understand the influence of altitude on the SCA.

The ASTER 30 m DEM of Beas catchment was classified into 10 elevation zones based on the sensitivity of SCA to altitude. The zones were distributed as E1 to E10, with elevation ranging from 947 to 6619 m, respectively. All the zones have three different intervals, such as E1, E2 to E9, and E10, with an altitude range of 553, 500, and 1119 m, respectively. The slope was classified into four major classes: S1 (0–15°), S2 (15–25°), S3 (25–35°), and S4 (>35°). The catchment was categorized into four aspect zones: A1 (North-East), A2 (South-East), A3 (North-West), and A4 (South-West) directions. The zonal map of the elevation, slope, and aspect is shown in Figure 19.3.

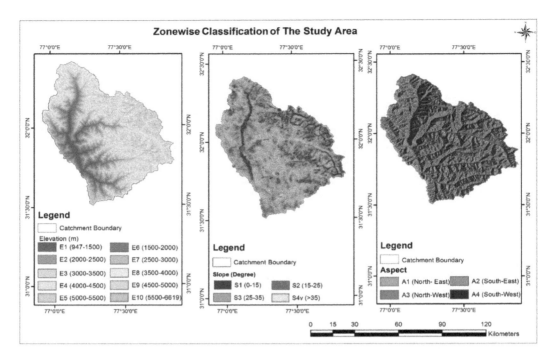

FIGURE 19.3 Zone-wise classification of Beas catchment.

TABLE 19.1
Details of Elevation Zones Derived from ASTER DEM

Sl. No.	Lower Elevation (m)	Upper Elevation (m)	Elevation Zone	Area (km²)	Area Covered (%)
1	947	1500	E1	211.6	4.25
2	1500	2000	E2	462.4	9.29
3	2000	2500	E3	615.9	12.37
4	2500	3000	E4	698.9	14.04
5	3000	3500	E5	597.6	12.01
6	3500	4000	E6	530.4	10.66
7	4000	4500	E7	650.7	13.09
8	4500	5000	E8	672.3	13.51
9	5000	5500	E9	468.1	9.41
10	5500	6619	E10	68.1	1.37

The percentage SCA in each elevation zone was estimated using a spatial analysis approach in ArcGIS 10.1 software for the years 2001–2016. Details of the area covered in km² and percentage for each elevation zone used for the SCA calculation are shown in Table 19.1.

Elevation is among the influencing parameters for snow accumulation. The snow cover extent increases with altitude, but cloud cover restricts the snow accumulation in higher-elevation zones. The clouds are particularly significant in the lower-elevation zones (3000–4000 m) during the monsoon and in the higher-elevation zones (>7000 m) during winter (Jost et al., 2007; Maskey et al., 2011). Snow cover extent decreases in altitude zones higher than 4000 m due to the presence of cloud cover in the winter season. Cloud-covered area was identified in all images, especially in the winter season, for 2001–2016. In winter season, most of the area is covered by cloud in high altitude zones. January showed maximum cloudy days in 2002, 2003, and 2008, whereas February showed maximum cloud cover in 2005–2007 and 2010–2013. The pattern of cloudy pixels shifted from January to February in the last few years. The seasonal SCA for elevation zones was calculated using an 8-day snow cover product for the years 2001–2016. The snow cover product was reclassified into snow and non-snow classes for all the years. The season-wise average for every year was calculated for 16 years. The results showed variation in snow cover for spring, summer, autumn, and winter seasons in all elevation zones. The average SCA in all four seasons (spring, summer, autumn, and winter) in the different elevation zones is shown in Figure 19.4. A decrement in SCA is observed, with major dips in 2003, 2006, and 2009. In the spring season, snow cover starts to ablate, and by the end of the summer season, it reaches minimum SCA. The SCA starts to increase again from autumn onwards, with a maximum SCA extent in the peak winter season. Low-elevation zones get a small amount of snow cover; thus zones E1 and E2 show the lowest snow cover in all seasons. The region with elevation above 3000 m receives snowfall, while the low-elevation region receives rainfall. As the altitude increases, the snow cover extent increases. Elevation zones from E6 onwards have annual snow cover. These zones are covered by permanent snow.

SCA melts with rising temperature. The resultant maximum and minimum snow cover days over 16 years have shown a shift in the snowfall and peak winter season from December–January to January–February. The seasonal snow cover extent in different elevation zones is shown in Figure 19.5. A noticeable change in river flow may result if this trend is found during the spring and autumn periods. Elevations ranging between 5000 and 7000 m are sensitive to climate change. The majority of glaciers exist in the 5000–7000 m altitude range. A significant amount of snow cover remains in this altitude range. The accumulation/ablation of snowmelt particularly depends on small variation of threshold temperature value.

An increase in snow cover at higher elevation zones in winter seasons has been reported in similar findings (Immerzeel et al., 2009; Maskey et al., 2011). In the present study, a falling trend in snow cover in January and an increasing trend in November have been observed. This trend may be attributed to a decrease in temperature in November and a gradual increase in December and January or to the presence of cloud cover in a particular month. The annual maximum snow extent is shifting forwards by about 6 to 7 days per year (Gurung et al., 2011; Paudel and Andersen, 2011; Ronco et al., 2016).

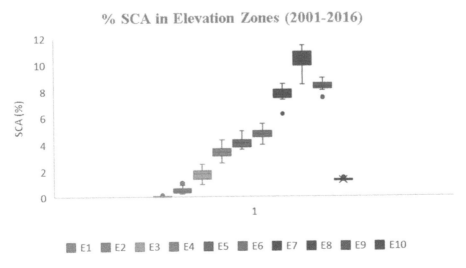

FIGURE 19.4 Seasonal (Spring, Summer, Autumn, and Winter) snow cover extent in elevation zones.

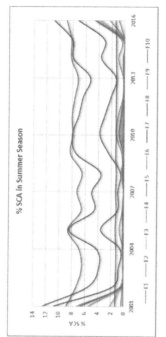
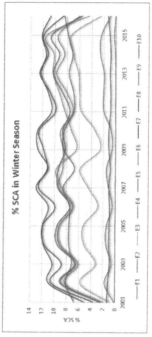
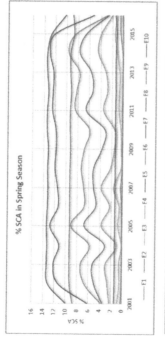
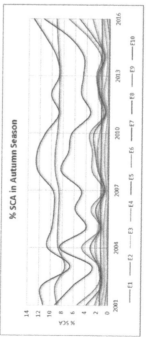

FIGURE 19.5 Seasonal (December–January and January–February) snow cover extent in elevation zones.

TABLE 19.2
Details of the Area Covered by Slope Classes

Sl. No.	Lower Value (Degrees)	Upper Class (Degrees)	Zone	Area (km²)	Percentage Area Covered
1	0	15	S1	772.77	15.53
2	15	25	S2	1186.42	23.85
3	25	35	S3	1396.78	28.08
4	35	>35	S4	1618.74	32.54

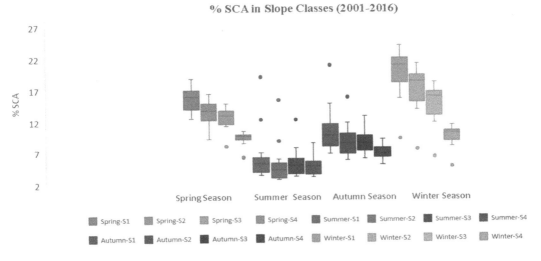

FIGURE 19.6 Seasonal SCA in different slope zones.

The direction of terrain slope has a strong influence on the SCA and snow accumulation. Slope and aspect affect incoming solar energy and moisture (Jain et al., 2009). In the present study, the slope was categorized into four classes, 0–15°, 15–25°, 25–35°, and >35°, as shown in Figure 19.3, and the area covered by each slope class is shown in Table 19.2.

Snow cover increases with a gradual rise in altitude. However, during the summer season, the sunset and sunrise are far from the northerly direction; thus, north-oriented slopes get similar solar radiation to south-oriented slopes (Jain et al., 2009). The resultant %SCA in different slope classes is shown in Figure 19.6. The summer season shows minimum SCA, while the winter season receives the maximum SCA in the region. Snow starts to melt in the spring season and has a minimum extent in the summer season. Snow accumulates in the autumn and winter seasons, and has a maximum extent in the peak winter season. The trend of snow cover extent in spring, summer, autumn, and winter can be observed from Figure 19.6. Slopes ranging from 0 to 10° have a significant impact on snow cover distribution (She et al., 2015). Maximum snow cover is observed in slope zone S1 for all the seasons. The autumn and winter seasons show maximum snow cover extent in the S1, S2, and S3 slope zones. Low slope classes contain more snow compared with high slope values. Slope class S4 (>35°) shows the minimum snow extent in all the seasons.

In the present study, the total basin area was distributed into four aspect zones: North-East, South-East, North-West, and South-West. The area covered by the North-East aspect is 24.14%, followed by South-East (23.59%), North-West (25.97%), and South-West (26.28%), as shown in Table 19.3.

TABLE 19.3
Area Covered by Aspect Class

Sl. No.	Lower Range (Degrees)	Upper Range (Degrees)	Zone	Area Covered (%)	Area (km²)
1	<315	<45	A1 (North-East)	24.11	1199.80
2	45	135	A2 (South-East)	23.65	1176.74
3	135	225	A3 (North-West)	25.90	1288.9
4	225	315	A4 (South-West)	26.34	1310.58

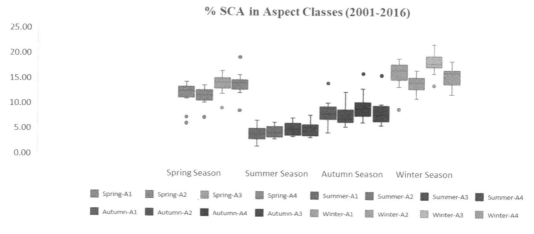

FIGURE 19.7 Seasonal SCA in different aspect zones.

During the winter season, the snow accumulation varies according to the direction of the slope. Aspect-wise SCA gives the influence of aspect on snow ablation and accumulation in (spring/summer season) March to June (Sharma et al., 2014). Snow starts ablating in March and reaches its lowest level in August. Less exposure to sunshine makes north-facing slopes more favorable for snow accumulations and SCD. This results in the development of glaciers in sheltered valleys on northern slope regions. The North-West and South-East directions of slopes accumulate more snow in low-altitude regions. Terrain with south-facing slopes is sunnier; thus, snow accumulation and SCD are lower for such regions. Snow ablates faster in southerly-sloping regions due to the significant reception of radiant energy. South-facing slopes have a higher snowline and receive more rainfall. This is one of the reasons why the snowline and rainfall are higher in the eastern Himalayan region than in the western Himalayan. The resultant snow cover shows maximum snow in A1 (North-East) and A3 (North-West) zones in all the seasons. The snow remains for a longer time in north-facing areas during the ablation period. The accumulation and ablation of snow depend on the amount of sunshine, insulation, and warm winds in the local region. Terrain having north-oriented slopes accumulates more snow than south-oriented slopes, as the northern slopes come under shadow during winter. The direction of mountain slopes regulates the local sunrise and sunset (Pratap Singh and Quick, 1993) (Figure 19.7).

The resultant SCA in aspect zones has shown higher snow cover in north-facing A1 (North-East) and A3 (North-West) aspect zones in the present study (She et al., 2015). The spatio-temporal analysis has reported the impact of topographical elements on the ablation and accumulation processes. Snow durations are longer on north-facing slopes and at lower snowline altitudes for all seasons (Ronco et al., 2016).

19.4 CONCLUSIONS

The spatio-temporal analysis has revealed the impact of elevation, slope, and aspect on snow accumulation and snowmelt processes. The OSS QGIS runs on different operating systems. It provides almost all GIS tools for spatial analysis, geoprocessing, cartography, modeling, and database management. It also supports different data and file formats, increasing data interoperability. The snow water equivalent, SCA, and snow depth influence the ablation, which affects the river flow regimes (Maskey et al., 2011). In the present study, an elevation above 3000 m has annual snow cover. A decreasing pattern in SCA has been seen in December and January. November has been reported to have an increase in snow cover extent. The slope classes S1 and S2 have shown significant snow extent, while S4 has been observed to have the lowest SCA. The maximum cloud cover period is also undergoing a shift towards February/March. February has started to receive peak winter snow durations instead of January. Similar findings have shown a shift in peak snowfall and snow period months in different regions (Gurung et al., 2011; Paudel and Andersen, 2011; Ronco et al., 2016). The study can be further correlated with temperature and rainfall data for climate change studies in the study area, which may help to obtain more information about the study area. The changes in peak snowfall period and peak winter seasons reveal the influence of climate change at a regional level.

REFERENCES

Abrams, M. B. R. (n.d.). *Aster User Handbook*. Version 2, Michael Abrams, Simon Hook, Jet Propulsion Laboratory, Pasadena CA.

Aizen, V. B., Aizen, E. M., Melack, J. M., & Dozier, J. (1997). Climatic and hydrologic changes in the Tien Shan, central Asia. *Journal of Climate*, *10*(6), 1393–1404. doi: doi: 10.1175/1520-0442(1997)010<1393:CAHCIT>2.0.CO;2

Anderton, S. P., White, S. M., & Alvera, B. (2004). Evaluation of spatial variability in snow water equivalent for a high mountain catchment. *Hydrological Processes*, *18*(3), 435–453. doi: 10.1002/hyp.1319

Antoine, G., Theurillat, J.-P., & Kienast, F. (1998). Predicting the potential distribution of plant species in an alpine environment. *Journal of Vegetation Science*, *9*(1998), 65–74. doi: 10.2307/3237224

Badea, A. C., & Badea, G. (2016). Considerations on open source GIS software Vs proprietary GIS software. *RevCAD Journal of Geodesy and Cadastre*, January, 15–27. www.researchgate.net/profile/Ana_Badea2/publication/322338186_CONSIDERATIONS_ON_OPEN_SOURCE_GIS_SOFTWARE_VS_PROPRIETARY_GIS_SOFTWARE/links/5a54d934aca2726c0ff1fea2/CONSIDERATIONS-ON-OPEN-SOURCE-GIS-SOFTWARE-VS-PROPRIETARY-GIS-SOFTWARE.pdf

Bergeron, J., Royer, A., Turcotte, R., & Roy, A. (2014). Snow cover estimation using blended MODIS and AMSR-E data for improved watershed-scale spring streamflow simulation in Quebec, Canada. *Hydrological Processes*, *28*(16), 4626–4639. doi: 10.1002/hyp.10123

Bocher, S., & Erwan, S. (2008). An overview on current free and open source desktop GIS developments. *International Journal of Geographical Information Science*, 1–24. www.dpi.inpe.br/cursos/ser300/Referencias/sstein_foss_desktop_gis_overview:IJGIS_2008.pdf

Brown, R. D. (2000). Northern hemisphere snow cover variability and change, 1915–97. *Journal of Climate*, *13*(2000), 2339–2355. doi: 10.1175/1520-0442(2000)013<2339:NHSCVA>2.0.CO;2

Changchun, X., Yaning, C., Weihong, L., Yapeng, C., & Hongtao, G. (2008). Potential impact of climate change on snow cover area in the Tarim River basin. *Environmental Geology*, *53*(7), 1465–1474. doi: 10.1007/s00254-007-0755-1

Dunn, S. M., & Colohan, R. J. E. (1999). Developing the snow component of a distributed hydrological model: A step-wise approach based on multi-objective analysis. *Journal of Hydrology*, *223*(1–2), 1–16. doi: 10.1016/S0022-1694(99)00095-5

Foster, J., Owe, M., & Rango, A. (1983). Snow cover and temperature relationships in North America and Eurasia. In *Journal of Climate and Applied Meteorology* (Vol. 22, Issue 3, pp. 460–469). doi: 10.1175/1520-0450(1983)022<0460:SCATRI>2.0.CO;2

Frei, A., & Robinson, D. A. (1999). Northern hemisphere snow extent: Regional variability 1972–1994. *International Journal of Climatology*, *19*(14), 1535–1560. doi: 10.1002/(SICI)1097-0088(19991130)19:14<1535::AID-JOC438>3.0.CO;2-J

Gafurov, A., & Bárdossy, A. (2009). Cloud removal methodology from MODIS snow cover product. *Hydrology and Earth System Sciences, 13*(7), 1361–1373. doi: 10.5194/hess-13-1361-2009

Gurung, D. R., Kulkarni, A. V., Giriraj, A., Aung, K. S., Shrestha, B., & Srinivasan, J. (2011). Changes in seasonal snow cover in Hindu Kush-Himalayan region. *The Cryosphere Discussions, 5*(2), 755–777. doi: 10.5194/tcd-5-755-2011

Hall, D. K., Riggs, G. A., & Salomonson, V. V. (1995). Development of methods for mapping global snow cover using moderate resolution imaging spectroradiometer data. *Remote Sensing of Environment, 54*(2), 127–140. doi: 10.1016/0034-4257(95)00137-P

Hall, D. K., & Riggs, G. A. (2007). Accuracy assessment of the MODIS snow products. *Hydrological Processes, 21*, 1534–1547.

Hewitt, K. (2005). The Karakoram anomaly? Glacier expansion and the "elevation effect", Karakoram Himalaya. *Mountain Research and Development, 25*(4), 332–340. doi: 10.1659/0276-4741(2005)025[0332:TKAGEA]2.0.CO;2

Hörsch, B. (2003). Modelling the spatial distribution of montane and subalpine forests in the central Alps using digital elevation models. *Ecological Modelling, 168*(3), 267–282. doi: 10.1016/S0304-3800(03)00141-8

Immerzeel, W. W., Droogers, P., deJong, S. M., & Bierkens, M. F. P. (2009). Large-scale monitoring of snow cover and runoff simulation in Himalayan river basins using remote sensing. *Remote Sensing of Environment, 113*(1), 40–49. doi: 10.1016/j.rse.2008.08.010

Jain, S. K., Goswami, A., & Saraf, A. K. (2009). Role of elevation and aspect in snow distribution in Western Himalaya. *Water Resources Management, 23*(1), 71–83. doi: 10.1007/s11269-008-9265-5

Jain, S. K., Goswami, A., & Saraf, A. K. (2010). Snowmelt runoff modelling in a Himalayan basin with the aid of satellite data. *International Journal of Remote Sensing, 31*(24), 6603–6618. doi: 10.1080/01431160903433893

Jost, G., Weiler, M., Gluns, D. R., & Alila, Y. (2007). The influence of forest and topography on snow accumulation and melt at the watershed-scale. *Journal of Hydrology, 347*(1–2), 101–115. doi: 10.1016/j.jhydrol.2007.09.006

Kaur, M., Mishra, V. D., & Sharma, J. K. (2012). Effects of topographic corrections on snow cover monitoring in Himalayan terrain using MODIS data. *International Journal of Engineering Research and Technlogy, 1*(5), 1–8.

Keller, F., Goyette, S., & Beniston, M. (2005). Sensitivity analysis of snow cover to climate change scenarios and their impact on plant habitats in alpine terrain. *Climatic Change, 72*(3), 299–319. doi: 10.1007/s10584-005-5360-2

Kour, R., Patel, N., & Krishna, A. P. (2015a). *Assessment of relationship between snow cover characteristics (SGI and SCI) and snow cover indices (NDSI and S3)*. April. doi: 10.1007/s12145-015-0216-4

Kour, R., Patel, N., & Krishna, A. P. (2015b). *Effects of terrain attributes on snow-cover dynamics in parts of Chenab basin, western Himalayas effects of terrain attributes on snow-cover dynamics in parts of Chenab basin, west … Effects of terrain attributes on snow-cover dynamics in parts of Chenab basin, western Himalayas*. October 2016. doi: 10.1080/02626667.2015.1052815

Kour, R., Patel, N., & Krishna, A. P. (2016). Climate and hydrological models to assess the impact of climate change on Climate and hydrological models to assess the impact of climate change on hydrological regime: A review. *Arabian Journal of Geosciences, 9*, 44. doi: 10.1007/s12517-016-2561-0

Krajčí, P., Holko, L., & Parajka, J. (2016). Variability of snow line elevation, snow cover area and depletion in the main Slovak basins in winters 2001–2014. *Journal of Hydrology and Hydromechanics, 64*(1), 12–22. doi: 10.1515/johh-2016-0011

Krishna, A. P., & Sharma, A. (2013). Snow cover and land surface temperature assessment of Gangotri basin in the Indian Himalayan Region (IHR) using MODIS satellite data for climate change inferences. *Earth Resources and Environmental Remote Sensing/GIS Applications Proceedings Volume 8893*, IV. doi: 10.1117/12.2029084

Kulkarni, A. V., Singh, S. K., Mathur, P., & Mishra, V. D. (2006). Algorithm to monitor snow cover using AWiFS data of RESOURCESAT-1 for the Himalayan region. *International Journal of Remote Sensing, 27*(June 2016), 2449–2457. doi: 10.1080/01431160500497820

Marchane, A., Jarlan, L., Hanich, L., Boudhar, A., Gascoin, S., Tavernier, A., Filali, N., LePage, M., Hagolle, O., & Berjamy, B. (2015). Assessment of daily MODIS snow cover products to monitor snow cover dynamics over the Moroccan Atlas mountain range. *Remote Sensing of Environment, 160*(February 2015), 72–86. doi: 10.1016/j.rse.2015.01.002

Maskey, S., Uhlenbrook, S., & Ojha, S. (2011). An analysis of snow cover changes in the Himalayan region using MODIS snow products and in-situ temperature data. *Climatic Change, 108*(1), 391–400. doi: 10.1007/s10584-011-0181-y

Menon, S., Koch, D., Beig, G., Sahu, S., Fasullo, J., & Orlikowski, D. (2010). Black carbon aerosols and the third polar ice cap. *Atmospheric Chemistry and Physics, 10*(10), 4559–4571. doi: 10.5194/acp-10-4559-2010

Mishra, B., Babel, M. S., & Tripathi, N. K. (2014). Analysis of climatic variability and snow cover in the Kaligandaki River Basin, Himalaya, Nepal. *Theoretical and Applied Climatology, 116*(3–4), 681–694. doi: 10.1007/s00704-013-0966-1

Mohammed, W. E. (2014). Free and open source GIS: An overview on the recent evolution of projects, standards and communities. *Proceeding of the 9th National GIS Symposium*, Saudi Arabia, KSA, April, 1–13.

Parajka, J., & Blöschl, G. (2008). Spatio-temporal combination of MODIS images - Potential for snow cover mapping. *Water Resources Research, 44*(3), 1–13. doi: 10.1029/2007WR006204

Paudel, K. P., & Andersen, P. (2011). Monitoring snow cover variability in an agropastoral area in the Trans Himalayan region of Nepal using MODIS data with improved cloud removal methodology. *Remote Sensing of Environment, 115*(5), 1234–1246. doi: 10.1016/j.rse.2011.01.006

Pimentel, R., Pérez-Palazón, M. J., Herrero, J., & José Polo, M. (2016). Using different remote sensing data to improve snow cover area representation. April, 2–3. *Geophysical Research Abstracts, 18*, EGU2016-16214. doi: 10.13140/RG.2.1.1721.0484

Pomeroy, J., Shook, K., Fang, X., Brown, T., & Marsh, C. (2013). Development of a snowmelt runoff model for the lower smoky river. *Centre for Hydrology Report No. 13, 13*, 89.

Pomeroy, J. W., Gray, D. M., Shook, K. R., Toth, B., Essery, R. L. H., Pietroniro, A., & Hedstrom, N. (1998). An evaluation of snow accumulation and ablation processes for land surface modelling. *Hydrological Processes, 12*(15), 2339–2367. doi: 10.1002/(SICI)1099-1085(199812)12:15<2339::AID-HYP800>3.0.CO;2-L

Pu, Z., Xu, L., & Salomonson, V. V. (2007). MODIS/Terra observed seasonal variations of snow cover over the Tibetan Plateau. *Geophysical Research Letters, 34*(6), 1–6. doi: 10.1029/2007GL029262

Riggs, G. A., & Hall, D. K. (2002). Reduction of cloud obscuration in the MODIS snow data product. *59th Eastern Snow Conference*, 205–212.

Ronco, P., Da Piperno, P., Avanzi, F., & Michele, C., De. (2016). Investigating the role of topography on snow cover duration and distribution in the Italian Apennines by means of MODIS data. *Geophysical Research Abstracts, 18*, 11656.

Salcedo, A. P., & Cogliati, M. G. (2014). Snow cover area estimation using Radar and optical satellite information. *Atmospheric and Climate Science, 4*, 514–523.

Sharma, V., Mishra, V. D., & Joshi, P. K. (2014). Topographic controls on spatio-temporal snow cover distribution in Northwest Himalaya. *International Journal of Remote Sensing, 35*(9), 3036–3056. doi: 10.1080/01431161.2014.894665

She, J., Zhang, Y., Li, X., & Feng, X. (2015). Spatial and temporal characteristics of snow cover in the Tizinafu watershed of the Western Kunlun Mountains. *Remote Sensing, 7*(4), 3426–3445. doi: 10.3390/rs70403426

Shen, S. S. P., Yao, R., Ngo, J., Basist, A. M., Thomas, N., & Yao, T. (2014). Characteristics of the Tibetan Plateau snow cover variations based on daily data during 1997–2011. *Theoretical and Applied Climatology, 120*, 445–453.

Singh, P., Jain, S. K., & Kumar, N. (1997). Estimation of snow and glacier-melt contribution to the Chenab River, Western Himalaya. *Mountain Research and Development, 17*(1), 49–56. doi: 10.2307/3673913

Singh, P., & Quick, M. C. (1993). Streamflow simulation of Satluj River in the western Himalayas. *Snow and Glacier Hydrology: Proceedings of an International Symposium Held at Kathmandu*, Nepal, 16–21 November 1992, *218*, 261–271.

Thakur, P. K., Garg, P. K., Aggarwal, S. P., Garg, R. D., & Mani, S. (2013). Snow cover area mapping using synthetic aperture Radar in Manali watershed of Beas river in the northwest Himalayas. *Journal of the Indian Society of Remote Sensing, 41*(4), 933–945. doi: 10.1007/s12524-012-0236-1

Thayyen, R. J., & Gergan, J. T. (2010). Role of glaciers in watershed hydrology: A preliminary study of a "Himalayan catchment." *Cryosphere, 4*(May), 115–128. doi: 10.5194/tcd-3-443-2009

Wang, J., & Li, S. (2006). Effect of climatic change on snowmelt runoffs in mountainous regions of inland rivers in Northwestern China. *Science in China, Series D: Earth Sciences, 49*(8), 881–888. doi: 10.1007/s11430-006-0881-8

Weng, Q., & Hu, X. (2008). Medium spatial resolution satellite imagery for estimating and mapping urban impervious surfaces using LSMA and ANN. *IEEE Transactions on Geoscience and Remote Sensing, 46*(8), 2397–2406. doi: 10.1109/TGRS.2008.917601

Ye, B. S., Ding, Y. J., Kang, E. S., Li, G., & Han, T. D. (1999). Response of the snowmelt and glacier runoff to the climate warming-up in the last 40 years in Xinjiang autonomous region, China. *Science in China, Series D: Earth Sciences, 42*(96), 44–51. doi: 10.1007/BF02878852

20 Land Surface Water Resource Monitoring and Climate Change

P. Das, V. Pandey, and Dipanwita Dutta

CONTENTS

20.1 Introduction	311
20.2 Surface Water Monitoring	313
20.2.1 Multispectral Optical Data	313
20.2.2 Microwave Remote Sensing	314
20.3 Water Quality Monitoring Using Remote Sensing	316
20.4 Open Source Image Processing and Geographic Information System (GIS) Software and Platforms	318
20.5 Climate Alteration Impacts	319
20.6 Conclusion	321
20.7 Supplementary	321
References	323

20.1 INTRODUCTION

Monitoring the impact of climate change on water resources is essential because of water's multi-dimensional uses, interactions, and influence on various land surface processes. The availability of solar radiation and seasonal precipitation, along with the soil type and topographic condition, regulates the hydrological settings and determines ecosystem productivity and structure, providing a suitable niche for various species (Tripathi et al., 2019), determines the dominant life forms (Das and Behera, 2019), and regulates the ecosystem services (Singh, 2007). On the contrary, recurrent dry spells over an area cause land degradation and productivity loss (Matin and Behera, 2019). The hydroclimatic condition of an area plays an important role in agricultural practices; specifically, it determines the cropping period, productivity, and choice of crops. Over the past decades, the increased greenhouse gas (GHG) concentration and average global temperature have posed threats to water resources as well as leading to decreased snow cover, increased sea level, and enhanced evaporation, and triggered a significant shift in the geographical distribution of precipitation (Ramesh and Yadava, 2005). Moreover, the projected threats of climate change simulated by the global climate models (GCM) have predicted significant alteration in climate conditions, which may lead to potential changes in species distribution, dependent land surface processes, water resources, and agricultural practices (Aggarwal, 2008).

Water resource management is considered an integral part of agriculture and water resource planning, and moreover, it is crucial for the development of a country (Bhatt and Mall, 2015). The effects of natural variability and alterations in climatic conditions on various hydrologic processes have been identified as dominant contributors to recurrent extreme weather events, such as drought (Pandey and Srivastava, 2019; Chanda and Maity, 2016), flood occurrences (Guhathakurta et al., 2011; Ghosh and Dutta, 2012), groundwater deficit (Kumar, 2012), alteration in soil moisture condition (Pandey et al., 2020), and river runoff (Singh et al., 2006), and such processes may indirectly alter local and regional climate processes (Vinayachandran et al., 2015). Furthermore, contributions from human disturbances, such as deforestation, urbanization, change in land use practices, etc., have exaggerated such processes, which may intensify in coming decades (Das et al., 2018; Venkatesh et al., 2020). The vulnerability to climate change is interlinked with other stressors, e.g., food security, an important parameter that controls the socio-political and economic conditions of an area (Leichenko and O'brien, 2002). The surface and sub-surface water availability and soil moisture content are essential for direct use in household, agriculture, industry, power generation, etc., and thus, it plays a major role in socio-economic and cultural practices. More than two-thirds of global freshwater is consumed in agricultural practices (Misra et al., 2013). In India, about 70% of rural households are primarily dependent upon agriculture and allied sectors, wherein 82% people are engaged as small and marginal farmers (FAO). It is worth noting that more than 60% of the total agricultural land in India is rain-fed, while only 35% of land has access to irrigation water (https://ccafs.cgiar.org/india#.XzWC7igzZPY). Alteration in climate conditions and an increase in extreme weather events will have adverse effects on agricultural productivity, food security, and the rural economy. The agricultural vulnerability index map generated for the present and projected climate scenarios in India indicates comparatively higher vulnerability for the states of Rajasthan, Gujarat, Maharashtra, Karnataka, and Madhya Pradesh (Rao et al., 2013). The water stress map also indicates an increasing vulnerability in the country, and around 54% of the area is experiencing high to extreme stress and facing the consequences of reduced groundwater level (www.indiawatertool.in/) (Figures 20.1 and 20.2).

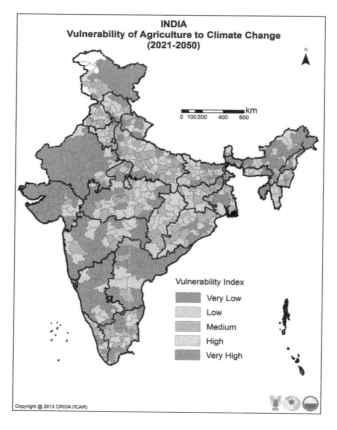

FIGURE 20.1 Agriculture Vulnerability Map under the Projected Climate during 2021–2050. (From Rao, C.R., et al., *Atlas on Vulnerability of Indian Agriculture to Climate Change*, Central Research Institute for Dryland Agriculture, Hyderabad, 2013.)

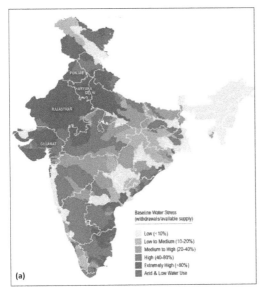 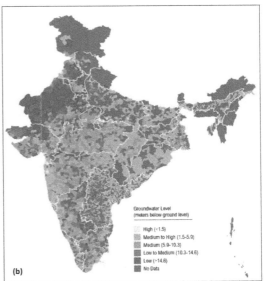

FIGURE 20.2 Water stress map (a) and groundwater level (b). (From World Resources Institute; www.indiawatertool.in/.)

20.2 SURFACE WATER MONITORING

20.2.1 MULTISPECTRAL OPTICAL DATA

The two major rivers, Ganga and Brahmaputra, in combination share about 50% of all the river water resources in India (Mujumdar, 2008). Storage of river runoff by constructing reservoirs and natural lakes has multiple benefits in mitigating the effects of extreme weather events such as drought and flood, providing freshwater, improving irrigation facilities, power generation, etc. Moreover, small household ponds are also beneficial for small-scale uses. Mapping and monitoring of the surface water area are essential for studying land surface processes and improving water resource management activities, flood and drought monitoring, and decision-making processes. Satellite remote sensing has emerged as a promising technology for assessing water resources and can be used as a complement to conventional field-based information (Jha et al., 2007). The optical data collected in a cloud-free environment allow the identification of water bodies through analyzing the surface reflectance (SR) pattern (Figure 20.3) and employing visual image interpretation, band thresholding, spectral signature matching, and spectral enhancement techniques (Behera et al., 2018), whereas the analysis of microwave (synthetic aperture radar [SAR]) data depends on the dielectric constant and surface scattering properties. In comparison to broadband multispectral data, the minute spectral deviations in the narrow spectral bands of hyperspectral data allow the mapping of constituents such as chlorophyll, algae, dissolved organic matter, sediment load, etc. (Brando and Dekker, 2003; Mishra et al., 2019; Jensen et al., 2019). The spectral enhancement uses two or more spectral bands, and the choice of bands depends on the differential SR properties of the land surface features. For example, the normalized difference vegetation index (NDVI) uses the normalized difference of the near-infrared (NIR) and red bands [(NIR − Red)/(NIR + Red)] and highlights the vegetation (higher positive value) due to contrasting SR in the NIR and red bands (Figure 20.3). In the case of a waterbody, low SR is observed in the red band and extremely low SR in the NIR and short-wave infrared (SWIR) bands due to its high absorption; which results in negative NDVI values for the water body. However, the presence of suspended sediment in the water body causes a higher SR compared with a clear water body.

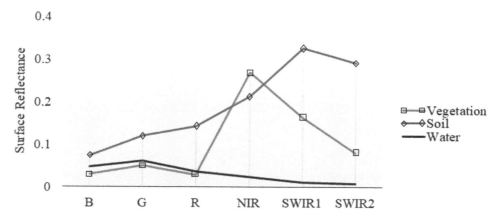

FIGURE 20.3 Surface reflectance curves of vegetation, soil, and water derived from Landsat-8 data.

Employing the highest and lowest SR bands of water body in the green and NIR bands, McFeeters (1996) developed the normalized difference water index (NDWI) [(Green − NIR)/(Green + NIR)]. A sample NDWI map for the Rengali dam, Odisha, India, shown in Figure 20.4, clearly discriminates the surface water area from other land surface features. The difference in the mean NDWI images of post-monsoon January and pre-monsoon May identifies the seasonal change in the surface water area. The NDWI values for pure water body pixels were observed to be >0.3, whereas lower NDWI values mainly represent mixed pixels with varying features. To reduce the similarity in NDWI index values over built-up lands and water bodies, Xu (2006) modified the index by replacing the NIR band with the SWIR band [MNDWI = (Green − SWIR)/(Green + SWIR)] and observed a significant adjustment. Feature space mapping, decomposition, and correlation analysis employing the spectral bands of multispectral or hyperspectral data enable effective band selection for spectral enhancement (Tang and Xu, 2017; Zhang et al., 2019). Moreover, statistical image classification techniques, i.e., supervised (maximum likelihood classifier, decision tree based classifier, and logistic regression), machine learning approaches (support vector machine, random forest, and artificial neural network), and time-series analysis are also widely used for mapping surface water and associated land surface features (Das and Pandey, 2019; Ghosh et al., 2017).

20.2.2 Microwave Remote Sensing

Microwave remote sensing relies on the dielectric constant and scattering properties of the land surface features. Due to its higher wavelength, the microwave signal can penetrate the clouds and map the surface water area even in the monsoon season. It is widely utilized for mapping the inundation area of floods and helps in developing effective mitigation planning, collecting data on hydrological parameters, etc. The side-looking system of a SAR captures the backscatter signals along the perpendicular direction of flight direction instead of vertical imaging. The incident microwave signal is reflected (specular reflection/ mirror effect) from the smooth water surface, while the diffused scattering from the other land surface features, such as soil, vegetation, buildings, etc., induces high backscatter values because of their roughness parameter and geometry. This leads to a lower return signal for the water body, which appears black in microwave data compared with the other land surface features. Moreover, the backscatter value for a water body may increase with waves or ripples, or a lower incidence angle, and depending on the polarization. Although the SAR backscatter (σ^0) provides an accurate water spread area, the presence of vegetation or built-up areas causes errors when mapping the flood inundation area. The presence of a subcanopy water layer may induce higher backscatter compared with subcanopy soil or other vegetation layers depending

Water Resource Monitoring and Climate Change 315

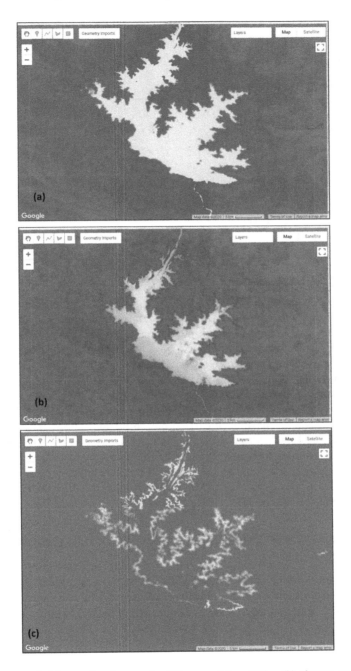

FIGURE 20.4 Mean NDWI map for the month of January (a) and May (b) change area (c) derived from Sentinel-2 MSI data using the Google Earth Engine (GEE) platform (code [1] is given in supplementary).

on the operating wavelength and following the canopy penetration capacity (Solbø and Solheim, 2005). Similarly, the presence of buildings in a city induces high error in flood inundation area mapping with microwave data. To assess the impact of climate alteration, studies have been carried out on regional- to global-scale surface water area mapping employing microwave remote sensing (Schroeder et al., 2015). Alternatively, a passive microwave radiometer such as the Advanced Microwave Scanning Radiometer (AMSR) records the brightness temperatures and enables

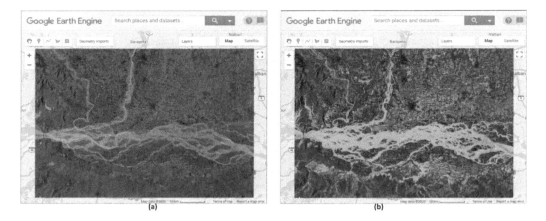

FIGURE 20.5 Water area, indicated by low values, exhibits the change in water area during pre-monsoon (a) and monsoon (b).

assessment of sea surface temperature (SST), surface wind speeds, atmospheric water vapor, cloud liquid water, and rain rate (www.remss.com/missions/amsr/). The high-frequency channel data of AMSR is also used for fractional open water inundation, flood and drought mapping by generating the look-up table of microwave emissivities for various features, and indicates the change in brightness temperature due to change in surface cover (Du et al., 2016; De Groeve, 2010; Galantowicz, 2002). While microwave data such as the Advanced Land Observing Satellite-Phased Array type L-band Synthetic Aperture Radar (ALOS-PALSAR), RADARSAT-1/2, TerraSAR-X, etc. are commercial, with the advent of the latest freely accessible high-resolution Sentinel-1 C-band SAR data, a significant number of studies have been carried out on mapping flood inundation (Borah et al., 2018; Anusha and Bharathi, 2020; Vanama et al., 2020). A sample water area change map is shown in Figure 20.5, which indicates the water spread area caused by the latest flood in the Brahmaputra River basin. The freely available Sentinel-1 C-band SAR data were employed in the study, and the VV backscatter values were examined for March and August 2020 in the Google Earth Engine platform (Supplementary GEE code 2).

20.3 WATER QUALITY MONITORING USING REMOTE SENSING

Increasing erosion of surface soil caused by intense rainfall, surface runoff, and human disturbances is leading to sediment load and turbidity in river water. Overburden of sediments transported by various streams has a detrimental effect on agriculture and the working life of a reservoir, as it reduces the effective water storage capacity (Dutta, 2016). A substantial number of mathematical and hydrological models – the Revised Universal Soil Loss Equation (RUSLE) (Dutta et al., 2015), Revised Universal Soil Loss Equation and Sediment Yield (RUSLE-SY) (Anees et al., 2018; Magesh and Chandrasekar, 2016), Modified Universal Soil Loss Equation (MUSLE) (Pandey et al., 2009; Arekhi et al., 2012), Soil and Water Assessment Tool (SWAT) (Dutta and Sen, 2018), multilayer perceptron (MLP) artificial neural network model (Singh et al., 2012), and Water Erosion Prediction Project (WEPP) (Pandey et al., 2008; Singh et al., 2011) – have been used successfully in sediment yield estimation at the river basin scale. It is worth noting that satellite remote sensing integrated with field measurements has been widely used for measuring the spatio-temporal variation of sediment yield with good accuracy. In order to assess water quality parameters such as turbidity, Secchi disk depth (SDD), chlorophyll content, surface temperature, suspended sediment (SS), etc., a substantial number of studies have employed moderate-resolution broadband Landsat imagery (El-Zeiny and El-Kafrawy, 2017; Nas et al., 2010). Previous research has demonstrated strong statistical interrelationships between satellite-derived indices and field information in the estimation of water constituents (El-Zeiny and El-Kafrawy, 2017; Kumar et al., 2016; Nas et al., 2010; Prasad et al., 2017). Studies have also been carried out on estimating the turbidity, SS, and heavy metals, integrating the moderate-resolution Landsat and MODIS (Moderate Resolution Imaging Spectroradiometer) data by using the spatial and temporal adaptive reflectance fusion model (STARFM), regression, and genetic algorithm (Swain and Sahoo, 2017).

In comparison to broadband multispectral data, the narrow spectral bands of hyperspectral data facilitate differential SR due to minute alterations in water quality parameters. Case studies identified the narrow spectral bands in the visible and red-edge region of the Airborne Visible InfraRed Imaging Spectrometer - Next Generation (AVIRIS-NG) data as suitable wavelength regions for mapping chlorophyll-a (Chl-a), turbidity, suspended sediment, and phosphorus in the Ganga River basin and Chilika lake, India (Bansod et al., 2018; Chander et al., 2019). Field-based spectral analysis reveals a significant positive correlation between the turbidity and total suspended solids in the spectral range between 700 and 900 nm (Wu et al., 2014). Such approaches have also been tested with broadband multispectral datasets (Murugan et al., 2016). Velloth et al. (2014) used radiative transfer models (RTM) for removing the attenuation of the water column and proved the potential of hyperspectral remote sensing for mapping coral reefs. The integration of the finer-resolution hyperspectral data and high-resolution WorldView-3 data overcomes the sun glint effects of water waves and enables mapping of submerged features such as seagrasses, drift algae, and algal bloom (Bostater et al., 2017). Datasets available from the Hyperspectral Imager for the Coastal Ocean (HICO), Hyperion, Medium Resolution Imaging Spectrometer (MERIS), Sentinel-2 and 3, and MODIS sensors, along with remote sensing–derived indices such as normalized difference chlorophyll index (NDCI), slope-based model ($Slope_{red}$ and $Slope_{NIR}$), etc., have been used widely in mapping the seagrass, algae, and chlorophyll concentration (Cho et al., 2014; Mishra et al., 2014; Mishra and Mishra, 2010). In comparison to the regression-based models, which require field measurements for parameterization, the optimized indices approximate the water constituents without field observation and enable comparative or time-series analysis. The NDCI was developed employing the two sensitive channels, 665 and 780 nm ([Rrs(708) − Rrs(665))/(Rrs(708)+Rrs(665)], wherein Rrs indicates remote sensing reflectance and values within parentheses indicate the central wavelength). It provides higher accuracy even for turbid water conditions. Mishra and Mishra (2012) generated a chart of NDCI value range and corresponding Chl-a concentration, which helps to approximate the Chl-a based on the NDCI value. The NDCI equation for the Sentinel-2 data can be expressed as (band5 − band4)/(band5 + band4). The Sentinel-2 derived sample NDCI image generated in the GEE and shown in Figure 20.6 indicates an NDCI value ranging between −0.1 and 0.03 in March and the corresponding Chl-a concentration, which varies between 7.5 and 20 mg.m^{-3}. A substantial number of remote sensing–based indices have been developed for assessing various

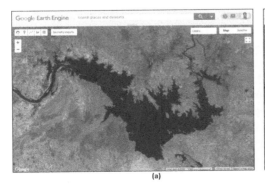
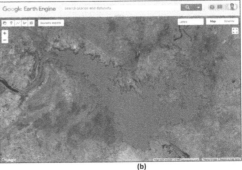

FIGURE 20.6 Sentinel-2 image SFCC (a) and derived NDCI image (b) for the Hirakud dam, Odisha, India.

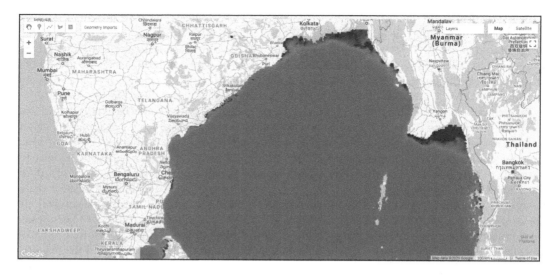

FIGURE 20.7 MODIS data–derived mean chlorophyll-a concentration map for the period of January to May 2019, plotted in GEE.

water quality parameters. For instance, the normalized difference turbidity index (NDTI: [(Red − Green)/(Red + Green)]), uses the red and green bands to estimate turbidity, especially in inland waters (Lacaux et al., 2007). The MODIS-derived mean Chl-a concentration of the Bay of Bengal region from January to May 2019 is shown in Figure 20.7, which indicates a maximum concentration value of 7 mg.m^{-3}.

20.4 OPEN SOURCE IMAGE PROCESSING AND GEOGRAPHIC INFORMATION SYSTEM (GIS) SOFTWARE AND PLATFORMS

In the past decade, a considerable number of open source software and advanced programming platforms have evolved for satellite image data processing and GIS data analysis. Among them, Quantum GIS (Q-GIS), Geographic Resources Analysis Support System (GRASS) GIS, ILWIS, Sentinel Application Platform (SNAP), and programming platforms like R and Python have gained popularity for their efficiency in image processing and modeling and their user-friendly graphical interface. Also, web-based GIS (WebGIS) platforms such as Google Earth Engine and ArcGIS online have been widely used in recent times. Q-GIS uses the Qt toolkit and C++, which supports vector, raster, and database formats; whereas the Python console allows the user for scripting. A large number of Python plugins have been developed for Q-GIS, which can be added for data processing according to the requirements of the user. The GRASS GIS was developed by the U.S. Army Corps of Engineers' Construction Engineering Research Laboratory (USA/CERL) in 1982. It is written in C, C++, and Python, having a graphical user interface (GUI) and command line syntax for user-defined analysis and automation. This software allows the processing of various spatial data (raster and vector), modeling, three-dimensional analysis, and database management with dedicated tools for processing LiDAR point cloud data, satellite images, aerial photographs, and unmanned aerial vehicle (UAV) images. The SNAP toolbox, primarily developed for the processing of various Sentinel data, is also able to perform several image processing tasks.

It is also worth mentioning that a substantial number of packages have been developed to perform geospatial data analysis. Python and R, two high-level programming platforms, are popularly used for analyzing the various types of spatial data as well as non-spatial data. On the other hand, the Google Earth Engine (earthengine.google.com/), one of the latest online programming platforms,

TABLE 20.1
Freely Available Satellite Data and Applications to Water Resources

Data Type	Satellite Data	Applications
Optical	Landsat, Sentinel-2, ASTER, MODIS, Advanced Very High Resolution Radiometer (AVHRR), Hyperion (hyperspectral till 2015)	• Surface water area/spatial extent • Water quality • Evapotranspiration • Glaciers • Crop water stress • Sea surface temperature
Microwave	Sentinel-1, Radarsat-1 (1996–2008 data), ALOS-PALSAR (2006-2011 data), SMAP, SMOS (Soil Moisture and Ocean Salinity), GRACE (Gravity Recovery and Climate Experiment)	• Flood or water inundation area • Soil moisture • Ground water • River discharge • Rainfall
LiDAR	IceSAT, GEDI (Global Ecosystem Dynamics Investigation)	• Sea surface height

offers a range of techniques using Python and JavaScript and provides accessibility to a large dataset comprising various satellite images (e.g., Landsat and Sentinel), image products (e.g., land use land cover map, digital elevation model, NDVI, EVI, fire, and sea surface temperature), spatial data (e.g., climate and atmosphere), GIS data (e.g., national boundaries), etc. This platform allows users to access the data, upload various data layers, perform processing and visualization, download the input, and generate the output, whereas ArcGIS online, developed by the Environmental Systems Research Institute (ESRI) (esri.com/en-us/arcgis/products/arcgis-online/overview), is an interactive online GIS data processing platform, which provides accessibility to a large number of GIS tools and allows users to create maps and visualize (Table 20.1).

20.5 CLIMATE ALTERATION IMPACTS

The slowly altering climatic conditions and extreme weather events will adversely affect various components of water resources, i.e., availability, quality, and hydrological balances, and intensive human activities such as deforestation, conversion of land, increasing artificial surfaces, urbanization, industrialization, etc., may exacerbate these processes. The shift in precipitation regime may lead to significant changes and detrimental effects on biodiversity, niche or habitat alteration, water availability, water quality, soil erosion, agricultural practices, etc. Such conflicts may further exacerbate the shifting or displacement of species as well as human inhabitants. The predicted water stress vulnerability indicates that more than 60% of the global population may experience significant stress by 2025, and climate change may further fuel the stress (Arnell, 1999). The projected change in climate conditions simulated by various GCMs under the Representative Concentrative Pathways (RCP) has predicted significant changes in the south Asian monsoon, with higher temperature and rainfall over India (Bal et al., 2016; Huo and Peltier, 2020). O'Brien (2004) studied vulnerability to multiple stressors, including climate change and socio-economic conditions, at the district level in India. Their results highlighted that the majority of the districts in the drier central and western parts of India are highly climate sensitive in the present climate conditions, which may expand towards the east and south under the projected climate change scenarios.

The projected temperature rise and drought conditions indicate an increase in soil salinization, aquifer depletion, and land degradation, which may also lead to desertification in a few regions, while the recurrent flooding will cause greater soil erosion and sedimentation (Misra et al., 2013). The impact of climate alteration on water quality can be evaluated through the monitoring of physico-chemical and biological parameters and micropollutants, which could be linked with anthropogenic disturbances and climatic factors, i.e., ambient temperature, extreme weather events, soil moisture condition, and solar radiation (Delpla et al., 2009). The combined effect of warming and human disturbances may significantly change nutrient loads such as nitrogen concentration in rivers and nitrogen, phosphorus, and harmful algae in lake water (Arheimer et al., 2005; Molina-Navarro et al., 2014; Michalak, 2016). Resilience refers to the power of a system to restore itself after experiencing a change in its normal conditions due to external forces. According to Murdoch et al. (2000), the water ecosystem resilience may experience significant degradation or frequent deviation from its tolerance limits under the projected climate change. The long-term monitoring of water availability and quality parameters, quantitative and comparative analysis of the variables, and understanding the combined effects of climate change and human disturbances are important for managing lake ecosystems (Murdoch et al., 2000; Michalak, 2016; Taylor et al., 2013). In their study, Rehana and Mujumdar (2011) discussed the impact of climate change on hydrological and water quality parameters. They observed a reduction in streamflow, water temperature, and dissolved oxygen in the Tunga-Bhadra River in recent years. Whitehead et al. (2018) highlighted the combined impact of climate and socioeconomic change on the Ganga–Brahmaputra–Meghna (GBM) River along with the Hooghly and Mahanadi River system. Their study showed increased streamflow in both monsoon and dry periods and alteration of nutrients as a consequence of various factors: diversion of water, population growth, urbanization and industrial development, agricultural practices, atmospheric deposition, and changes in water balance and fluxes.

The changing climate will lead to significant modifications in water and energy balances, atmospheric processes, glacier melting, river discharges, soil moisture, and groundwater level. Simulation studies have indicated a substantial increase in CO_2 concentration that will lead to methane emissions from the wetlands and water inundation regions (Shindell et al., 2004). The decadal analysis of satellite remote sensing data indicates an expansion of more than 43% in the total surface area of moraine dammed glacial lakes (MDGLs) in the central and eastern Himalayas from 1990 to 2015, which could be attributed to the impact of warming climate in this region (Begam and Sen, 2019). The human-modified changes in land use practices and projected change in climatic conditions will significantly modify the river basin hydrological balances and will lead to increasing surface runoff, base-flow, and sediment yield and decreasing evapotranspiration (Das et al., 2018; Molina-Navarro et al., 2014; Nilawar and Waikar, 2019). The possible shift in climate conditions may increase flood and drought intensity and agricultural water demand, which may require much higher supplemental irrigation, suitable water harvesting structures, micro-storage facilities in watersheds, etc. (Pathak et al., 2014). Burney and Ramanathan (2014) studied the combined influence of climate (precipitation and temperature) and two short-lived climate pollutants (SLCPs), tropospheric ozone and black carbon, on wheat and rice yield. They observed a statistically significant reduction in yield for wheat (36% lower) and rice (20%, statistically insignificant). The combined impacts of global climate and GHG concentration are likely to exert pressure on the agricultural system and socioeconomic conditions. Measures to mitigate the projected influences on agriculture and crop yield require improved climate projections, accurate and real-time crop monitoring, robust modeling or simulation platforms, climate-adaptive management strategies and practices, and alternative and modern farming systems (Vermeulen et al., 2012; Mall et al., 2006). To develop appropriate mitigation planning and frame suitable policies, Aggarwal et al. (2008) suggested a few key activities: (i) strengthening the research to develop effective planning and mitigation measures, integrating biotechnology, advanced information, and space technology; (ii) reliable weather forecasting and early warning systems for drought, flood, heat and cold-waves, and disease outbreaks, and providing advisory services to farmers, (iii) facilitating financial incentives to farmers for resource conservation (water, nutrients, and energy) and regulating the market prices of various crops; (iv) deploying or developing suitable water harvesting structures or proper irrigation facilities; and (v) mobilizing national and international opinion on food security and poverty alleviation, addressing vulnerability under the increased GHGs and climate change.

20.6 CONCLUSION

Land utilization, agricultural practices, livelihood, and socio-economic conditions differ greatly among landscapes. Also, the projected changes in climate conditions may have varying impacts on the diverse hydrological regimes of various landscapes. Thus, a landscape-level approach is essential to develop suitable strategies and planning to optimize water resource utilization and to mitigate the impact of climate change. Additionally, suitable policies can be framed for landscape management, ensuring sustainable livelihood and socio-economic developments. The Landsat satellite data available for the past five decades can be useful to understand long-term changes, i.e., climate change impacts on various land surface processes, including surface water resources. On the other hand, the latest Sentinel-2 optical data have become popular for monitoring heterogeneous landscapes, whereas the free microwave Sentinel-1 data are being widely used for flood and water resources mapping under cloudy conditions. However, the unavailability of uninterrupted moderate- to high-resolution hyperspectral data is one of the major constraints on studying water resources and water quality measurement. Considering the limited access to proprietary commercial software, the potentiality of open source software and programming platforms needs to be explored for better management of water resources and addressing the key issues related to the global water crisis. Besides, suitable platforms can be developed employing open source tools and codes for improved near-real-time water resource monitoring, which will be useful for water resource managers and all stakeholders.

20.7 SUPPLEMENTARY

Google Earth Engine (GEE) code 1: to derive mean NDVI for a period [*the geometry is added using the polygon tool in GEE]

```
//Defining Map Center
Map.centerObject(geometry, 13);

//Visualization Colour
var SFCCvis = {bands: ['B8', 'B4', 'B3'], min: 0, max: 3000};
var Indexvis = {min: -1, max: 0.3, palette: ['0D1CA6', '0DA63D',
                'A6A60D', 'blue']};

//Cloud Masking Function
function maskS2clouds(image) {
  var qa = image.select('QA60');
  var cloudBitMask = 1 << 10;
  var cirrusBitMask = 1 << 11;
  var mask = qa.bitwiseAnd(cloudBitMask).eq(0)
    .and(qa.bitwiseAnd(cirrusBitMask).eq(0));
  return image.updateMask(mask);
}

//Dataset Selection and Data Visualization
var S2dataset = ee.ImageCollection('COPERNICUS/S2')
  .filterDate('2017-01-01', '2017-01-31')
  .filter(ee.Filter.lt('CLOUDY_PIXEL_PERCENTAGE', 20))
  .map(maskS2clouds)
```

```
.filterBounds(geometry);
Map.addLayer(S2dataset, SFCCvis, 'SFCC Jan');

var S2dataset1 = ee.ImageCollection('COPERNICUS/S2')
.filterDate('2017-05-01', '2017-05-31')
.filter(ee.Filter.lt('CLOUDY_PIXEL_PERCENTAGE', 20))
.map(maskS2clouds)
.filterBounds(geometry);
Map.addLayer(S2dataset1, SFCCvis, 'SFCC May');

//NDWIFunction
var addNDWI = function(image){
 return image
.addBands(image.normalizedDifference(["B3", "B8"])
.rename('NDWI'))
.float();
};

var NDWIS2 = S2dataset.map(addNDWI);
var selectNDWI = NDWIS2.select('NDWI');
var mean_NDWI=selectNDWI.mean().clip(geometry);
Map.addLayer(mean_NDWI, Indexvis,'mean NDWI Jan');

var NDWIS21 = S2dataset1.map(addNDWI);
var selectNDWI1 = NDWIS21.select('NDWI');
var mean_NDWI1=selectNDWI1.mean().clip(geometry);
Map.addLayer(mean_NDWI1, Indexvis,'mean NDWI May');

//NDWI Change Mapping
var NDWI_change=mean_NDWI.subtract(mean_NDWI1);
var NDWI_change_mask=NDWI_change.gt(0.4);
Map.addLayer(NDWI_change_mask, {min: 0,max: 1,
palette: ['grey', 'blue']}, 'NDWI Change');
```

Google Earth Engine (GEE) code 2: Sentinel-1 SAR data visualization [*the geometry is added using the polygon tool in GEE]

```
//Defining Map Center
Map.centerObject(geometry, 10);
//Visualization Colour
var color={min: -25, max: 0, palette: ['aqua','grey','black']};
//Sentinel-1 SAR Data Selection
var imgVV = ee.ImageCollection('COPERNICUS/S1_GRD')
.filter(ee.Filter.listContains('transmitterReceiverPolarisation', 'VV'))
.filter(ee.Filter.eq('instrumentMode', 'IW'))
.filterDate('2020-03-01','2020-08-15')
.filterBounds(geometry)
.select('VV');

var S1 = imgVV.filter(ee.Filter.eq('orbitProperties_pass',
'DESCENDING'));
//Data List Creation
var data_list= S1.toList(S1.size());
```

```
//Data Visualization
var img_pm = ee.Image(data_list.get(0)).clip(geometry);
Map.addLayer(img_pm, color, 'Pre-monsoon data');
var img_m = ee.Image(data_list.get(23)).clip(geometry);
Map.addLayer(img_m, color, 'monsoon data');
```

REFERENCES

Aggarwal, P. K. (2008). Global climate change and Indian agriculture: impacts, adaptation and mitigation. *Indian Journal of Agricultural Sciences*, 78(11), 911.

Anees, M. T., Abdullah, K., Nawawi, M. N. M., Norulaini, N. A. N., Syakir, M. I., & Omar, A. K. M. (2018). Soil erosion analysis by RUSLE and sediment yield models using remote sensing and GIS in Kelantan state, Peninsular Malaysia. *Soil Research*, 56(4), 356–372.

Anusha, N., & Bharathi, B. (2020). Flood detection and flood mapping using multi-temporal synthetic aperture radar and optical data. *The Egyptian Journal of Remote Sensing and Space Science*, 23, 207–219.

Arekhi, S., Shabani, A., & Rostamizad, G. (2012). Application of the modified universal soil loss equation (MUSLE) in prediction of sediment yield (Case study: Kengir Watershed, Iran). *Arabian Journal of Geosciences*, 5(6), 1259–1267.

Arheimer, B., Andréasson, J., Fogelberg, S., Johnsson, H., Pers, C. B., & Persson, K. (2005). Climate change impact on water quality: model results from southern Sweden. *AMBIO: A Journal of the Human Environment*, 34(7), 559–566.

Arnell, N. W. (1999). Climate change and global water resources. *Global Environmental Change*, 9, S31–S49.

Bal, P. K., Ramachandran, A., Palanivelu, K., Thirumurugan, P., Geetha, R., & Bhaskaran, B. (2016). Climate change projections over India by a downscaling approach using PRECIS. *Asia-Pacific Journal of Atmospheric Sciences*, 52(4), 353–369.

Bansod, B., Singh, R., & Thakur, R. (2018). Analysis of water quality parameters by hyperspectral imaging in Ganges River. *Spatial Information Research*, 26(2), 203–211.

Begam, S., & Sen, D. (2019). Mapping of moraine dammed glacial lakes and assessment of their areal changes in the central and eastern Himalayas using satellite data. *Journal of Mountain Science*, 16(1), 77–94.

Behera, M. D., Tripathi, P., Das, P., Srivastava, S. K., Roy, P. S., Joshi, C., Behera, P. R., Deka, J., Kumar, P., Khan, M. L., Tripathi, O. P., Dash, T., & Krishnamurthy, Y. V. N. (2018). Remote sensing-based deforestation analysis in Mahanadi and Brahmaputra river basin in India since 1985. *J. Environ. Manage.*, 206, 1192–1203. 10.1016/j.jenvman.2017.10.015

Bhatt, D., & Mall, R. K. (2015). Surface water resources, climate change and simulation modeling. *Aquatic Procedia*, 4(0), 730–738.

Borah, S. B., Sivasankar, T., Ramya, M. N. S., & Raju, P. L. N. (2018). Flood inundation mapping and monitoring in Kaziranga National Park, Assam using Sentinel-1 SAR data. *Environmental Monitoring and Assessment*, 190(9), 520.

Bostater Jr, C. R., Oney, T. S., Rotkiske, T., Aziz, S., Morrisette, C., Callahan, K., & Mcallister, D. (2017). Hyperspectral signatures and WorldView-3 imagery of Indian River Lagoon and Banana River Estuarine water and bottom types. In *Remote Sensing of the Ocean, Sea Ice, Coastal Waters, and Large Water Regions 2017* (Vol. 10422, p. 104220E). International Society for Optics and Photonics.

Brando, V. E., & Dekker, A. G. (2003). Satellite hyperspectral remote sensing for estimating estuarine and coastal water quality. *IEEE Transactions on Geoscience and Remote Sensing*, 41(6), 1378–1387.

Burney, J., & Ramanathan, V. (2014). Recent climate and air pollution impacts on Indian agriculture. *Proceedings of the National Academy of Sciences of the United States of America*, 111(46), 16319–16324.

Chanda, K, Maity, R. (2016). Uncovering global climate fields causing local precipitation extremes. *Hydrological Sciences Journal* 61(7), 1227–1237.

Chander, S., Gujrati, A., Hakeem, K. A., Garg, V., Issac, A. M., Dhote, P. R., ... & Sahay, A. (2019). Water quality assessment of River Ganga and Chilika lagoon using AVIRIS-NG hyperspectral data. *Current Science*, 116(7), 1172.

Cho, H. J., Ogashawara, I., Mishra, D., White, J., Kamerosky, A., Morris, L., ... & Banisakher, D. (2014). Evaluating hyperspectral imager for the coastal ocean (HICO) data for seagrass mapping in Indian River Lagoon, FL. *GIScience & Remote Sensing*, 51(2), 120–138.

Das, P., & Behera, M. D. (2019). Can the forest cover in India withstand large climate alterations? *Biodivers Conserv*, 28, 2017–2033.

Das, B., Ghosh, T. S., Kedia, S., Rampal, R., Saxena, S., Bag, S., Mitra, R., Dayal, M., et al. (2018). Analysis of the gut microbiome of rural and urban healthy Indians living in sea level and high altitude areas. *Sci. Rep.*, 8, 10104.

De Groeve, T. (2010). Flood monitoring and mapping using passive microwave remote sensing in Namibia. *Geomatics, Natural Hazards and Risk*, 1(1), 19–35.

Delpla, I., Jung, A. V., Baures, E., Clement, M., & Thomas, O. (2009). Impacts of climate change on surface water quality in relation to drinking water production. *Environment International*, 35(8), 1225–1233.

Du, J., Kimball, J. S., Jones, L. A., & Watts, J. D. (2016). Satellite passive microwave detection of surface water inundation changes over the pan-Arctic from AMSR. *AGUFM*, 2016, B52C–05.

Dutta, D., Das, S., Kundu, A., & Taj, A. (2015). Soil erosion risk assessment in Sanjal watershed, Jharkhand (India) using geo-informatics, RUSLE model and TRMM data. *Modeling Earth Systems and Environment*, 1:37.

Dutta, S. (2016). Soil erosion, sediment yield and sedimentation of reservoir: a review. *Modeling Earth Systems and Environment*, 2(3), 123.

Dutta, S., & Sen, D. (2018). Application of SWAT model for predicting soil erosion and sediment yield. *Sustainable Water Resources Management*, 4(3), 447–468.

El-Zeiny, A., & El-Kafrawy, S. (2017). Assessment of water pollution induced by human activities in Burullus Lake using Landsat 8 operational land imager and GIS. *Egyptian Journal of Remote Sensing and Space Sciences*, 20, S49–S56.

FAO. www.fao.org/india/fao-in-india/india-at-a-glance/en/#:~:text=Agriculture%2C%20with%20its%20allied%20sectors,275%20million%20tonnes%20(MT) [Visited on 13-08-2020]

Galantowicz, J. F. (2002). High-resolution flood mapping from low-resolution passive microwave data. In *IEEE International Geoscience and Remote Sensing Symposium* (Vol. 3, pp. 1499–1502). IEEE.

Ghosh, S., & Dutta, S. (2012). Impact of climate change on flood characteristics in Brahmaputra basin using a macro-scale distributed hydrological model. *Journal of Earth System Science*, 121(3), 637–657.

Ghosh, S. M., Saraf, S., Behera, M. D., & Biradar, C. (2017). Estimating agricultural crop types and fallow lands using multi temporal Sentinel-2A imageries. *Proceedings of the National Academy of Sciences, India Section A: Physical Sciences*, 87(4), 769–779.

Guhathakurta, P., Sreejith, O. P., & Menon, P. A. (2011). Impact of climate change on extreme rainfall events and flood risk in India. *Journal of Earth System Science*, 120(3), 359.

Huo, Y., & Peltier, W. R. (2020). Dynamically downscaled climate change projections for the South Asian monsoon: mean and extreme precipitation changes and physics parameterization impacts. *Journal of Climate*, 33(6), 2311–2331.

Jensen, D., Simard, M., Cavanaugh, K., Sheng, Y., Fichot, C. G., Pavelsky, T., & Twilley, R. (2019). Improving the transferability of suspended solid estimation in wetland and deltaic waters with an empirical hyperspectral approach. *Remote Sensing*, 11(13), 1629.

Jha, M. K., Chowdhury, A., Chowdary, V. M., Peiffer, S. (2007). Groundwater management and development by integrated remote sensing and geographic information systems: prospects and constraints. *Water Resources Management* 21(2), 427–467.

Kumar, C. P. (2012). Climate change and its impact on groundwater resources. *International Journal of Engineering Science*, 1(5), 43–60.

Kumar, V., Sharma, A., Chawla, A., Bhardwaj, R., & Thukral, A. K. (2016). Water quality assessment of river Beas, India, using multivariate and remote sensing techniques. *Environmental Monitoring and Assessment*, 188(3), 137.

Lacaux, J. P., Tourre, Y. M., Vignolles, C., Ndione, J. A., & Lafaye, M. (2007). Classification of ponds from high-spatial resolution remote sensing: application to Rift Valley fever epidemics in Senegal. *Remote Sensing of Environment*, 106(1), 66–74.

Leichenko, R. M., & O'brien, K. L. (2002). The dynamics of rural vulnerability to global change: the case of southern Africa. *Mitigation and Adaptation Strategies for Global Change*, 7(1), 1–18.

Magesh, N. S., & Chandrasekar, N. (2016). Assessment of soil erosion and sediment yield in the Tamiraparani sub-basin, South India, using an automated RUSLE-SY model. *Environmental Earth Sciences*, 75(16), 1208.

Mall, R. K., Singh, R., Gupta, A., Srinivasan, G., & Rathore, L. S. (2006). Impact of climate change on Indian agriculture: a review. *Climatic Change*, 78(2–4), 445–478.

Matin, S., & Behera, M. D. (2019). Studying evidence of land degradation in the Indian Ganga River Basin- a Geoinformatics approach. *Environmental Monitoring and Assessment*, 191, 803. https://doi.org/10.1007/s10661-019-7694-7.

McFeeters, S. K. (1996). The use of the normalized difference water index (NDWI) in the delineation of open water features. *International Journal of Remote Sensing*, 17(7), 1425–1432.

Michalak, A. M. (2016). Study role of climate change in extreme threats to water quality. *Nature*, 535(7612), 349–350.

Mishra, D. R., & Mishra, S. (2010). Plume and bloom: effect of the Mississippi River diversion on the water quality of Lake Pontchartrain. *Geocarto International*, 25(7), 555–568.

Mishra, D. R., Schaeffer, B. A., & Keith, D. (2014). Performance evaluation of normalized difference chlorophyll index in northern Gulf of Mexico estuaries using the Hyperspectral Imager for the Coastal Ocean. *GIScience & Remote Sensing*, 51(2), 175–198.

Mishra, S., & Mishra, D. R. (2012). Normalized difference chlorophyll index: a novel model for remote estimation of chlorophyll-a concentration in turbid productive waters. *Remote Sensing of Environment*, 117, 394–406.

Mishra, S., Stumpf, R. P., Schaeffer, B. A., Werdell, P. J., Loftin, K. A., & Meredith, A. (2019). Measurement of cyanobacterial bloom magnitude using satellite remote sensing. *Scientific Reports*, 9(1), 1–17.

Misra, A. K. (2013). Climate change impact, mitigation and adaptation strategies for agricultural and water resources, in Ganga Plain (India). *Mitigation and Adaptation Strategies for Global Change*, 18(5), 673–689.

Molina-Navarro, E., Trolle, D., Martínez-Pérez, S., Sastre-Merlín, A., & Jeppesen, E. (2014). Hydrological and water quality impact assessment of a Mediterranean limno-reservoir under climate change and land use management scenarios. *Journal of Hydrology*, 509, 354–366.

Mujumdar, P. P. (2008). Implications of climate change for sustainable water resources management in India. *Physics and Chemistry of the Earth, Parts A/B/C*, 33(5), 354–358.

Murdoch, P. S., Baron, J. S., & Miller, T. L. (2000). Potential effects of climate change on surface-water quality in North America. *Journal of the American Water Resources Association*, 36(2), 347–366.

Murugan, P., Sivakumar, R., Pandiyan, R., & Annadurai, M. (2016). Comparison of in-situ hyperspectral and Landsat ETM+ data for chlorophyll-a mapping in case-II water (Krishnarajapuram Lake, Bangalore). *Journal of the Indian Society of Remote Sensing*, 44(6), 949–957.

Nas, B., Ekercin, S., Karabörk, H., Berktay, A., & Mulla, D. J. (2010). An application of Landsat-5TM image data for water quality mapping in Lake Beysehir, Turkey. *Water, Air, & Soil Pollution*, 212(1–4), 183–197.

Nilawar, A. P., & Waikar, M. L. (2019). Impacts of climate change on streamflow and sediment concentration under RCP 4.5 and 8.5: a case study in Purna river basin, India. *Science of the Total Environment*, 650, 2685–2696.

O'Brien, K., Leichenko, R., Kelkar, U., Venema, H., Aandahl, G., Tompkins, H., ... & West, J. (2004). Mapping vulnerability to multiple stressors: climate change and globalization in India. *Global Environmental Change*, 14(4), 303–313.

Pandey, A., Chowdary, V. M., & Mal, B. C. (2009). Sediment yield modelling of an agricultural watershed using MUSLE, remote sensing and GIS. *Paddy and Water Environment*, 7(2), 105–113.

Pandey, A., Chowdary, V. M., Mal, B. C., & Billib, M. (2008). Runoff and sediment yield modeling from a small agricultural watershed in India using the WEPP model. *Journal of Hydrology*, 348(3–4), 305–319.

Pandey, V., & Srivastava, P. K. (2019). Integration of microwave and optical/infrared derived datasets for a drought hazard inventory in a sub-tropical region of India. *Remote Sensing*, 11(4), 439.

Pandey, V., Srivastava, P. K., Mall, R. K., Munoz-Arriola, F., & Han, D. (2020). Multi-satellite precipitation products for meteorological drought assessment and forecasting in Bundelkhand region of Central India. *Geocarto International*, 1–17. https://doi.org/10.1080/10106049.2020.1801862.

Pathak, H., Pramanik, P., Khanna, M., & Kumar, A. (2014). Climate change and water availability in Indian agriculture: impacts and adaptation. *Indian Journal of Agriculture and Science*, 84, 671–679.

Prasad, S., Saluja, R., & Garg, J. K. (2017, October). Modeling chlorophyll-a and turbidity concentrations in river Ganga (India) using Landsat-8 OLI imagery. In *Earth Resources and Environmental Remote Sensing/GIS Applications* VIII (Vol. 10428, p. 1042814). International Society for Optics and Photonics.

Ramesh, R., & Yadava, M. G. (2005). Climate and water resources of India. *Current Science*, 89, 818–824.

Rao, C. R., Raju, B. M. K., Rao, A. S., Rao, K. V., Ramachandran, V. R. K., Venkateswarlu, B., & Sikka, A. K. (2013). *Atlas on Vulnerability of Indian Agriculture to Climate Change*. Central Research Institute for Dryland Agriculture, Hyderabad.

Rehana, S., & Mujumdar, P. P. (2011). River water quality response under hypothetical climate change scenarios in Tunga-Bhadra river, India. *Hydrological Processes*, 25(22), 3373–3386.

Schroeder, R., McDonald, K. C., Chapman, B. D., Jensen, K., Podest, E., Tessler, Z. D., ... & Zimmermann, R. (2015). Development and evaluation of a multi-year fractional surface water data set derived from active/passive microwave remote sensing data. *Remote Sensing*, 7(12), 16688–16732.

Shindell, D. T., Walter, B. P., & Faluvegi, G. (2004). Impacts of climate change on methane emissions from wetlands. *Geophysical Research Letters*, 31(21), 1–5.

Singh, A., Imtiyaz, M., Isaac, R. K., & Denis, D. M. (2012). Comparison of soil and water assessment tool (SWAT) and multilayer perceptron (MLP) artificial neural network for predicting sediment yield in the Nagwa agricultural watershed in Jharkhand, India. *Agricultural Water Management*, 104, 113–120.

Singh, P., Arora, M., & Goel, N. K. (2006). Effect of climate change on runoff of a glacierized Himalayan basin. *Hydrological Processes. An International Journal*, 20(9), 1979–1992.

Singh, R. K., Panda, R. K., Satapathy, K. K., & Ngachan, S. V. (2011). Simulation of runoff and sediment yield from a hilly watershed in the eastern Himalaya, India using the WEPP model. *Journal of Hydrology*, 405(3–4), 261–276.

Singh, S. P. (2007). *Himalayan forest ecosystem services*. Central Himalayan Environment Association, Nainital, Uttarakhand, India, 53.

Solbø, S., & Solheim, I. (2005). Towards operational flood mapping with satellite SAR. In *Envisat & ERS Symposium, Salzburg, Austria, 6–10 September 2004* (Vol. 572).

Swain, R., & Sahoo, B. (2017). Mapping of heavy metal pollution in river water at daily time-scale using spatio-temporal fusion of MODIS-aqua and Landsat satellite imageries. *Journal of Environmental Management*, 192, 1–14.

Tang, F., & Xu, H. (2017). Impervious surface information extraction based on hyperspectral remote sensing imagery. *Remote Sensing*, 9(6), 550.

Taylor, R. G., Scanlon, B., Döll, P., Rodell, M., Van Beek, R., Wada, Y., ... & Konikow, L. (2013). Ground water and climate change. *Nature Climate Change*, 3(4), 322–329.

Tripathi, R., Moharana, K. C., Nayak, A. D., Dhal, B., Shahid, M., Mondal, B., Mohapatra, S. D., Bhattacharyya, P., Fitton, N., Smith, P., Shukla, A. K., Pathak, H., & Nayak, A. K. (2019) Ecosystem services in different agro-climatic zones in eastern India: impact of land use and land cover change. *Environ Monit Assess*, 191(2), 1–16.

Vanama, V. S. K., Mandal, D., & Rao, Y. S. (2020). GEE4FLOOD: rapid mapping of flood areas using temporal Sentinel-1 SAR images with Google Earth Engine cloud platform. *Journal of Applied Remote Sensing*, 14(3), 034505.

Velloth, S., Mupparthy, R. S., Raghavan, B. R., & Nayak, S. (2014). Coupled correction and classification of hyperspectral imagery for mapping coral reefs of Agatti and Flat Islands, India. *International Journal of Remote Sensing*, 35(14), 5544–5561.

Venkatesh, K., Ramesh, H., & Das, P. (2020). Modelling stream flow and soil erosion response considering varied land practices in a cascading river basin. *Journal of Environmental Management*, 264, 110448.

Vermeulen, S. J., Aggarwal, P. K., Ainslie, A., Angelone, C., Campbell, B. M., Challinor, A. J., ... & Lau, C. (2012). Options for support to agriculture and food security under climate change. *Environmental Science & Policy*, 15(1), 136–144.

Vinayachandran, P. N., Jahfer, S., & Nanjundiah, R. S. (2015). Impact of river runoff into the ocean on Indian summer monsoon. *Environmental Research Letters*, 10(5), 054008.

Whitehead, P. G., Jin, L., Macadam, I., Janes, T., Sarkar, S., Rodda, H. J., ... & Nicholls, R. J. (2018). Modelling impacts of climate change and socio-economic change on the Ganga, Brahmaputra, Meghna, Hooghly and Mahanadi river systems in India and Bangladesh. *Science of the Total Environment*, 636, 1362–1372.

Wu, J. L., Ho, C. R., Huang, C. C., Srivastav, A. L., Tzeng, J. H., & Lin, Y. T. (2014). Hyperspectral sensing for turbid water quality monitoring in freshwater rivers: empirical relationship between reflectance and turbidity and total solids. *Sensors*, 14(12), 22670–22688.

Xu, H. (2006). Modification of normalised difference water index (NDWI) to enhance open water features in remotely sensed imagery. *International Journal of Remote Sensing*, 27(14), 3025–3033.

Zhang, J., Zhang, Q., Bao, A., & Wang, Y. (2019). A new remote sensing dryness index based on the near-infrared and red spectral space. *Remote Sensing*, 11(4), 456.

Index

A

Accuracy assessment, 54, 201
Advanced Microwave Scanning Radiometer (AMSR), 315–316
Analytical Hierarchical Approach (AHP), 22, 24, 186, 227
Artificial neural network (ANN), 152–153, 156
Assessment and monitoring of vegetative drought, 88
Automated Water Extraction Index (AWEI), 200

B

Big BSVD Data, 127
Big (Geo) Data, 123

C

Cartosat-1 DEM, 168
Categories of droughts, 251–256
Cellular Automata, 1
Cement nala bund (CNB), 189
Change detection analysis, 54, 109
Check dams, 175
Cloud computing, 136
Coarse resolution satellite data, 125
Community-formulated plantation guidelines, 116
Computation of top of atmosphere (TOA) reflectance, 73–74
Confusion matrix, 54–55
Consistent ratio (CR), 26
Crowdsource platform, 131–135
Cyclone-induced changes in vegetation cover, 113

D

Desertification, 33
Digital elevation model (DEM), 6, 17
Drought categories, 84
Drought frequency, 250
Drought progression, 262–263
Dynamic surface water extent (DSWE), 150

E

Earthen nala bund (ENB), 189
Ecosystem services value (ESV), 215–217
Error matrix, 10
Estimation of resource degradation, 55

F

False color composite (FCC), 8
Fani, 111
FOSS for geographic information (FOSSGIS), 166–167
Fragmentation analysis, 79
FRAGSTAT, 3, 72, 74
Free and open source software (FOSS), 122, 130, 165–167
Free satellite imagery, 128–129
FREEWAT, 166
Fuzzy set modeling, 24

G

Ganges–Brahmaputra–Meghna (GBM) Delta, 212
Geographic information systems (GIS), 2, 6, 122
Geographic Resources Analysis Support System (GRASS), 85
GeoServer, 171, 172
Global Assessment of Land Degradation and Improvement (GLADA), 35

H

High resolution satellite data, 126
Himachal Pradesh, 35–36
Historical cyclonic events along eastern coast of India, 106

I

Image fusion, 152, 154
Impact of mining activity, 6
Integrated groundwater management (IGM), 226
Integrated Land and Water Information System (ILWIS), 3, 86, 182
Interactive web GIS application, 172

K

Kappa statistics, 10, 64
Kendall correlation, 153

L

Land capability, 184
Land evaluation, 185
Land information systems (LIS), 122–123
Land resource development (LRD), 2
Land resource management (LRM), 122, 135
Land resource planning, 121
Landsat TM, 51–52, 73
Landscape metrics, 74–75
Land suitability analysis, 2
Land use/land cover (LULC), 3, 8, 10, 52, 74, 107, 185, 214
Lentil suitability, 30
Loose bolder structure (LBS), 189

M

MapWindowGIS, 1, 170
Maximum likelihood (MXL) algorithm, 52–53
Medium resolution satellite data, 125
Moderate Resolution Imaging Spectroradiometer (MODIS), 24, 36

Modified Normalized Difference Water Index (MNDWI), 200
Modis Terra snow, 299
Multi-attribute utility theory (MAUT), 183
Multi-criteria evaluation (MCE), 22, 27
Multiple-criteria decision-making (MCDM), 26
Multi-spectral Instrument (MSI), 3

N

National Oceanic and Atmospheric Administration (NOAA), 34
Net primary productivity (NPP), 34
Normalized difference built-up index (NDBI), 3
Normalized difference chlorophyll index (NDCI), 317
Normalized difference vegetation index (NDVI), 3, 9, 16, 39, 108
Normalized difference water index (NDWI), 199

O

Open Geospatial Consortium (OGC), 2
Open source software (OSS), 165, 170
Open source vector data, 127
Ordered weighted averaging (OWA), 2
Overall accuracy, 16

P

Pair-wise comparison matrix, 26
Palmer Drought Severity Index (PDSI), 245
Participatory approach for integrated WRDP, 181
POSTGIS, 171
Pre-processing of satellite image, 52, 72–74, 106
Producer accuracy, 16

Q

Quantum geographical information system (QGIS), 3, 85, 171

R

RAI, 282
Rain use efficiency (RUE), 34
Remote sensing (RS), 2, 6
Residual NDVI, 38
Resource use efficiency (RUE), 34, 37
RESTREND, 35
Revised Universal Soil Loss Equation (RUSLE), 89–91
Root mean square error (RMSE), 157

S

Saraikela, Jharkhand, 167
Search engines for accessing satellite data, 127
Shuttle Radar Topography Mission (SRTM), 24
Significance of land use types For ESV, 220
Soil erodibility, 90
Soil erosion, 84
Soil erosion modelling, 88
Soil Moisture Active Passive (SMAP), 23–24
Soil suitability, 29–30
Spearman's correlation, 153
Standardised Precipitation-Evapotranspiration Index (SPEI), 85, 248–249
Standardized precipitation index (SPI), 280–281
Sub-pixel water fraction (SWF), 150
Suitability index (SI), 27
Surface water monitoring, 313–316

T

Temperature condition index, 88
Top of atmosphere, 73
Trend analysis of drought, 251

U

User accuracy, 16

V

Vegetation condition index, 88, 280
Vegetation Health Index (VHI), 88–89

W

Water quality, 316
Water resource action plan generation, 185
Water resource development (WRD), 2
Water resource development plan (WRDP), 166–167
Water resource management, 194
Web GIS, 171
Weighted linear combination (WLC), 27
Weighted overlay, 186
Weighted overlay techniques, 186
West Bengal, 23
WRDP using open source GIS, 170–171

Y

Yield anomaly index (YAI), 282